A to Z GIS

An Illustrated Dictionary of Geographic Information Systems

Third edition

By Esri
Illustrated by Kelly Brownlee

Esri Press
Redlands, California

Esri Press, 380 New York Street, Redlands, California 92373-8100

Printed in the United States of America.

ISBN: 9781589488113
Library of Congress Control Number: 2024948461

For purchasing and distribution options (both domestic and international), please visit esripress.esri.com.

Preface

The previous edition of this dictionary came out in 2006, nearly 20 years ago, and much has changed in that time. Geographic information systems (GIS) are now indispensable tools that have transcended their initial applications in government, utilities, and natural resource management. Today, GIS technology is used across diverse sectors such as urban planning, health care, transportation, and others. This broad adoption underscores the need for a clear and comprehensive understanding of the terminology associated with GIS. Because of its interdisciplinary nature, GIS has borrowed terms from related fields. In many cases, the meanings of these words have evolved and shifted over time. Over the last 20 years, numerous new terms have been added to the dictionary, and many existing terms have been refined or redefined.

With nearly 3,000 terms, *A to Z GIS: An Illustrated Dictionary of Geographic Information Systems* was designed to be a comprehensive technical dictionary for GIS students and professionals alike. Our goal for this dictionary was to create definitions that are technically accurate yet not intimidating to the GIS novice.

Choosing terms

The first step in creating a dictionary is choosing which terms to include and which to leave out. This process was a challenge, since there are surprisingly few "pure" GIS terms. Most terms used in GIS have been adopted from such related fields as cartography, computing, geodesy, mathematics, remote sensing, statistics, and surveying.

Our solution was to identify two classes of terms within the lexicon:

- Core GIS terms that describe GIS concepts, processes, and operations. Although many of these terms may have origins in other fields, they are more strongly correlated with GIS than with any related field. We included as many of these terms as possible.
- Terms from related fields such as cartography, computing, geodesy, geography, GPS, and remote sensing. These can be thought of as source fields that have lent methods, data, theory, and metaphors to GIS. We included the terms from related fields that GIS practitioners or students are mostly likely to come across in the course of their activities but left out terms that lie outside the GIS context.

Writing the definitions

GIS is an evolving science, and we work to remain aware of and define terms based on current developments in numerous interconnected disciplines. In 2006, when this book was first published, we chose the original list from a database of thousands of potential terms and definitions used in glossaries and dictionaries across Esri. Our final list was carefully reviewed by subject matter experts and editors at Esri as well as an external panel of academic reviewers for consistency and accuracy, and their insights were invaluable. Since that time, the definitions we add continue to follow the growth of GIS and its partner fields, helping new generations of users improve their understanding.

Creating the illustrations

The dictionary's illustrations were created by Kelly Brownlee, featuring some by Jennifer Jennings from the previous edition. Brownlee's challenge was to reimagine Jennings' illustrations and pair clarity and technical accuracy with extreme simplicity. For the maps in this dictionary, the goal of conveying the main idea of a concept in a tight space often meant a departure from traditional cartographic methods.

The future

This dictionary has truly been a collaborative effort, and we are grateful to all the subject matter experts, reviewers, and other staff and contributors who have been involved in its compilation. The expertise that has gone into this dictionary is broad, but we welcome feedback and alternative interpretations. Those who wish to provide feedback may do so through the online GIS Dictionary at support.esri.com/gisdictionary or by email at gisdictionary@esri.com. We will consider all viewpoints and submissions for future printings and editions. GIS has a strong future, and as applications for GIS multiply, GIS terminology will continue to evolve and require definition.

Acknowledgments

We extend our gratitude to Jack Dangermond, cofounder and president of Esri, for his support of this book, and for providing the kind of environment that fosters creativity and collaboration. More than 200 dedicated individuals at Esri have contributed definitions, edits, or comments for *A to Z GIS*. Without their collaboration and expertise, this book would not exist. We sincerely appreciate everyone who took time out of their busy schedules to write and review terms and definitions and answer our questions. Jeff Liedtke, John Nelson, Kevin M. Kelly, and Tim Ormsby in particular were instrumental. We are especially thankful to the dictionary editing team, Rebekah Folsom, Sarah David, Shanon Sims, and Molly Zurn, who maintain the dictionary online and helped us immensely with this new edition. Big thanks to the previous edition's editors Tasha Wade and Shelly Sommer, without whom there would be no *A to Z GIS*. Lastly, this new edition shines with the wonderful new illustrations by Kelly Brownlee. Special thanks to Steve Frizzell, Craig Carpenter, and Victoria Roberts for their guidance in illustrating each term effectively.

Contributors within Esri from 2005 to 2024 include Eric Akin, Rebeka Alvarez-Heck, Jamil Alvi, Christine Anthony, David Arctur, David Austin, Jonathan Bailey, Scott Ball, Peter Becker, Cody Benkelman, Mrinmayee Bharadwaj, Victor Bhattacharyya, Suzanne Boden, Bob Booth, Hal Bowman, Judy Boyd, Dave Boyles, Steve Bratt, Joe Breman, Patrick Brennan, Pat Breslin, Evan Brinton, Lee Brinton, Clint Brown, Aileen Buckley, Rob Burke, Kathy Capelli-Breier, Krista Carlson, Tarun Chandrasekhar, Colin Childs, Dan Clark, Kristin Clark, Shane Clarke, Dan Cobb, Amy Collins, Clayton Crawford, David Crawford, Scott Crosier, Brian Cross, Matt Crowder, Katy Dalton, Dave Danko, Jane Darbyshire, Marilyn Daum, Eleanor Davies, David Davis, Nana Y. Dei, Jon DeRose, Mara Dolan, Thomas Dunn, Sara Eddy, Cory Eicher, Gregory Emmanuel, Rupert Essinger, Mark Feduska, Katherine Fitzgerald, Derek Foll, John Foster, Witold Fraczek, Steve Frizzell, Charlie Frye, Ignacia Galvan, Peng Gao, Phoebe Gelbard, Sophia Giebler, Shelly Gill, Craig Gillgrass, Rhonda Glennon, Lisa Godin, Craig Greene, Craig Greenwald, Lauren Scott Griffin, Michael Grossman, Sarah Hanson, Paul Hardy, Melanie Harlow, Alan Hatakeyama, Mark Henry, Jim Herries, Catherine Hill, Vicki Hill, Tim Hodson, Jennifer Itatani, Jen Jennings, Robert Jensen, Ann Johnson, Karen Johnston, Kevin Johnston, Catherine Jones, Rob Jordan, Gary Kabot, Peter Kasianchuk, Amy Kastrinos, Tim Kearns, Anita Kemp, Melita Kennedy, Jon Kimerling, Steve Kopp, Kory Kramer, Konstantin Krivoruchko, Al Laframboise, Juan Laguna, Derek Law,

Christine Leslie, Will Lewis, Shing Lin, Adrien Litton, Mike Livingston, Robin Lovell, Clint Loveman, Steve Lynch, Kevin M. Kelly, Andy MacDonald, Gary MacDougall, Alan MacEachren, Keith Mann, Michael Mannion, Frank Martin, Jim Mason, Sean McCarron, Jill McCoy, Heather McCracken, Matt McGrath, Ginger McKay, Elizabeth Mezenes, Christopher Moore, Scott Morehouse, Bill Moreland, Doug Morgenthaler, Joel Morrison, Makram Murad, Diana Muresan, Jonathan Murphy, Scott Murray, Claudia Naber, Brad Niemand, Nawajish Noman, Serene Ong, Sarah Osborne, Krista Page, Brian Parr, Jamie Parrish, Christopher Patterson, Meredith Payne, Andrew Perencsik, Kim Peter, Rhonda Pfaff, Morakot Pilouk, Christie Pleiss, Ghislain Prince, Edie Punt, Sterling Quinn, Amir Razavi, Jeff Reinhart, Jeff Rogers, Rick Rossi, Sara Sanchez, Bojan Šavrič, Sandi Schaeffer, Frederic Schettini, Charles Serafy, Jeff Shaner, Nathan Shephard, Gillian Silvertand, Shanon Sims, Daryl Smith, Damian Spangrud, Cathy Spisszak, Marc St. Onge, Bjorn Svensson, Jeff Swain, Sally Swenson, Agatha Tang, Corey Tucker, Patty Turner, Sarah VanHoy, Marika Vertzonis, Aleta Vienneau, Nathan Warmerdam, David Watkins, Kyle Watson, Lindsay Weitz, Tiffany Wilkerson, Craig Williams, Jason Willison, Jill Williston, Niki Wong, Simon Woo, Randy Worch, Dawn Wright, David Wynne, Lingtao Xie, Hong Xu, Rebecca Yarbrough, and Larry Young.

We are also indebted to the Esri Copyediting team and the following academic and industry reviewers and contributors:

Dr. Barbara Buttenfield, Department of Geography, University of Colorado, Boulder; Jeffrey Danielson, U.S. Geological Survey; Dr. Michael N. DeMers, Department of Geography, New Mexico State University; Mr. Jeffrey D. Hammerlinck, University of Wyoming, Laramie; Dr. Duane F. Marble, The Ohio State University; Professor W. Andrew Marcus, University of Oregon; Dr. Nicholas Nagle, Department of Geography, University of Colorado, Boulder; and Roland Viger, U.S. Geological Survey.

Image and data credits

Digital orthophoto quadrangle: Image from U.S. Geological Survey.
Draping: Cartography by Craig Carpenter, Esri; image from U.S. Geological Survey.
Hillshading: Cartography by Daniel Coe, Washington Geological Survey.
Landsat: Image from U.S. Geological Survey.
Quadrangle: Image from U.S. Geological Survey.
Voxel: Cartography by Nathan Shephard, Esri; data courtesy of the World Ocean Database (WOD), from NOAA.

How to use this dictionary

Parts of an entry

- *Headword*—The term being defined is in bold type at the beginning of the entry.
- *Taxonomy*—This is used to classify definitions by subject area and is set in brackets.
- *Definition*—The meaning of the headword is explained in the definition. For terms that have more than one definition, the definitions are numbered and placed in order. GIS-relevant definitions typically appear first; definitions from related fields follow.
- *Cross-references*—Listed after the last definition in an entry with "See also," these terms are related in some way to the headword. They might be synonyms, antonyms, broader terms, or narrower terms.
- *Illustration*—Illustrations appear after the definition they refer to and are labeled with the headword.

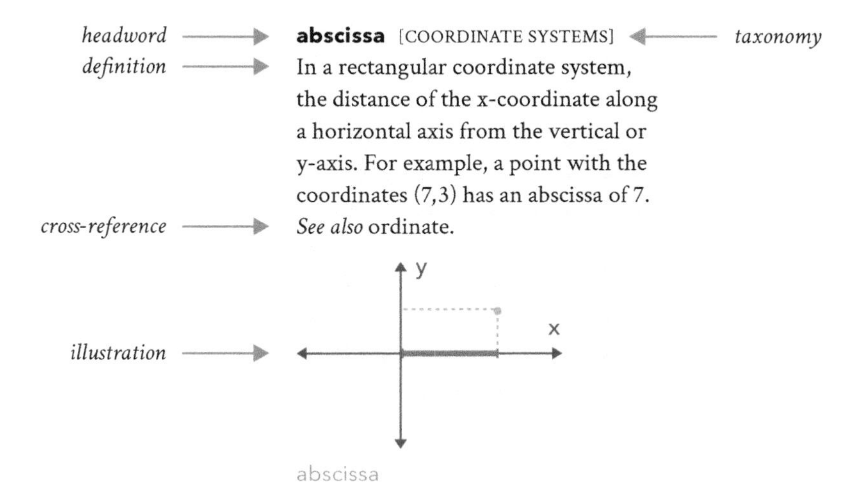

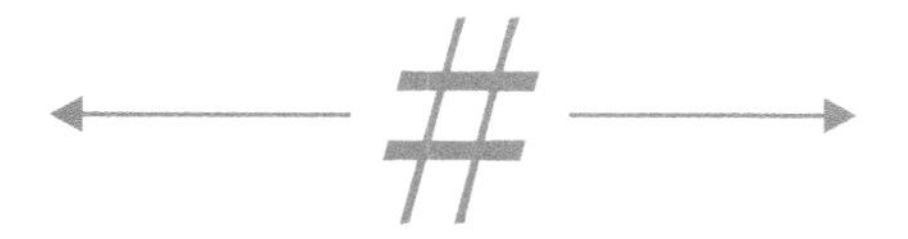

# #

1NF *See* first normal form.

24-bit color [VISUALIZATION] An 8-bit byte used for the red, green, and blue (RGB) components of a pixel color. Also called true color. *See also* pixel, RGB.

2NF *See* second normal form.

3D feature [3D GIS] A representation of a three-dimensional, real-world object in a map or scene, with elevation values (z-values) stored within the feature's geometry and optionally, attributes stored in a feature table. In applications such as CAD, 3D features are often referred to as 3D models. *See also* 3D model, elevation, feature, real world, z-value.

3D graphic [3D GIS] A representation of a three-dimensional, real-world object in a map or scene, with elevation values (z-values) stored within the feature's geometry. Unlike 3D features, 3D graphics do not have attributes. *See also* 3D model, graphic, real world, z-value.

3D mesh [3D ANALYSIS] A digital 3D textured model where the ground and aboveground feature surfaces are densely and accurately reconstructed. *See also* edge, face, surface.

3D model [3D GIS] A construct used to portray an object in three dimensions. In GIS, 3D models are often referred to as 3D features.

3D multipatch *See* multipatch.

3D object [3D ANALYSIS] A digital representation of a real-world feature with three-dimensional geometric characteristics, such as length, width, and height. Within GIS, 3D objects are used for spatial modeling and analysis. *See also* geometry, real world.

3D polygon *See* polyhedron.

3D scene [3D ANALYSIS, ESRI SOFTWARE] A method used to display three-dimensional data. The viewer can adjust the display of 3D data to identify and select features. *See also* scene.

3D shape [3D GIS] A point, line, or polygon that stores x-, y-, and z-coordinates as part of its geometry. A point has one set of z-coordinates; lines and polygons have z-coordinates for each vertex in a shape. *See also* polyhedron, sphere, spheroid.

3D surface area [CARTOGRAPHY] An area calculation that takes the slope of the landscape into account. *See also* slope.

3D symbol [SYMBOLOGY] A symbol with properties that allow it to be rendered in three dimensions. *See also* symbol.

3D tile [3D ANALYSIS] A data format used to stream and visualize large-scale 3D geospatial datasets. 3D tiles are specifically designed for displaying complex 3D models, terrain, and other spatial data with smooth rendering and interaction. *See also* 3D model, smoothing, terrain surface.

3D-perspective surface map [MAP DESIGN] A map that shows a continuous surface in oblique perspective. *See also* continuous surface, oblique-perspective map.

3NF *See* third normal form.

A

abscissa [COORDINATE SYSTEMS] In a rectangular coordinate system, the distance of the x-coordinate along a horizontal axis from the vertical or y-axis. For example, a point with the coordinates (7,3) has an abscissa of 7. *See also* ordinate.

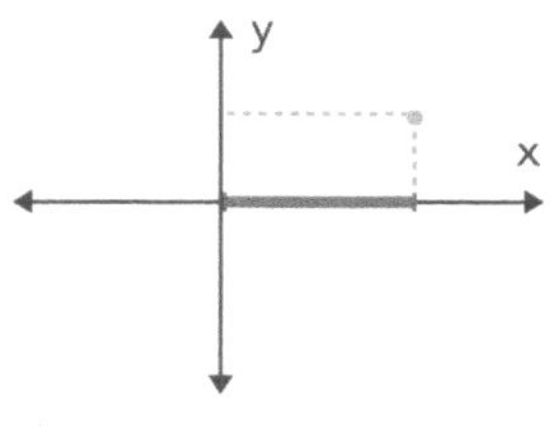

abscissa

absolute accuracy [COORDINATE SYSTEMS, DATA QUALITY] The degree to which the position of an object on a map conforms to its correct location on the earth according to an accepted coordinate system. *See also* coordinate system, relative accuracy.

absolute coordinates [COORDINATE SYSTEMS] Coordinates that are referenced to the origin of a given coordinate system. *See also* coordinates.

absolute relief [CARTOGRAPHY] The elevation values at locations in the landscape given in reference to a specified datum. *See also* datum, elevation.

absolute relief mapping method [CARTOGRAPHY] A mapping method for showing numeric elevation information.

acetate [ESRI SOFTWARE] Circles, lines, polygons, points, or markers that become transparent when not active. Acetate features are overlaid on other map layers and can be independently annotated.

across-track scanner [REMOTE SENSING] Also known as a whisk broom sensor. A remote-sensing tool with an oscillating mirror that moves back and forth across a satellite's direction of travel, creating scan line strips that are contiguous or that overlap slightly to produce an image. *See also* along-track scanner, sensor.

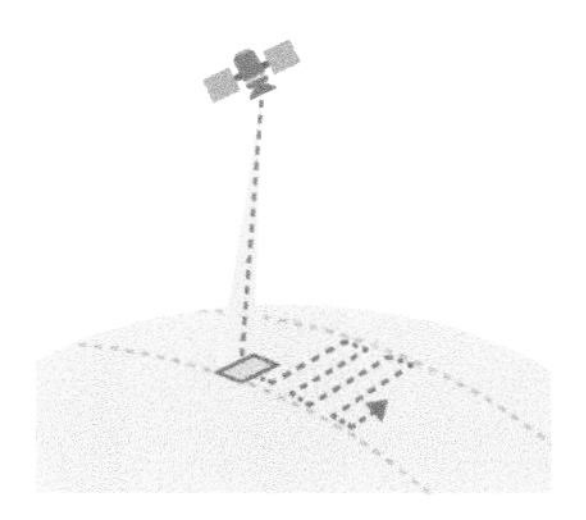
across-track scanner

active remote sensing [REMOTE SENSING] A remote sensing system, such as radar, that produces electromagnetic radiation and measures its reflection back from a surface. *See also* passive remote sensing, remote sensing.

A

actual scale [CARTOGRAPHY] The local measurement of scale at any point on a map; actual scale can vary from one location on a map to another. *See also* principal scale, scale factor.

acutance [PHOTOGRAMMETRY] A measure, using a microdensitometer or other instrument, of how well a photographic system shows sharp edges between contiguous bright and dark areas. *See also* remote sensing, microdensitometer, contiguous.

address [CADASTRAL AND LAND RECORDS] The designation of a location, typically used for navigation or postal service purposes, which consists of text and numeric elements, such as a street number, street name, city, and postal code arranged in a specific format. *See also* geocoding, location, matching.

address data [GEOCODING] Data that contains address information used for geocoding. Address data may consist of one individual address or a table containing many addresses. *See also* geocoding.

address data format [GEOCODING] The arrangement of address information in a database, most often consisting of such address elements as house number, street direction, street name, street type, city, and postal code. *See also* geocoding.

address data model [GEOCODING] The rules of a geodatabase that specifically accommodate address-related material, such as streets, zones, and ranges. Relevant properties include the allowed address elements, their attribute values, and the relationships between these. An address data model also facilitates address data storage. *See also* geocoding.

address element [GEOCODING] One of the components that compose an address. House numbers, street names, street types, and street directions are examples of address elements. *See also* address, address data, locator, geocoding.

address event [ADDRESS MATCHING, ESRI SOFTWARE] Features that can be located based on address matching with a street network or other address identifier, such as postal codes or lot numbers. *See also* address matching, event, postal code.

address event table [GEOCODING, ESRI SOFTWARE] A document object containing addresses but no spatial reference information. Using GIS software, address event tables can be geocoded to create a spatial data layer. *See also* address, address event, geocoding, table.

address field [GEOCODING, ESRI SOFTWARE] A column in a table that stores one or more address elements. An address field can be present in reference data, address data, or both. *See also* field, geocoding.

address format [GEOCODING] The particular structure and arrangement of address elements and a corresponding method of matching that can be used for a specific application. The address format may vary based on locale or country. *See also* geocoding, locale.

address geocoding *See* geocoding.

address locator *See* locator.

address locator property *See* locator property.

address locator style *See* locator style.

address matching [ADDRESS MATCHING, ESRI SOFTWARE] A process that compares an address or a table of addresses to the address attributes of a reference dataset to determine whether a particular address falls within an address range associated with a feature in the reference dataset. If an address falls within a feature's address range, it is considered a match and a location can be returned. *See also* geocoding, matching, match score.

address range [GEOCODING, ESRI SOFTWARE] Street numbers running from lowest to highest along a street or street segment. Address ranges often indicate ranges on the left and right sides of streets. *See also* address, geocoding.

address standardization [GEOCODING] The process of breaking down an address into elements and converting those elements with standard abbreviations or spellings. For best practices, this process applies to preparing the reference data and address data for matching. *See also* address, geocoding.

adjacency

1. [GEOGRAPHY] A type of spatial relationship in which two or more polygons share a side or boundary.

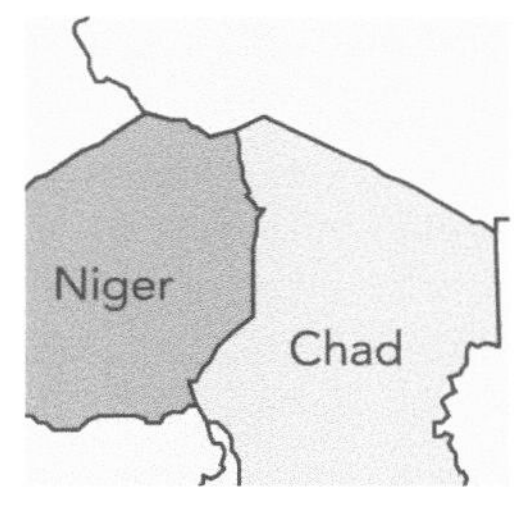

adjacency 1

2. [EUCLIDEAN GEOMETRY] The state or quality of lying close or contiguous.

See also boundary, conterminous, contiguous.

adjacency query [SPATIAL ANALYSIS] A statement or logical expression used to select geographic features that share a boundary. *See also* adjacency.

Advanced Very High Resolution Radiometer [REMOTE SENSING] Also known by the acronym *AVHRR*. A sensor on National Oceanic and Atmospheric Administration (NOAA) polar-orbiting satellites for measuring visible, shortwave, mediumwave, and longwave (thermal) radiation reflected or emitted from vegetation, cloud cover, shorelines, water, snow, and ice. AVHRR data is often used

A

for weather prediction, sea surface, and vegetation mapping. The highest AVHRR resolution is 1.1 km per pixel at nadir; the swath is 2399.5 km. *See also* electromagnetic radiation, remote sensing, sensor.

aerial [PHYSICS] Refers to the air or atmosphere. *See also* aerial photograph, areal.

aerial photograph [AERIAL PHOTOGRAPHY] A photograph of the earth's surface taken from an aircraft, satellite, or other remote platform. Aerial photography is often used as a cartographic data source for basemaps, to locate geographic features, and to interpret environmental conditions. *See also* orthophoto, oblique photograph, vertical photograph.

aeronautical chart [CARTOGRAPHY] A map created specifically for air navigation. *See also* chart, marine navigation, navigation chart, sectional aeronautical chart.

affine transformation

1. [COORDINATE SYSTEMS, SPATIAL ANALYSIS] A geometric transformation that scales, rotates, skews, or translates images or coordinates between any two Euclidean spaces. It is commonly used in GIS to transform maps between coordinate systems. Affine transformations preserve parallel lines, ratios of distances between points, and all points on a straight line remain on a straight line; affine transformations do not necessarily preserve angles or lengths.

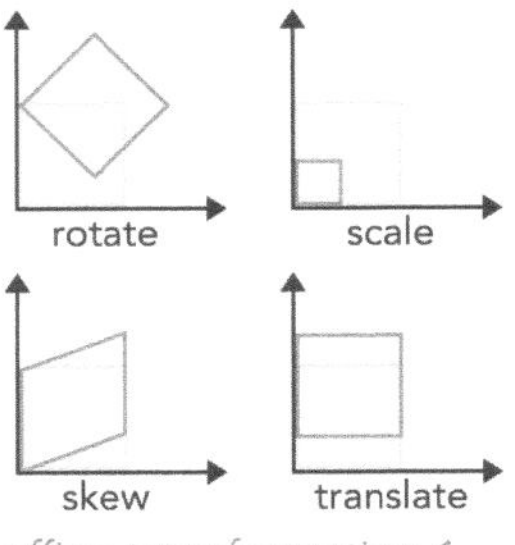

affine transformation 1

2. [DATA CONVERSION] In imagery, a six-parameter coordinate transformation that defines the relationship between pixels and an image coordinate system for interior orientation or other geometric transformations. Affine transformation can include translation, rotation, scale change, and shearing.

See also coordinate transformation, geometric transformation, Helmert transformation, reprojection, transformation.

aggregation [DATA EDITING] The process of collecting a set of similar, usually adjacent, polygons (with their associated attributes) to form a single, larger entity.

agile sensor [SATELLITE IMAGING] A sensor that can be moved—in altitude or position—using its platform. Agile movement can include pointing it over a target, remaining above a targeted area, or spreading across the target area. *See also* sensor.

agonic line [CARTOGRAPHY] The line of zero magnetic declination along which the true and magnetic north poles are aligned. *See also* magnetic declination, magnetic north, true north.

AI [PROGRAMMING] Acronym for *artificial intelligence.* Programming that continually adapts, infers patterns, generalizes, and improves output over time. Within GIS, AI can be used to create optimal routes based on learned user preferences. Or, using natural language processing (NLP)—a branch of AI focused on communication, such as text and speech—a GIS system can improve the response speed to spoken or text location queries with spatially correct results. *See also* deep learning, machine learning, neural network, semantic search.

air navigation [NAVIGATION] The process of planning, recording, and controlling the movement of an aircraft. *See also* marine navigation, navigation.

air photo *See* aerial photograph.

air station *See* exposure station.

airbase vector [PHOTOGRAMMETRY] A line that connects the two exposure stations of a stereopair. *See also* exposure station, stereopair.

Airy 1830 ellipsoid [GEODESY] A mathematical model of the earth defined by English mathematician George Biddell Airy (1801–1892) to best fit the regions of Great Britain and Ireland. *See also* ellipsoid.

AIS *See* automatic identification system.

AIXM [NAVIGATION] Acronym for *Aeronautical Information Exchange Format.* A model and XML schema used to describe aeronautical data transactions created and maintained by the U.S. Federal Aviation Administration (FAA), the U.S. National Geospatial Intelligence Agency (NGA), and the European Organisation for the Safety of Air Navigation (EUROCONTROL). *See also* schema.

albedo [PHYSICS] A measure of the reflectivity of an object or surface; the ratio of the amount of radiation reflected by a body to the amount of energy striking it. *See also* electromagnetic radiation, radiation, reflectance.

Albers equal-area conic projection [CARTOGRAPHY] A conformal, conic map projection designed to preserve the relative sizes of areas on a map. The Albers equal-area conic projection is particularly useful when mapping regions with significant variations in latitude, such as countries or continents, as it minimizes area distortion within the selected region. Devised in 1805 by German cartographer Heinrich C. Albers, it is also known as the Albers equal-area projection. *See also* conformal projection, conic projection, Gall-Peters projection.

A

alert [USABILITY] A message that calls attention to a notable situation or informs users of changes in the state of a monitored situation.

algorithm [COMPUTING] A mathematical procedure used to solve problems with a series of steps. Algorithms are usually encoded as a sequence of computer commands. *See also* heuristic, soundex.

alias [COMPUTING] An alternative name specified for fields, tables, files, or datasets that is typically more descriptive and user-friendly than the actual name. On computer networks, a single email alias may refer to a group of email addresses.

aliasing [GRAPHICS (COMPUTING)] The jagged appearance of curves and diagonal lines in a raster image. Aliasing becomes more apparent as the size of the raster pixels is increased or the resolution of the image is decreased. *See also* pixel, raster.

alidade [SURVEYING]

1. A peep sight mounted on a straightedge and used to measure direction.
2. The part of a theodolite containing the telescope and attachments.

See also theodolite.

aligned dimension [SURVEYING] A drafting symbol that runs parallel to the baseline and indicates the true distance between beginning and ending dimension points. *See also* linear dimension.

allocation [NETWORK ANALYSIS] The process of assigning entities or edges and junctions to features until the feature's capacity or limit of impedance is reached. For example, streets may be assigned to the most accessible fire station within a six-minute radius, or students may be assigned to the nearest school until it is full. *See also* cost-weighted allocation, network analysis, straight-line allocation.

almanac

1. [GPS] A file transmitted from a satellite to a receiver that contains information about the orbits of all satellites included in the satellite network. Receivers refer to the almanac to determine which satellite to track.
2. [ASTRONOMY, METEOROLOGY] An annual publication containing weather forecasts, information on astronomical events, and miscellaneous facts, arranged according to the calendar of a given year.

See also satellite constellation.

along-track scanner [REMOTE SENSING] Also known as a push broom sensor. In satellite imagery, a sensor that uses a linear array in the focal plane to create an image by moving that array through space. The linear, projected footprint of the array is perpendicular to the sensor's movement. *See also* across-track scanner, sensor.

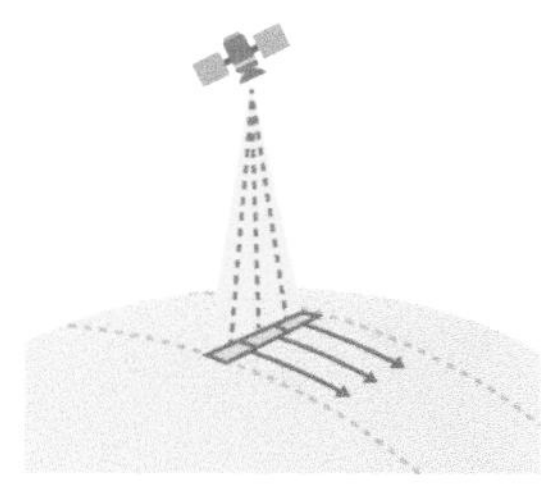
along-track scanner

alphanumeric grid [CARTOGRAPHY] A grid of numbered rows and lettered columns (or vice versa) superimposed on a map that is used to find and identify features. Alphanumeric grids are commonly used as a reference system on local street maps.

alternate key [DATABASE STRUCTURES] An attribute or set of attributes in a relational database that provides a unique identifier for each record and could be used as an alternative to the primary key. *See also* key.

alternate name [GEOCODING] A name for an address element, usually a street name, that is different from the official or most common name. For example, a highway number might be an alternate name for a street name. *See also* alias.

altimeter [CARTOGRAPHY] A reference instrument used to measure the height above a vertical datum, usually mean sea level. *See also* mean sea level, vertical datum.

altitude [CARTOGRAPHY] The height or vertical elevation of a point above a reference surface. Altitude measurements are usually based on a given reference datum, such as mean sea level. *See also* azimuth, elevation.

A

AM/FM

1. [UTILITIES] Acronym for *automated mapping/facilities management.* GIS or CAD-based systems used by utilities and public works organizations for storing, manipulating, and mapping facility information, such as the location of geographically dispersed assets.
2. [PUBLIC SAFETY] Acronym for *amplitude modulation/frequency modulation.* The methods of radio broadcasting.

ambiguity [UNCERTAINTY] In GIS, a state of uncertainty in data classification that exists when an object may appropriately be assigned two or more values for a given attribute. For example, coastal areas experiencing tidal fluctuations may be dry land at some times and underwater at other times. Ambiguity may be caused by changeable conditions in reality, by incomplete or conflicting definitions of attributes, or by subjective differences in the evaluation of data. It may also be caused by disputes, as when two parties claim ownership of the same tract of land. *See also* vagueness, uncertainty.

American National Standards Institute [STANDARDS] Also known by the acronym *ANSI.* The private, nonprofit organization that develops U.S. industry standards for products, services, processes, systems, and personnel through consensus and public review. The organization also aligns U.S. standards with comparable international standards.

American Society of Photogrammetry and Remote Sensing *See* ASPRS.

American Standard Code for Information Interchange [STANDARDS] Also known by the acronym *ASCII.* The de facto standard for formatting internet text files that assigns a 7-bit binary number to each alphanumeric or special character. ASCII defines 128 possible characters. *See also* American National Standards Institute, binary.

amoeba *See* complex market area.

anaglyph [MAP DISPLAY] A stereo image that is created by superimposing two images of the same area. The images are displayed in complementary colors, usually red and blue or green. When viewed through filters of corresponding colors, the images appear as one 3D image. Anaglyphs are useful for visualizing terrain, elevation models, or other spatial data in 3D or virtual reality (VR) applications; they can provide a better understanding of the topography, spatial relationships, and features of the landscape. *See also* image, photogrammetry.

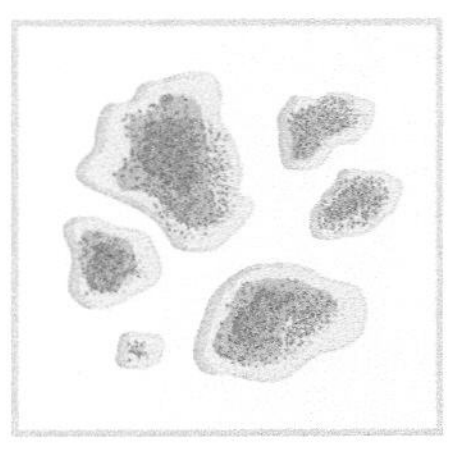

anaglyph

analog [MATHEMATICS] Represented continuously rather than in discrete steps; having value at any degree of precision. *See also* encoding, signal.

analog image [GRAPHICS (COMPUTING)] An image represented by continuous variation in tone, such as a photograph. *See also* aerial photograph, digital image, image.

analysis [ANALYSIS/GEOPROCESSING] A systematic examination of a problem or complex entity to provide new information from what is already known. *See also* autocorrelation.

analysis extent [SPATIAL ANALYSIS] The geographic bounding area within which spatial analysis will occur. The bounding area is set by defining the x,y coordinates of opposite corners, usually the bottom left and top right corners of results.

analysis mask *See* mask.

analysis of variance [STATISTICS] A statistical procedure used to evaluate the variance of the mean values for two or more datasets to assess the probability that the data comes from the same sample or statistical population. *See also* probability.

ancillary data

1. [DIGITAL IMAGE PROCESSING] Data from sources other than remote sensing, used to assist in analysis and classification or to populate metadata.
2. [DATA EDITING] Supplementary data.

See also digital elevation model.

ancillary source [DATABASE STRUCTURES] A supplementary source of information.

Anderson schema [REMOTE SENSING] The land cover or land use classification schema used by the U.S. Geological Survey specifically for remotely sensed imagery. The basis for the U.S. National Land Cover Database (NLCD). *See also* remote sensing, U.S. Geological Survey.

angle scale [CARTOGRAPHY] The system of marks that show angle measurements along the edge of a protractor.

angular deformation [CARTOGRAPHY] The minimum average distortion in shape for the area mapped.

angular range [CARTOGRAPHY] The difference between the maximum and minimum azimuth angles. *See also* azimuth.

angular shift [CARTOGRAPHY] The movement in position that would occur if the center of a shift line was plotted relative to an x,y axis related to compass directions. *See also* shift line.

angular unit [GEODESY] The unit of measurement on a sphere or a spheroid, usually degrees. Some map projection parameters, such as the central meridian and standard parallel, are defined in angular units. *See also* linear unit.

animated map [MAP DISPLAY] A map that has had some form of digital animation added, typically to add a sense of change over time, space, or both. A common use is in meteorology.

animation [3D ANALYSIS, ESRI SOFTWARE] Data that defines dynamic property changes to associated objects. Animation enables dynamic or temporal changes or navigation through the display, or the modification of layer and scene properties, such as layer transparency or the scene background. *See also* keyframe.

anisotropic [MODELING] Having nonuniform spatial distribution of movement or properties, usually across a surface. *See also* anisotropy, isotropic.

anisotropy [SPATIAL STATISTICS (USE FOR GEOSTATISTICS)] A property of a spatial process or data in which spatial dependence (autocorrelation) changes with both the distance and the direction between two locations. *See also* anisotropic, isotropy, autocorrelation.

annotated orthophotomap [MAP DESIGN] An orthophoto on which conventional map symbols are overlaid. *See also* orthophoto.

annotation

1. [MAP DESIGN] In cartography, text or graphics on a map that provide information for the map reader. An annotation may identify or describe a specific map entity, provide general information about an area on the map, or supply information about the map itself.

annotation 1

2. [ESRI SOFTWARE] A part of a web map (text or graphics) that can be individually selected, positioned, and modified. An annotation may be manually entered or generated from labels. Annotations can be stored as features in a geodatabase or as map annotation.

See also label, annotation class, feature-linked annotation, map annotation.

annotation class [MAP DESIGN] A subset of annotation in a standard or feature-linked geodatabase annotation feature class that contains properties that determine how the subset of annotation will display on a web map. A standard or feature-linked geodatabase annotation feature class may contain one or more annotation classes. *See also* annotation, annotation feature class.

annotation construction method [MAP DESIGN] One of several procedures that dictate what type of annotation feature is created and the number of points required to create annotation features on a web map. Construction methods include horizontal, straight, curved, leader line, and follow feature. *See also* annotation, annotation feature class.

annotation feature class [MAP DESIGN] A geodatabase feature class that stores text or graphics that provide information about features or general areas of a web map (annotation). An annotation feature class may be linked to another feature class, so that edits to the features are reflected in the corresponding annotation (feature-linked annotation). *See also* annotation group.

annotation group [MAP DESIGN] A container within a map document for organizing and managing text or graphics that provide additional information about features or general areas of a web map. Annotation groups assist with managing the display of different sets of annotation. *See also* annotation feature class.

annotation layer [MAP DESIGN] A web map layer that references annotation. Information stored for annotation includes a text string, a position at which it can be displayed, and display characteristics. *See also* layer.

annotation target [MAP DESIGN, ESRI SOFTWARE] The annotation group or feature class in a map document where new or pasted annotation is stored. *See also* annotation, annotation feature class.

annual increase or decrease [CARTOGRAPHY] The annual change in compass variation and direction of magnetic declination that is shown on charts in minutes of a degree per year. *See also* compass variation, magnetic declination.

annular drainage pattern [CARTOGRAPHY] A ringlike drainage pattern associated with maturely dissected dome or basin structures.

annulus [MATHEMATICS] A ring-shaped object formed by the area bounded by two concentric circles, often used to determine a neighborhood for statistical calculations.

ANOVA *See* analysis of variance.

ANSI *See* American National Standards Institute.

anthropogenic [HUMAN FACTORS] Caused by or resulting from human activity.

antialiasing [DIGITAL IMAGE PROCESSING] Adding pixels of intermediate color between the object and the background to smooth a jagged edge in an image.

antipodal meridian [CARTOGRAPHY] The meridian at the opposite side of the earth from the one being considered.

antipode [GEODESY] Any point on the surface of a sphere that lies 180 degrees (opposite) from a given point on the same surface, so that a line drawn between the two points through the center of the sphere forms a true diameter. *See also* sphere, zenith.

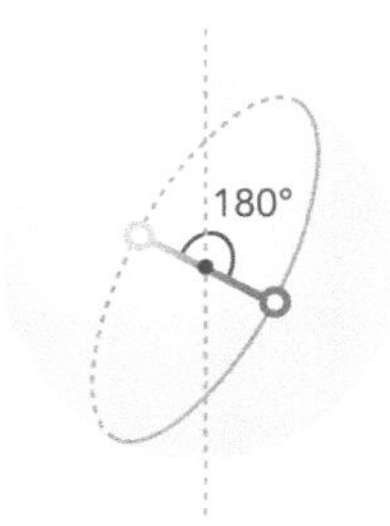

antipode

any-vertex connectivity [NETWORK ANALYSIS] In network datasets, a type of edge connectivity policy that states that an edge may connect to another edge or junction where they have coincident vertices. *See also* endpoint connectivity, edge connectivity policy.

anywhere fix [GPS] A position that a GPS receiver can calculate without knowing its own location or the local time.

AOI [MAP DESIGN] Acronym for *area of interest*. The extent used to define a focus area for a map or database.

aphylactic projection
See compromise projection.

A

API *See* application programming interface.

API key [PROGRAMMING] API is an acronym for *application programming interface*. An encrypted access token that defines the scope and permissions for accessing private digital content. *See also* application programming interface, authentication.

apogee [ASTRONOMY] In an orbit path, the point at which the object in orbit is farthest from the center of the body being orbited. *See also* perigee.

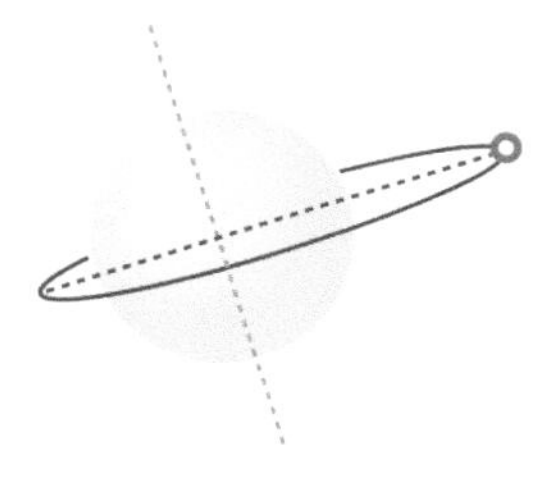

apogee

apparent reflectance

1. [DIGITAL IMAGE PROCESSING] The ratio of upwelling irradiance to downwelling irradiance, resulting in a dimensionless unit.
2. [ESRI SOFTWARE] A function that calibrates imagery values using sun elevation, sensor gain/bias, and acquisition date to derive top of atmosphere (TOA) reflectance and sun angle correction. Also known as albedo or surface brightness in planetary science.

See also albedo.

appending [ANALYSIS/GEOPROCESSING] Adding features from multiple data sources of the same data type into an existing dataset. *See also* merging.

application [ANALYSIS/GEOPROCESSING] The use of a GIS to solve problems, automate tasks, or generate information within a specific field of interest. For example, a common agricultural application of GIS is determining fertilization requirements based on field maps of soil chemistry and previous crop yields.

application programming interface [PROGRAMMING] Also known by the acronym *API*. A set of interfaces, methods, protocols, and tools that application developers use to build or customize a software program. *See also* SDK.

application web service [INTERNET] A web service that solves a particular problem—for example, a web service that finds all the hospitals within a certain distance of an address. *See also* web service.

approximate contour [CARTOGRAPHY] A contour used to indicate the approximate location of contours where information isn't reliable, usually shown in areas where the vegetative surface cover precludes economically contouring the ground so that the contours meet National Map Accuracy Standards. Also called an indefinite contour. *See also* contour, National Map Accuracy Standards.

A

Aqua [SATELLITE IMAGING] A satellite, launched in 2002, in orbit around the earth to study the precipitation, evaporation, and cycling of water. It carries the Moderate Resolution Imaging Spectroradiometer (MODIS) sensor on board. *See also* MODIS.

arbitrary grid cell location system [CARTOGRAPHY] A grid cell location system composed of numbered columns and lettered rows, or conversely, lettered columns and numbered rows. *See also* grid, grid cell location system, row, column.

arbitrary symbol [SYMBOLOGY] A symbol that has no visual similarity to the feature it represents—for example, a circle used to represent a city or a triangle used to represent a school. *See also* mimetic symbol.

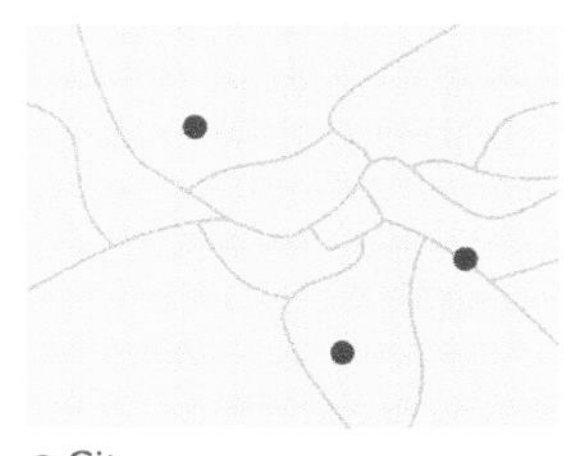

arbitrary symbol

arc

1. [DATA STRUCTURES] On a map, a shape defined by a connected series of unique x,y coordinate pairs. An arc may be straight or curved.

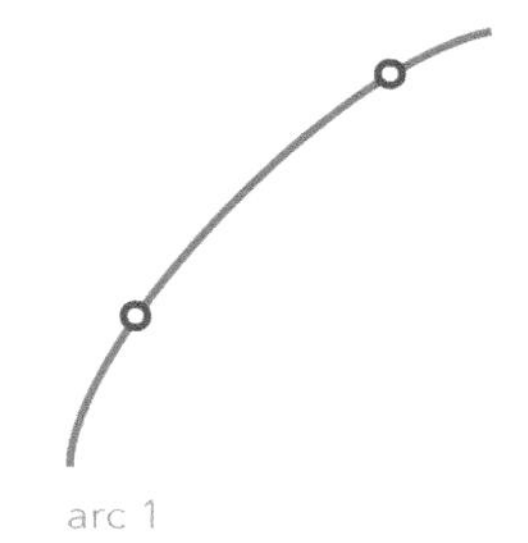

arc 1

2. [DATA STRUCTURES] A coverage feature class that represents lines and polygon boundaries. One line feature can contain many arcs. Arcs are topologically linked to nodes and to polygons.

See also line, node, path, x,y coordinates.

arc degree *See* degree. *See also* arcsecond, minute, second.

ArcGIS account [ESRI SOFTWARE] A user profile authorized to access Esri products, tools, SDKs, and services, typically through an organization and group. Each ArcGIS account is assigned unique credentials. The user type and role establish the available capabilities. *See also* authorization, credits, group, members, organization, share.

arcminute *See* minute. *See also* arcsecond, degree, degrees-minutes-seconds, second.

arc-node topology [DATA STRUCTURES] The data structure in a coverage used to represent linear features and polygon boundaries and to support analysis functions, such as network tracing. Nodes represent the beginning and ending vertices of each arc. Arcs that share a node are connected, and polygons are defined by a series of connected arcs. *See also* arc, node, polygon, topology.

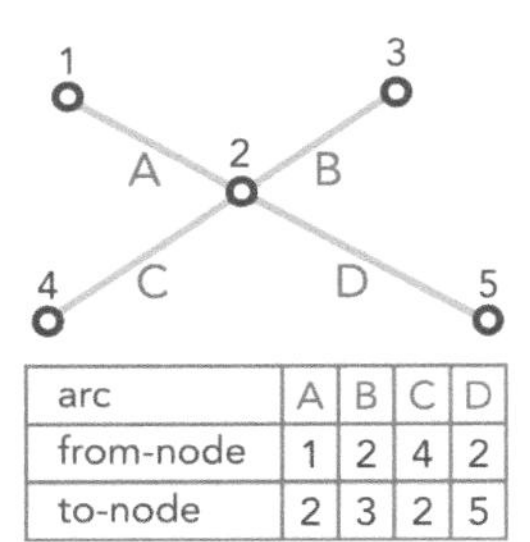

arc	A	B	C	D
from-node	1	2	4	2
to-node	2	3	2	5

arc-node topology

arcsecond

1. [ASTRONOMY] A unit of angular measurement that is 1/3600 of a degree. There are 60 arcminutes in a degree and 60 arcseconds in an arcminute. Arcseconds of latitude remain nearly constant, whereas arcseconds of longitude decrease mathematically as one moves toward the earth's poles.
2. [MATHEMATICS] The distance traversed on the earth's surface when traveling 1/3600th of a degree of latitude or longitude.

See also degree, degrees-minutes-seconds, minute, second.

arctan *See* arctangent.

arctangent [CARTOGRAPHY] The slope angle (an angle that has a tangent equal to a given number, calculated as tan–1). Sometimes shortened to arctan.

are [STANDARDS] A metric areal unit of measure equal to 100 square meters. One are is equal to 1,076.39 square feet, or 0.025 acres. *See also* areal, hectare, measurement, unit of measure.

area

1. [EUCLIDEAN GEOMETRY] A closed, two-dimensional shape defined by its boundary or by a contiguous set of raster cells.
2. [EUCLIDEAN GEOMETRY] A calculation of the size of a two-dimensional feature, measured in square units.
3. [CARTOGRAPHY] A polygon used to represent an area feature on a map

See also polygon.

area cartogram *See* cartogram.

area chart [GRAPHICS (MAP DISPLAY)] A chart that emphasizes the difference between two or more groups of data; for example, the changes in a population from one year to the next. The area of interest is usually shaded in a solid color.

area correspondence [CARTOGRAPHY] The similarity between a feature and a standard two-dimensional shape.

area count [CARTOGRAPHY]

1. A tally of the total area occupied by a certain type of feature within a data collection unit.
2. A measure of the abundance of area features in a data collection unit.

See also area feature, data collection unit.

area feature [CARTOGRAPHY] A two-dimensional feature, object, or region. Also called a polygon. *See also* feature, polygon.

area of interest *See* AOI.

area pattern [CARTOGRAPHY] A set of shapes or lines that create a pattern within an area.

area proportion [CARTOGRAPHY] A value obtained by dividing an area count by the data collection unit area. *See also* area count, data collection unit.

areal [GEOMORPHOLOGY] Pertains to surface or area. *See also* aerial, modifiable areal unit problem.

areal scale *See* scale.

arête [GEOGRAPHY] A sharp narrow ridge found in rugged mountains.

Argos [GPS] A global satellite constellation that collects, processes, and shares environmental data from fixed and mobile platforms. Argos was developed as a joint project by the French space agency, Centre National d'Études Spatiales (CNES), and two United States agencies: the National Aeronautics and Space Administration (NASA) and the National Oceanic and Atmospheric Administration (NOAA). *See also* DORIS, satellite constellation, SPOT.

argument

1. [COMPUTING] In computing, a value or expression passed to a function, command, or program.
2. [MATHEMATICS] In mathematics, an independent variable of a function.

See also function, variable.

arithmetic expression [MATHEMATICS] A number, variable, function, or combination of these, with operators or parentheses, or both, that can be evaluated to produce a single number. *See also* expression.

arithmetic function [SPATIAL ANALYSIS] A type of mathematical function that performs a calculation on the values of cells in an input raster.

arithmetic operator *See* operator.

arpent section *See* long lot.

array [GPS] A set of objects that are connected to function as a unit. In GPS technology, an array of satellites is used to pinpoint locations on the earth. *See also* matrix, variable.

artificial intelligence *See* AI.

artificial neural network *See* neural network.

ascending node [ASTRONOMY] The point at which a satellite traveling south to north crosses the equator. *See also* descending node.

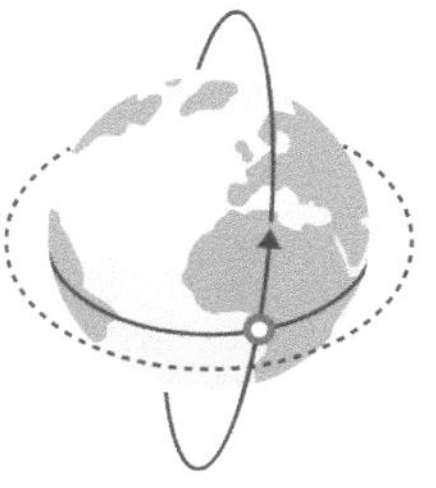

ascending node

A

ASCII *See* American Standard Code for Information Interchange.

aspatial data *See* nonspatial data.

aspatial query *See* attribute query.

aspect

1. [ANALYSIS/GEOPROCESSING] The compass direction that a topographic slope faces, usually measured in degrees from north. Aspect can be generated from continuous elevation surfaces. For example, the aspect recorded for a TIN face is the steepest downslope direction of the face, and the aspect of a cell in a raster is the steepest downslope direction of a plane defined by the cell and its eight surrounding neighbors.

   Can also be stated as the downslope direction of the maximum vertical change in the surface determined over a given horizontal distance.

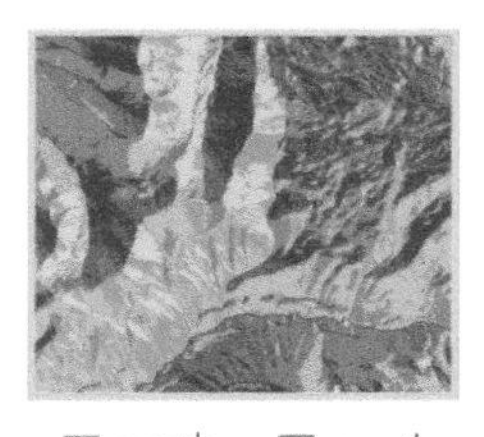

aspect 1

2. [MAP PROJECTIONS] The conceptual center of a projection system.
3. [CARTOGRAPHY] The location of the point or lines of tangency on a developable surface.

See also developable surface, slope.

aspect map [CARTOGRAPHY] A map constructed from contours or digital elevation model data to show specific aspect categories. *See also* aspect, digital elevation model.

aspect ratio

1. [VISUALIZATION] The width-to-height ratio of any quadrilateral.
2. [HARDWARE] The ratio of the width of an image to its height. The aspect ratio of a standard computer monitor is 4:3 (rectangular).

aspect-slope map [CARTOGRAPHY] A map on which aspect (direction) and slope (steepness) of an area's elevation are shown together.

ASPRS [CARTOGRAPHY] Acronym for *American Society of Photogrammetry and Remote Sensing.* A United States organization that, in 1990, developed a spatial-accuracy standard for large-scale topographic maps, engineering plans, and other detailed maps of the ground as an alternative to the National Map Accuracy Standards. *See also* National Map Accuracy Standards.

assumed bearing [SURVEYING] A bearing measured from an arbitrarily chosen reference line called an assumed meridian. *See also* bearing.

ASTER [REMOTE SENSING] Acronym for *Advanced Spaceborne Thermal Emission and Reflection Radiometer.* A Japanese sensor on board the Terra satellite.

astrolabe [ASTRONOMY] An instrument that measures the vertical angle between a celestial body and the horizontal plane at an observer's position. The astrolabe was replaced by the sextant in the fifteenth century for navigation, but modern versions are still used to determine local time and latitude. *See also* sextant.

astrolabe

asynchronous [DATA EDITING] Not synchronous; that is, not occurring together or at the same time. *See also* data transfer, synchronous.

asynchronous request [COMPUTING] A request from a client application that does not require a response from the server for the client application to continue its process.

atlas [CARTOGRAPHY] A collection of maps usually related to a particular area or theme and presented together. Examples of atlases include world atlases, historical atlases, and biodiversity atlases. *See also* chart, map.

atlas grid *See* alphanumeric grid.

atmospheric window [REMOTE SENSING] Parts of the electromagnetic spectrum that can be transmitted through the atmosphere with relatively little interference. *See also* backscatter, electromagnetic spectrum, remote sensing.

A

atomic clock [PHYSICS] A clock that keeps time by the radiation frequency associated with a particular atomic reaction. Atomic clocks are used in official timekeeping. *See also* chronometer, coordinated universal time.

ATR [NAVIGATION] Acronym for *Automatic Traffic Recorder*. *See also* ATR station.

ATR station [NAVIGATION] ATR is an acronym for *Automatic Traffic Recorder*. A traffic volume counter station permanently installed in a road network to record the number of vehicles that pass its location; it transmits the data with telemetry to computers at agency headquarters for immediate data processing and to make traffic flow maps. *See also* telemetry, traffic flow map.

attenuation [REMOTE SENSING] The dimming and blurring effects in remotely sensed images caused by the absorption and scattering of light or other radiation that passes through the earth's atmosphere. *See also* band-pass filter, remote sensing.

attractiveness [BUSINESS] A measure of the combined attributes of a center or site that are considered positive features or that draw in potential customers or tenants. *See also* gravity model.

A

attribute [DATA MODELS] Values that provide information about a feature, raster cell, or record in a GIS. For example, the attributes of a building feature might include its name, city, and label font when shown on a web map. *See also* feature, item, modifier, record.

attribute accuracy [ACCURACY] The degree to which the description of geographic feature characteristics can be demonstrated to conform to true or accepted values.

attribute change map [CARTOGRAPHY] A map that shows the change in an attribute over time by superimposing data for several dates on a single map or by using a series of choropleth, prism, or other maps, all of which are similar. *See also* choropleth map, stepped-surface map.

attribute data [DATA MODELS] Tabular or textual data describing the geographic characteristics of features. *See also* nonspatial data.

attribute domain [DATA STRUCTURES] In a geodatabase, a mechanism for enforcing data integrity. Attribute domains define what values are allowed in a field in a feature class or nonspatial attribute table. If the features or non-spatial objects have been grouped into subtypes, different attribute domains can be assigned to each of the subtypes. *See also* coded value domain, domain, range domain.

attribute error [CARTOGRAPHY] A type of error caused by misreporting the characteristics of a feature.

attribute precision [CARTOGRAPHY] The amount of detail used to report the characteristics of a feature. *See also* precision.

attribute query [DATA ANALYSIS] A request for records of features in a table based on their attribute values. *See also* attribute.

attribute table [DATA STRUCTURES] A database or tabular file containing information about a set of geographic features, usually arranged so that each row represents a feature, and each column represents one feature attribute. In raster datasets, each row of an attribute table corresponds to a certain zone of cells having the same value. In a GIS, attribute tables are often joined or related to spatial data layers, and the attribute values they contain can be used to find, query, and symbolize features or raster cells. *See also* text attribute table.

FID	Species	Color
1	oak	brown
2	pine	green
3	fir	green

attribute table

attribute transformation [COORDINATE SYSTEMS, SPATIAL ANALYSIS] The process of manipulating the nonspatial characteristics associated with spatial features (such as population, temperature, and other descriptive data) to provide an analytical framework. Common attribute transformations include aggregation, filtering, joining, calculation, and classification. *See also* aggregation, classification, feature, filter, joining, thematic map.

attribution [DATA EDITING] The process of assigning attributes to features. *See also* attribute.

authalic [MAP PROJECTIONS] A map projection that preserves the area of all regions on the map. The authalic projection is commonly used to create maps that maintain equal area properties. *See also* authalic sphere, equal-area projection.

authalic projection *See* authalic.

authalic sphere [CARTOGRAPHY] A sphere with the same surface area as a specified reference ellipsoid.

authentication [COMPUTING] The process of validating user identity. *See also* API key, authorization.

authorization [COMPUTING] Privileges granted related to Esri software.

autocorrelation [STATISTICS] The correlation or similarity of values, generally values that are nearby in a dataset. Temporal data is said to exhibit serial autocorrelation when values measured close together in time are more similar than values measured far apart in time. Spatial data is said to exhibit spatial autocorrelation when values measured nearby in space are more similar than values measured farther away from each other. *See also* correlation, spatial autocorrelation, Tobler's First Law of Geography.

automated cartography [GRAPHICS (MAP DISPLAY)] The process of making maps using computer systems that carry out many of the tasks associated with map production. *See also* GIS.

automated feature extraction [DATA CAPTURE, ESRI SOFTWARE] The identification of geographic features and their outlines in remote sensing imagery through postprocessing technology that enhances feature definition, often by increasing feature-to-background contrast or using pattern recognition software.

automated mapping/facilities management *See* AM/FM.

automated text placement [DATA EDITING] An operation in which text is automatically placed on or next to features on a digital map by a software application according to rules set by the software user.

automatic identification system [NAVIGATION] Also known by the acronym *AIS.* As a supplemental system to marine radar, AIS uses shipboard transceivers to provide the speed, identification, position, and course of nearby marine vessels to vessel traffic services (VTS).

A

automation [COMPUTING] The automatic functioning of a machine, system, or process, without the need for human interaction.

automation scale [DATA CAPTURE] The scale at which nondigital data is made digital; for example, a map digitized at a scale of 1:24,000 has an automation scale of 1:24,000. The data can be rendered at different display scales. *See also* display scale.

average diversity index [CARTOGRAPHY] The average diversity for all cells on a map. *See also* diversity.

average fragmentation index [CARTOGRAPHY] A measure used to assess the degree of feature spatial dispersion within a geographic area.

average point spacing
1. [3D GIS] The average distance separating sample points in a point dataset. A terrain dataset uses the average point spacing of a dataset to define a horizontal tiling system for dividing input source measurements.
2. [CARTOGRAPHY] The average distance between point features or centroids and their neighbors.

See also centroid, terrain tiles.

average spatial shift [CARTOGRAPHY] The difference in location for all features between two time periods, calculated by measuring the length of each shift line, adding their lengths together, and dividing the sum of their lengths by the number of features. *See also* shift line.

AVHRR *See* Advanced Very High Resolution Radiometer.

AVIRIS [REMOTE SENSING] Acronym for *Airborne Visible/Infrared Imaging Spectrometer*. An airborne hyperspectral sensor.

AVL [NAVIGATION] Acronym for *Automated Vehicle Location*. A means to automatically determine and transmit the geographic location of a vehicle.

axis
1. [COORDINATE SYSTEMS] A line along which measurements are made to determine the coordinates of a location.

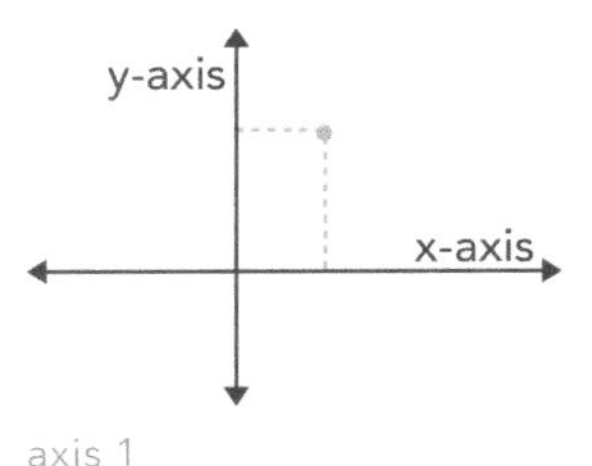

axis 1

2. [COORDINATE SYSTEMS] In a spherical coordinate system, the line that directions are related to and from which angles are measured.
3. [ASTRONOMY] In astronomy, the imaginary line through the poles about which a rotating body turns.

See also dimension, height, radius.

azimuth

1. [CARTOGRAPHY] A direction in degrees clockwise from north from a given point.

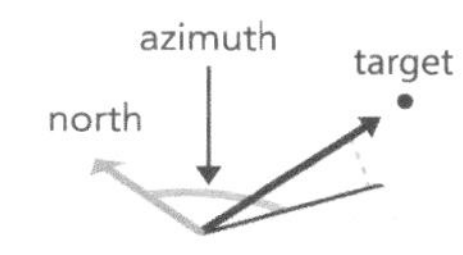

azimuth 1

2. [ANALYSIS/GEOPROCESSING] The direction from which a light source illuminates a surface. Azimuth is measured from north in clockwise degrees, from 0 to 359.9.
3. [NAVIGATION] In navigation, the horizontal angle, measured in degrees, between a reference line drawn from a point and another line drawn from the same point to a point on the celestial sphere. Normally, the reference line points true north, and the angle is measured clockwise from the reference line.
4. [SATELLITE IMAGING] In satellite image collection, the direction the sensor is pointing, which varies from 0° to 360°, starting with north at 0°.

See also nadir.

azimuthal equidistant projection

[MAP PROJECTIONS] A planar map projection with straight-line meridians (radiating outward from the pole) and equally spaced parallels; this arrangement of parallels and meridians results in all straight lines drawn from the point of tangency being great-circle routes. *See also* great circle, planar projection.

azimuthal projection [MAP PROJECTIONS]

1. A map projection that transforms points from a spheroid or sphere onto a tangent or secant plane. The azimuthal projection is also known as a zenithal projection.
2. A map projection that preserves global directions. Sometimes called a true-direction projection.

See also secant projection, planar projection.

B

Babylonian sexagesimal system [ONTOLOGIES] A system for describing degrees (°), minutes ('), and seconds ("), in which there are 60 minutes in a degree and 60 seconds in a minute. Commonly referred to as degrees, minutes, seconds (DMS). *See also* degrees-minutes-seconds.

back azimuth [CARTOGRAPHY] The direction line drawn on a map from the known position of a distant feature back to the viewer's position; a line from the viewer's current position backward along a route; the opposite direction from a given azimuth calculated by adding 180 degrees to azimuths less than 180 degrees or subtracting 180 degrees from azimuths greater than 180 degrees. Also known as a backsight or a direction line. *See also* azimuth.

back bearing [CARTOGRAPHY] The opposite direction of a bearing. A back bearing reverses the letters that indicate direction (N, S, E, W for north, south, east, west, respectively)—for example, N becomes S and E becomes W. *See also* bearing.

background [OUTPUT, ESRI SOFTWARE] The backdrop of the view. The color of the background can be set to suggest sky, empty space, or any color that improves visualization. *See also* view, visualization.

background image [MAP DISPLAY] A satellite image, aerial photograph, or scanned map over which vector data is displayed. Although a background image can be used to align coordinates, it is not linked to attribute information and is not part of the spatial analysis in a GIS.

backscatter [REMOTE SENSING] Electromagnetic energy that is reflected back toward its source by terrain or particles in the atmosphere. *See also* atmospheric window, electromagnetic radiation, remote sensing.

backsight *See* back azimuth.

backslope [CARTOGRAPHY] A hillslope element that is the linear portion of the hillslope below the shoulder descending to the concave footslope portion of the hillslope. *See also* hillslope element.

BAM *See* best available map.

band [REMOTE SENSING] A wavelength range in the spectrum of reflected or radiated electromagnetic (EM) energy to which a remote sensor is sensitive. For example, red, green, and blue are the bands used by a digital camera; the red band collects light from approximately 625–740 nanometers. Once a sensor collects data from a band, it stores the data in a file or portion of a file devoted to that range, which is also referred to as a band. Also commonly known as a spectral band. *See also* electromagnetic spectrum, raster dataset band, sensor.

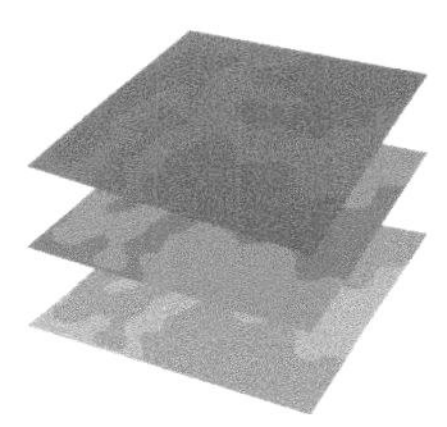

band

band ratio [DIGITAL IMAGE PROCESSING] A digital image-processing technique that enhances the spectral difference between electromagnetic bands by dividing the measure of reflectance for a pixel in one image band by its geometrically corresponding pixel from another image band. The resulting image is also called the band ratio. Commonly used to mitigate illumination effects in terrain variation or to enhance phenomena in objects of interest. *See also* band, electromagnetic spectrum, reflectance.

band separate [REMOTE SENSING] An image format that stores each band of data in a separate file. *See also* band.

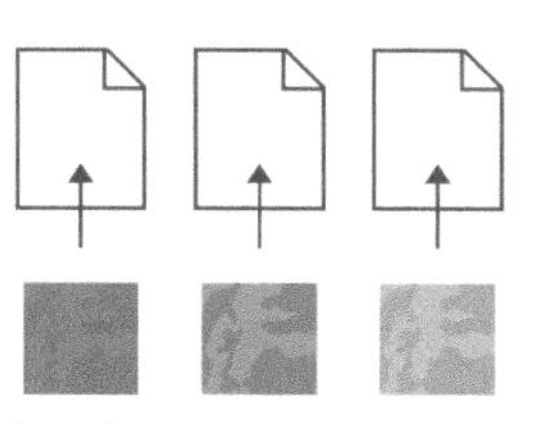

band separate

band-pass filter [REMOTE SENSING] A wave filter that allows signals in a certain frequency to pass through, while blocking or attenuating signals at other frequencies. *See also* high-pass filter, low-pass filter.

bar scale *See* scale bar.

barrier

1. [NETWORK ANALYSIS] An entity that prevents flow from traversing a network edge or junction.
2. [SPATIAL ANALYSIS] A line feature used to keep certain points from being used in the calculation of new values when a raster is interpolated. The line can represent a cliff, ridge, or some other interruption in the landscape. Only the sample points on the same side of the barrier as the current processing cell will be considered.

See also flag, impedance.

base data [DATA ANALYSIS] Map data over which other, thematic information is placed.

base height [AERIAL PHOTOGRAPHY] In aerial photography, the height or altitude from which a photograph is taken. *See also* aerial photograph.

B

base height ratio [AERIAL PHOTOGRAPHY] In aerial photography, the distance on the ground between the centers of overlapping photos, divided by sensor altitude. In a stereomodel, base height ratio is used to determine vertical exaggeration. *See also* base height, stereomodel.

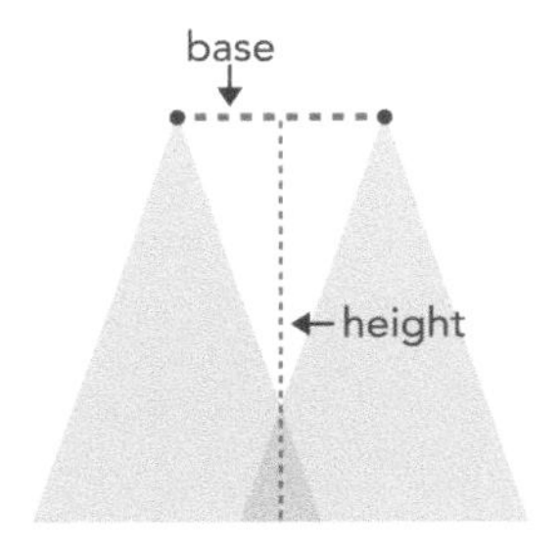

base height ratio

base layer [DATA ANALYSIS] A data layer in a GIS to which all other layers are geometrically referenced. *See also* basemap, layer.

base plate [CARTOGRAPHY] The part of an orienteering compass that is marked with a ruler and sometimes map scales. *See also* orienteering compass.

base station [GPS]

1. A GPS/GNSS receiver at a known location that broadcasts and collects correction information for roving GPS/GNSS receivers.
2. A GPS receiver at a stationary position on a precisely known point, typically a surveyed benchmark.

See also benchmark, CORS, differential correction, GNSS, GPS.

base symbol [ESRI SOFTWARE] The default symbol used to represent an event or a feature on a map. *See also* symbol.

baseline

1. [SURVEYING] An accurately surveyed line from which other lines or the angles between them are measured.
2. [SURVEYING] The parallel determined by government land surveyors that intersects the principal meridian to establish an initial point in the U.S. Public Land Survey System and Canada's Dominion Land Survey. Also called a geographer's line.

baseline 2

3. [GPS] In GPS, the physical distance between a base station and a rover.

See also base station, DLS, PLSS, reference line.

basemap

1. [VISUALIZATION] A foundational layer on a map that is the basis of GIS visual and geographic context. A basemap may include background reference information such as landforms, roads, landmarks, and administrative boundaries. It displays below any other layers and is typically excluded from the legend.

A basemap is used for locational or spatial data reference.

2. [DATA ANALYSIS] A map to which GIS data layers are registered and rescaled.
3. [MAP DESIGN] In imagery, a collection of geometrically correct imagery used as the backdrop for GIS layers.

See also cartography, map.

batch geocoding [GEOCODING] The process of geocoding many address records at the same time.

batch processing [ESRI SOFTWARE] A method for processing data automatically in which the data is grouped into batches and processed by the computer at one time, without user interaction.

batch vectorization [DATA CONVERSION] An automated process that converts raster data into vector features for an entire raster or a portion of it based on user-defined settings. *See also* vectorization.

bathymetric curve *See* depth contour.

bathymetric map [CARTOGRAPHY] A map representing the depths and contours of water bodies, such as a seafloor, river bottom, or lake bed. *See also* depth contour, hydrography, topography.

bathymetric map

bathymetry [CARTOGRAPHY] The measurement of underwater depths and landforms, such as a lake bed, river bottom, or seafloor. *See also* depth contour, hydrography, submerged contour.

battle dimension [DEFENSE, ESRI SOFTWARE] The primary area in which a military unit operates, such as air, space, ground, sea, surface, and subsurface. *See also* modifier.

battleships grid *See* alphanumeric grid.

Bayes' theorem [STATISTICS] A theorem that describes the conditional probability of a given event based on factors that are relevant to the event. Developed by English mathematician Thomas Bayes (1702–1761), Bayes' theorem is used in probability theory and statistics. *See also* probability.

Bayesian statistics [STATISTICS] A statistical approach to measuring likelihood based on the synthesis of a prior distribution and current sample data. Classical approaches to statistics estimate the probability of an event by averaging all possible data. The Bayesian approach, in contrast, weights probability according to actual data from a particular situation. It also factors in data from sources outside the statistical investigation, such as past experience, expert opinion, or prior belief. This outside information is described by a distribution that includes all possible values for the parameter. *See also* Bayes' theorem.

bearing

1. [SURVEYING] The horizontal direction of a point in relation to another point, expressed as an angle from a known direction, usually north, and usually measured from 0 degrees at the reference direction clockwise through 360 degrees. Bearings are differentiated based on whether the meridian is true, magnetic, or assumed (relative).

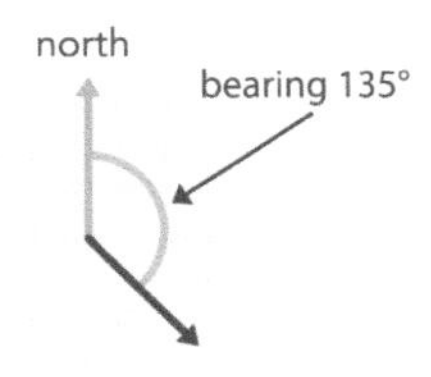

bearing 1

2. [GEOGRAPHY] In navigation, a horizontal angle given in degrees ranging from 0 to 90.

See also azimuth, heading.

bearing card [CARTOGRAPHY] The graphic face of a compass, graduated to show bearings rather than azimuths. *See also* bearing, compass.

bedrock [GEOGRAPHY] The solid rock that underlies loose surface materials, such as soil, sand, clay, or gravel.

BeiDou Navigation Satellite System [GPS] A global geopositioning satellite constellation used for positioning, navigation, and timing (PNT); operated by the Chinese government, it is part of the international GNSS constellations spread between several orbital planes. *See also* Galileo, geopositioning, GLONASS, GNSS, GPS, PNT, satellite constellation, satellite navigation.

benchmark [SURVEYING] A permanent survey monument or brass plate that establishes the exact elevation of a structure above or below an adopted vertical geodetic datum, typically using precise leveling methods. *See also* surveying, survey monument.

best available map [CARTOGRAPHY] Also known by the acronym *BAM*. The most suitable data source for a map. *See also* data source, map.

best route [ESRI SOFTWARE] The route of least impedance between two or more locations, considering connectivity and travel restrictions such as one-way streets and rush-hour traffic.

Bézier curve [EUCLIDEAN GEOMETRY] A curved line with a shape that is derived mathematically rather than by a series of connected vertices. In graphics software, a Bézier curve usually has two endpoints and two handles that can be moved to change the direction and the steepness of the curve. Named for the French engineer Pierre Bézier (1910–1999). *See also* parametric curve, segment.

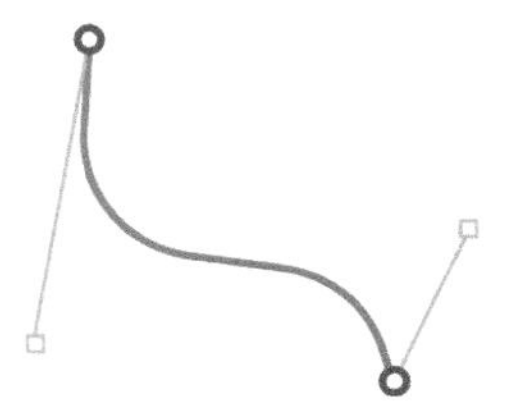

Bézier curve

Bhattacharyya distance [DIGITAL IMAGE PROCESSING] A measure of the theoretical distance between two normal distributions of spectral classes, which acts as an upper limit on the probability of error in a Bayesian estimate of correct classification. Named for the Indian mathematician Anil Kumar Bhattacharyya (1915–1996). *See also* Bayesian statistics, probability.

bifurcation ratio [CARTOGRAPHY] The ratio of links or stream segments of one order to those of the next higher order. *See also* drainage, watershed.

bilinear interpolation [VISUALIZATION] A resampling method that uses a weighted average of the four nearest cells to determine a new cell value. *See also* interpolation, resampling.

billboarding [GRAPHICS (MAP DISPLAY)] A method for displaying graphics associated with features in a three-dimensional map display by posting them vertically as two-dimensional symbols and orienting them to always face the user. *See also* symbology.

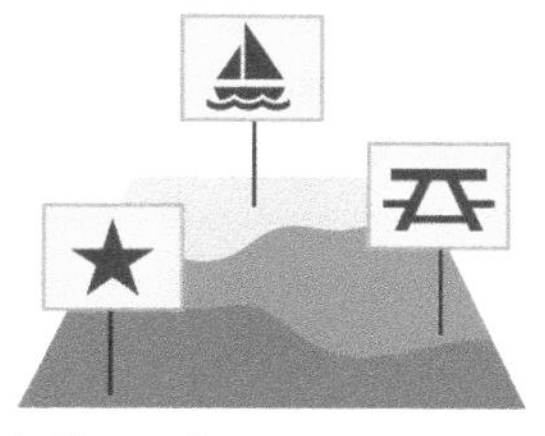

billboarding

BIM *See* building information model.

bin

1. [DATA ANALYSIS] In a variogram map, each cell that groups lags that share similar distance and direction. Bins are commonly formed by dividing the sample area into a grid of cells or sectors and are used to calculate the empirical semivariogram for kriging.
2. [OUTPUT] In a histogram, user-defined size classes for a variable.

See also histogram, kriging, lag, semivariogram.

binary [COMPUTING] In computer programming, having only two states (or values), such as yes or no, on or off, true or false, or 0 or 1. *See also* Boolean expression, expression, logical expression.

binary large object [DATABASE STRUCTURES] Also known by the acronym *BLOB.* A large block of data, such as an image, a sound file, or geometry, stored in a database. The database cannot read the BLOB's structure and only references it by its size and location. *See also* data, data type.

B

bingo grid *See* alphanumeric grid.

binomial distribution [STATISTICS] A distribution describing the probability of obtaining exactly K successes in N independent trials, where each trial results in either a success or a failure. *See also* probability.

biogeography [BIOLOGY] The study of the geographical distribution of living things. *See also* distribution, geography.

biomass [ENVIRONMENTAL GIS] The total amount of organic matter in a defined area; usually refers to vegetation. *See also* density.

bispectral plot [GEOREFERENCING] In imagery, a graph where the pixels from two electromagnetic bands are plotted, one band defined on the x-axis and the other on the y-axis. This method is used to evaluate the correlation and variance between two spectral bands and to analyze the separability and distribution of classes and features on the ground. *See also* band, chart.

bit depth [DATA STRUCTURES] The range of values that a particular raster format can store, based on the formula 2n. An 8-bit depth dataset can store 256 unique values. *See also* digital image, raster.

bitmap [DATA STRUCTURES] An image format in which one or more bits represent each pixel on the screen. The number of bits per pixel determines the shades of gray or number of colors that a bitmap can represent. *See also* pixel, PNG.

bivariate map [MAP DESIGN] A map that displays two variables on a single map by combining two different sets of symbols or colors. *See also* choropleth map.

bivariate mean [CARTOGRAPHY] The average of the horizontal and vertical coordinates for *n* features.

BLM [FEDERAL GOVERNMENT] Acronym for *Bureau of Land Management.* A United States federal government agency within the U.S. Department of the Interior that administers America's public lands. Prior to 1946, known as the U.S. General Land Office (GLO).

BLOB *See* binary large object.

block [ESRI SOFTWARE] A group of records in a compressed file geodatabase feature class or table that are stored together. The arrangement of compressed data into blocks helps optimize query performance. *See also* record.

block adjustment [DIGITAL IMAGE PROCESSING] A computational process that uses sensor models to adjust and determine optimal exterior orientation data for a collection of overlapping, remotely sensed images.

block diagram [CARTOGRAPHY] A map that portrays a portion of terrain as if it was cut out of the surface of the earth. The vertical sides of the block allow the subsurface geologic information to be shown.

block group [FEDERAL GOVERNMENT] A unit of U.S. census geography that is a combination of census blocks. A block group is the smallest unit for which the U.S. Census Bureau reports a full range of demographic statistics. There are approximately 700 residents per block group. A block group is a subdivision of a census tract. *See also* block, census tract, demographics.

block kriging [SPATIAL STATISTICS (USE FOR GEOSTATISTICS)] A kriging method in which the average expected value in an area around an unsampled point is generated rather than the estimated exact value of an unsampled point. Block kriging is commonly used to provide better variance estimates and smooth interpolated results. *See also* kriging.

blocking [ESRI SOFTWARE] A geocoding indexing process that reduces the number of potential matches that need to be checked. *See also* geocoding, index.

blunder [SURVEYING] In surveying, a defective measurement that can be detected by a statistical test. *See also* error, surveying.

bolometer [PHOTOGRAMMETRY] In imagery, a device that measures incident thermal electromagnetic radiation.

Boolean expression [MATHEMATICS] An expression that results in a true or false (logical) condition. For example, in the Boolean expression "height > 70 AND diameter = 100," all locations where the height is greater than 70 and the diameter is equal to 100 would be given a value of 1, or true, and all locations where this criterion is not met would be given a value of 0, or false. Named for the English mathematician George Boole (1815–1864). *See also* binary, expression, logical expression.

Boolean operation [DATA MANAGEMENT] A GIS operation that uses Boolean operators to combine input datasets into a single output dataset. *See also* Boolean operator.

Boolean operator [MATHEMATICS] A logical operator used in the formulation of a Boolean expression. Common Boolean operators include AND, which specifies a combination of conditions (A and B must be true); OR, which specifies a list of alternative conditions (A or B must be true); NOT, which negates a condition (A but not B must be true); and XOR (exclusive or), which makes conditions mutually exclusive (A or B may be true but not both A and B). *See also* operator, Boolean expression.

border arcs [DATA MODELS] The arcs that create the boundary line of a polygon coverage. *See also* arc, boundary, polygon.

B

boundary

1. [CARTOGRAPHY] A dividing line.
2. [SURVEYING] A line separating adjacent political entities, such as countries or districts; adjacent tracts of privately owned land, such as parcels; or adjacent geographic zones, such as ecosystems. A boundary is a line that may or may not follow physical features, such as rivers, mountains, or walls.

See also boundary line.

boundary effect [DATA QUALITY] A problem caused by arbitrary boundaries being imposed on spatial data that represents unbounded spatial phenomena. These can include edge effects, where patterns of interaction or interdependency across the borders of a bounded region are ignored or distorted, and shape effects, where the shape imposed on a bounded area affects perceived interactions between phenomena. *See also* modifiable areal unit problem, spatial data.

boundary line [CARTOGRAPHY] A division between adjacent political entities, tracts of private land, or geographic zones. Boundary lines may be imaginary lines, physical features that follow those lines, or the graphical representation of those lines on a map. Boundary lines between privately owned land parcels are usually called property lines. *See also* boundary.

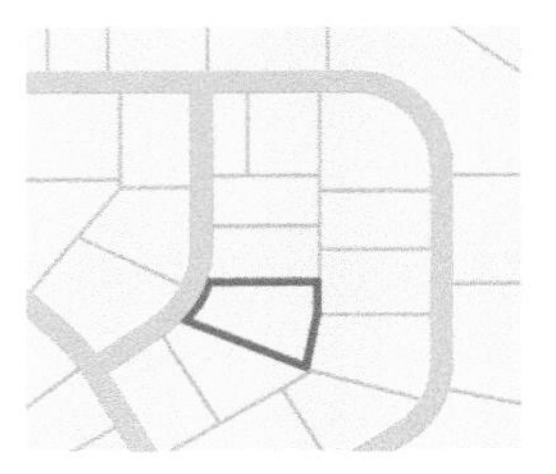

boundary line

boundary monument [SURVEYING] An object that marks an accurately surveyed position on or near a boundary. *See also* surveying.

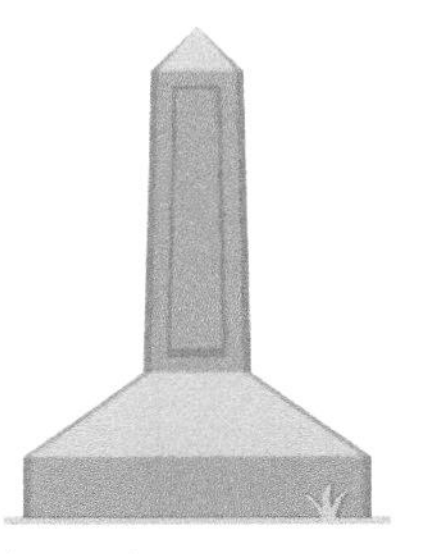

boundary monument

boundary survey [CADASTRAL AND LAND RECORDS]

1. A map that shows property lines and corner monuments of a parcel of land.
2. The survey taken to gather data for a map that shows property lines and corner monuments of a parcel of land.

See also cadastral survey, surveying.

bounded continuous color [MAP DESIGN] An approach to selecting class colors in which all values above or below user-selected bounds are represented by the same colors.

bounded continuous size [MAP DESIGN] An approach to selecting circle sizes in which all values above or below size bounds are represented by circles of the same size.

bounding rectangle [MAP DISPLAY] A rectangle—aligned with the coordinate axes and placed on a map display—that encompasses a geographic feature, group of features, or an area of interest. It is defined by minimum and maximum coordinates in the x and y directions and is used to represent, in a general way, the location of a geographic area. *See also* envelope.

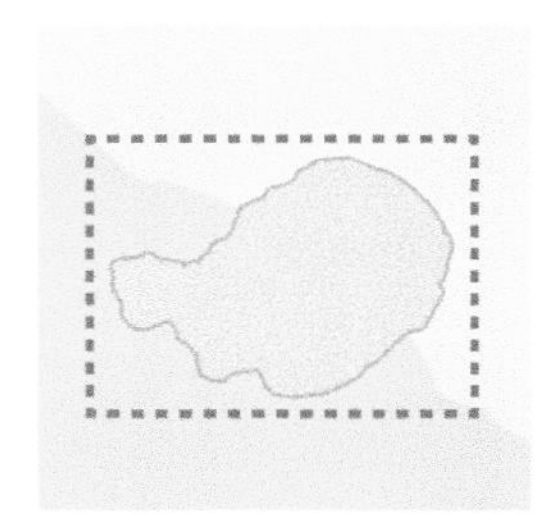

bounding rectangle

Bowditch rule *See* compass rule.

branching network [NETWORK ANALYSIS] A network that has only one possible path between pairs of places.

break [NETWORK ANALYSIS] An object used in vehicle routing problem (VRP) analysis. A break can be used to model a specified period of rest along a route within a VRP instance. *See also* network analysis, vehicle routing problem.

breakline [DATA STRUCTURES] A line in a TIN that represents a distinct interruption in the slope of a surface, such as a ridge, road, or stream. Breaklines function as edge constraints: No triangle in a TIN may cross a breakline. Z-values along a breakline can be constant or variable. *See also* slope, structure line, surface constraint, TIN.

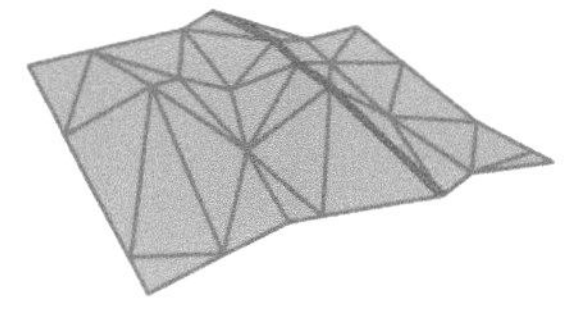

breakline

brightness difference [PHOTOGRAMMETRY] The difference in brightness between a foreground color and its background. *See also* brightness equation.

brightness equation [PHOTOGRAMMETRY] An equation used to compute the brightness difference. *See also* brightness difference.

brightness theme [DATA MODELS, SPATIAL ANALYSIS] A grid theme with cell values that are used to vary the brightness of another grid theme. The cell values in one grid can be visually plotted against those in another. Most commonly, hillshade grids are used as brightness themes for elevation grids to display the elevation surface in relief. *See also* data model, spatial analysis, theme.

B

browse graphic [VISUALIZATION] An image associated with data to provide a visual summary of the service. *See also* graphic, image.

buffer [SPATIAL ANALYSIS]

1. An area of specified distance or time around one or more features. For example, to protect an endangered species, construction activity is not allowed within a 500-foot buffer of its habitat.

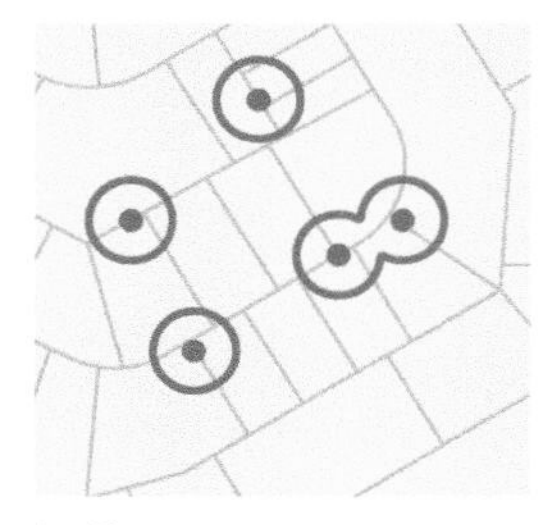

buffer 1

2. A polygon enclosing a point, line, or polygon at a specified distance.

See also proximity analysis, zone.

build parcel [GEODESY] A process in the parcel fabric that builds parcels from polygons or from lines. When parcels are built, missing parcel features (polygons, lines, and points) are created. *See also* construction line.

building information model [DEVELOPMENT] Also known by the acronym *BIM*. A virtual representation of a physical structure, typically used to simulate the functional and physical characteristics of a building throughout its life cycle. An incomplete list of industries that use BIM include construction, architecture, and infrastructure. *See also* digital twin.

building setback line [CARTOGRAPHY] The distance from a lot line beyond which buildings or improvements may not extend without permission from an authority. *See also* lot.

bump mapping [MAP DESIGN] A map visualization method used to indicate surface texture or land cover, achieved through modification of the original digital elevation model by adding elevation values around randomly scattered points that represent vegetation in the landscape. Also called texture shading. *See also* digital elevation model.

Bureau of Land Management *See* BLM.

bust *See* closure error.

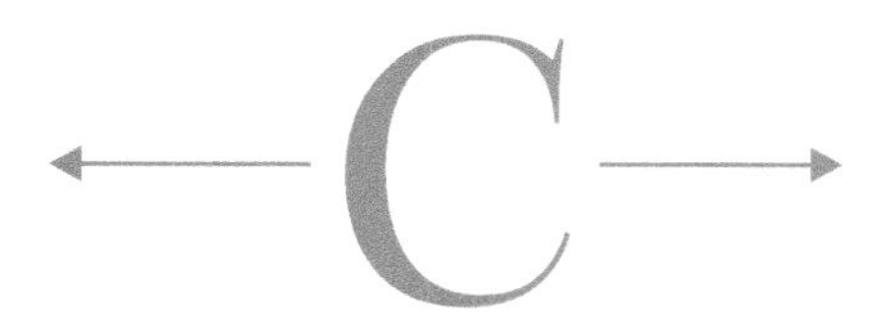

C/A code *See* civilian code.

C4I [DEFENSE] In defense, an acronym used to signify that a computer program or system supports command, control, communication, computers, and information.

CAC *See* coefficient of areal correspondence.

CAD *See* computer-aided design.

CAD dataset *See* CAD feature dataset.

CAD drawing [GRAPHICS (COMPUTING)] The digital equivalent of a drawing, figure, or schematic created using a CAD system. *See also* computer-aided design.

CAD feature dataset [ESRI SOFTWARE] The feature representation of a CAD file in a geodatabase-enforced schema. *See also* multipatch, point, polygon, polyline.

CAD file [ESRI SOFTWARE] The digital equivalent of a drawing, figure, or schematic created using a CAD system. *See also* computer-aided design.

CAD layer [ESRI SOFTWARE] A layer that references a set of CAD data. *See also* computer-aided design, layer.

CAD staging geodatabase [DATABASE STRUCTURES] A normalized, fixed set of feature classes and data tables of a predefined schema from a collection of input CAD drawings. *See also* computer-aided design.

cadastral map [CADASTRAL AND LAND RECORDS] A map containing detailed information about a property, often acquired from a cadastral survey.

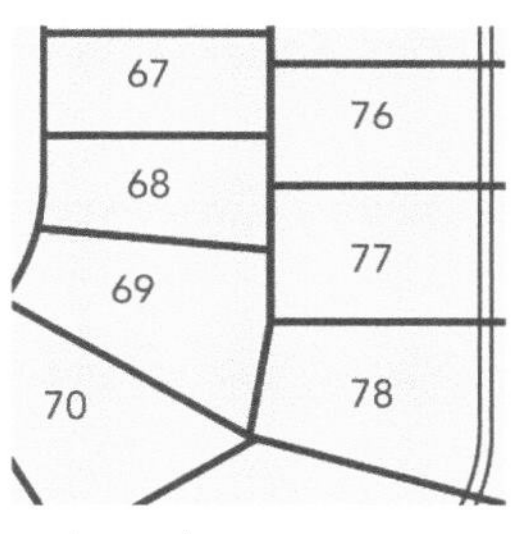

cadastral map

cadastral survey [CADASTRAL AND LAND RECORDS] A boundary survey taken for the purposes of ownership and taxation. *See also* cadastre.

cadastre [CADASTRAL AND LAND RECORDS] An official record of the dimensions and value of land parcels, used to record ownership and assist in calculating taxes. *See also* parcel.

calculated average mean [STATISTICS] A single number—typically represented as $\bar{x}$—that summarizes the center point or typical value of a dataset. Used as a measure of central tendency. The calculated average mean is the sum of the individual points in a dataset divided by the total number of individual points. *See also* standard deviation.

C

calibrated focal length [PHOTOGRAMMETRY] An adjusted value of the focal length of a camera or sensor that minimizes distortion and best approximates an ideal camera geometry and determines interior orientation in photogrammetric computations. Calculated to equalize the positive and negative values of distortion over the entire field used by the camera or sensor. *See also* ideal camera model.

calibration

1. [ACCURACY] The comparison of the accuracy of an instrument's measurements to a known standard.
2. [SPATIAL ANALYSIS] In spatial analysis, the selection of attribute values and computational parameters that will cause a model to properly represent the situation being analyzed. For example, in pathfinding and allocation, calibration generally refers to assigning or calculating impedance values.

callout line [GRAPHICS (MAP DISPLAY)] A line on a map extending between a feature's geographic position and its corresponding symbol or label, used in areas where there is not enough room to display a symbol or label in its correct location. *See also* label, symbol.

camera [MAP DISPLAY] A processing parameter that renders a point from which the observer's perspective is defined for maps and scenes. A camera's perspective has a location (x longitude, y latitude, and z elevation for scenes), a heading (an x-axis angle about which the camera is rotated, in degrees), a pitch (the angle at which the camera is rotated vertically relative to the y-axis, in degrees), and roll (the angle the camera is rotated about its center axis, in degrees). *See also* observer, scene.

camera station *See* exposure station.

camera tilt [AERIAL PHOTOGRAPHY] The movement of the camera within a vertical plane when an aerial photograph is taken. *See also* aerial photograph.

camouflage detection film *See* CIR film.

candidate [ESRI SOFTWARE] A record returned as a potential match for an address in the geocoding process.

candidate key [COMPUTING] In a relational database, any key that can be used as the primary key in a table. *See also* primary key.

capacity [ANALYSIS/GEOPROCESSING] In location-allocation, the maximum number of people or units that a center can service, contain, or have assigned to it. *See also* location-allocation.

caption [ESRI SOFTWARE] The descriptive text that accompanies an image. As part of the user interface, captions are usually editable and customizable.

carbonate bedrock [GEOGRAPHY] A class of bedrock composed primarily of carbonate minerals, usually limestone, dolomite, or marble.

cardinal direction *See* cardinal point.

cardinal point [NAVIGATION] One of the four compass directions on the earth's surface: north, south, east, or west. *See also* compass point.

cardinality [MATHEMATICS] The correspondence or equivalency between sets; how sets relate to each other. For example, if one row in a table is related to three rows in another table, the cardinality is one to many. *See also* edge-junction cardinality.

cardo maximus [CARTOGRAPHY] A north–south boundary line that began in the middle of the streets of Roman Berytus (now Beirut, Lebanon), that was used to survey centuria under the Roman Centuriation system. *See also* Centuriation system.

carriage [CARTOGRAPHY] The part of a polar planimeter that is fixed to or slides along the tracer arm and contains the measuring wheel. *See also* polar planimeter.

carrier [PHYSICS] An electromagnetic wave, such as radio, with modulations that are used as signals to transmit information. *See also* electromagnetic radiation, signal.

carrier-aided tracking [GPS] Signal processing that uses the GPS carrier signal to lock onto the PRN code generated by the satellite. *See also* PRN code.

carrier-phase GPS [GPS] GPS measurements that are calculated using the carrier signal of a satellite. *See also* code-phase GPS.

carrying contour [SYMBOLOGY] A single line representing multiple coincident contour lines, used to represent vertical or near-vertical topographic features such as cliffs, cuts, escarpments, and fills. *See also* contour line, contour interval.

Cartesian coordinate system

1. [COORDINATE SYSTEMS] A two-dimensional, planar coordinate system in which horizontal distance is measured along an x-axis and vertical distance is measured along a y-axis. Each point on the plane is defined by an x,y coordinate. Relative measures of distance, area, and direction are constant throughout the Cartesian coordinate plane. Named for the French mathematician and philosopher René Descartes (1596–1650).

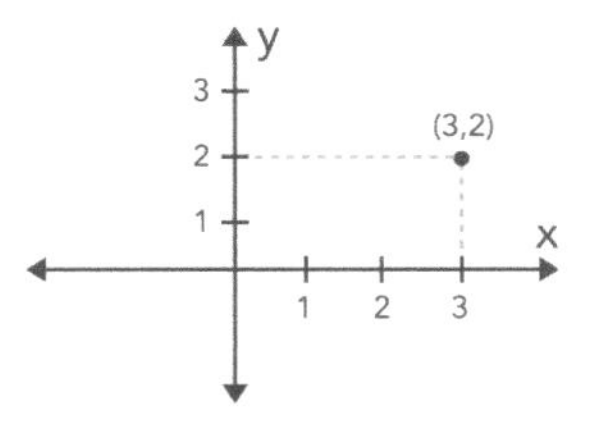

Cartesian coordinate system 1

2. [REMOTE SENSING] In imagery, a three-dimensional coordinate system where each axis is orthogonal, the scale in each direction is the same, and the reference for photogrammetry is earth stationary.

See also coordinate system.

cartogram [MAP DESIGN] A diagram or abstract map in which geographical areas, or data collection units, are distorted proportionally to the value of an attribute. *See also* data collection unit, distortion, map.

cartogram

cartographer [CARTOGRAPHY] One who practices the art and science of expressing graphically, usually through maps, the natural and cultural features of the earth. *See also* cartography, cultural feature, map, natural feature.

cartographic abstraction [GRAPHICS (MAP DISPLAY)] The process of transforming data that has been collected about the environment into a graphic representation of features and attributes that are relevant to the purpose of the map.

cartographic generalization [CARTOGRAPHY] The abstraction, reduction, and simplification of features so that a map is clear and uncluttered at a given scale. *See also* generalization.

cartographic modeling map *See* composite variable map.

cartographic selection [CARTOGRAPHY] The process of deciding which classes of features to show on the map.

cartographic symbolization [CARTOGRAPHY] The use of signs and graphic symbolism on maps.

cartography [CARTOGRAPHY] The art and science of expressing graphically, usually through maps, the natural and social features of the earth. *See also* map.

cartometrics [CARTOGRAPHY] Making measurements on maps.

cartouche [MAP DESIGN] An enclosed, ornamental area on the map that contains the title, author, general description, legend, scale bar, and so on. Cartouches are rarely used on modern maps.

cartouche

case [MAP PROJECTIONS] Used for map projections to describe whether the projection surface touches or intersects the generating globe. *See also* generating globe.

cased line symbol [CARTOGRAPHY] A symbol in which the interior line is bounded by a casing that is shown in a different color. *See also* casing.

casing [CARTOGRAPHY] The exterior edges of a cased line symbol. *See also* cased line symbol.

catchment *See* watershed.

categorical color scheme [CARTOGRAPHY] A set of colors used to show different categories of nominal data. *See also* nominal data.

categorical data *See* nominal data.

categorical map [CARTOGRAPHY] A map that has polygons enclosing areas that are assumed to be uniform or areas to which a single description can apply. *See also* polygon.

categorical raster *See* discrete raster.

category

1. [COGNITION] A distinct class within a conceptual system for grouping things with shared attributes.
2. [DATA STORAGE, ESRI SOFTWARE] A collection of related files, layers, data sources, or classes. Administrative, cultural, and elevation are examples of categories.

See also class, data.

CATRF2022 [MODELING] Acronym for *Caribbean Terrestrial Reference Frame of 2022.* A geometric reference frame of the National Spatial Reference System (NSRS) used to define the geodetic latitude, geodetic longitude, and ellipsoidal height of points on the Caribbean tectonic plate. The coordinates defined in CATRF2022 are time-dependent. *See also* NAPGD2022, National Spatial Reference System, reference frame.

CBD *See* central business district.

CBSA *See* core-based statistical area.

CCD [HARDWARE] Acronym for *charge-coupled device.* A light-sensitive integrated circuit that captures and stores image information in the form of localized electrical charges that vary with incident light intensity. CCDs are used for high-quality scientific imaging; CCD imaging chips are used as the sensitive focal planes of digital sensors.

celestial sphere [ASTRONOMY] The sky, considered as the inside of a sphere of infinitely large radius that surrounds the earth, on which all celestial bodies except the earth are imagined to be projected. *See also* horizon, nadir, zenith.

cell

1. [GRAPHICS (COMPUTING)] The smallest unit of information in raster data, usually square in shape. In a map or GIS dataset, each cell represents a portion of the earth, such as a square meter or square mile, and usually has

an attribute value associated with it, such as soil type or vegetation class.

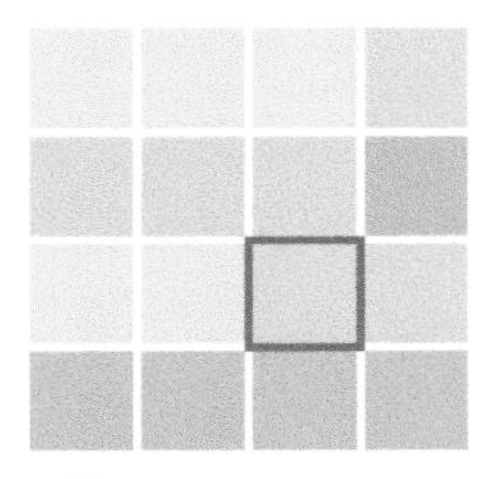

cell 1

2. [GRAPHICS (COMPUTING)] A pixel.
3. [NON-ESRI SOFTWARE] A small drawing, usually of a frequently used or complex symbol, notation, or detail. Cells are similar to blocks in AutoCAD drawings.

See also grid, pixel, raster.

cell size [DATA MODELS] The dimensions on the ground of a single cell in a raster, measured in map units. Cell size is often used synonymously with pixel size. *See also* cell.

cell statistics [SPATIAL ANALYSIS, ESRI SOFTWARE] A spatial analysis function that calculates a statistic for each cell of an output raster based on the values of each cell in the same location of multiple input rasters. *See also* cell, spatial analysis.

cellular automaton [MODELING] A mathematical construction consisting of a row or grid of cells in which each cell has an initial value—from a known and limited number of possible values—and all cells are simultaneously evaluated and updated according to their internal states and the values of their neighbors. The simplest cellular automaton is a row in which each cell has one of two values, such as red or green. In this case, there are eight possible value combinations for a cell and its neighbors. (If a green cell with two red neighbors is notated RGR, the eight combinations are RRR, RRG, RGR, GRR, RGG, GRG, GGR, GGG.) A set of rules determines whether a cell changes value when it is evaluated. A sample rule might be, "A green cell becomes red if it has a red neighbor on both sides." Successive updates, or generations, of a cellular automaton may produce complex patterns. Cellular automata are of interest in spatial modeling and are often used to model land-cover change.

census block [FEDERAL GOVERNMENT] The smallest geographic entity for which the U.S. Census Bureau tabulates decennial census data. Many blocks correspond to city blocks bounded by streets, whereas blocks in rural areas may include several square miles and have some boundaries that are not streets. *See also* block group, census tract, demography, enumeration district.

census block group [FEDERAL GOVERNMENT] A cluster of U.S. census blocks that have the same first digit within a four-digit identifying number of a census tract. *See also* census block, census tract, enumeration district.

census geography [FEDERAL GOVERNMENT] Any one of various types of precisely defined geographic areas used by the U.S. Census Bureau to collect and aggregate data. The largest unit of area is the entire United States, whereas the smallest is a census block. *See also* block group, census block.

census tract [FEDERAL GOVERNMENT] A small, relatively permanent statistical subdivision of a county or statistically equivalent entity delineated by local participants as part of the U.S. Census Bureau's Participant Statistical Areas Program. Tract boundaries normally follow physical features but may also follow administrative boundaries or other nonphysical features. *See also* census block, block group.

center

1. [EUCLIDEAN GEOMETRY] The point in a circle or in a sphere equidistant from all other points on the object.

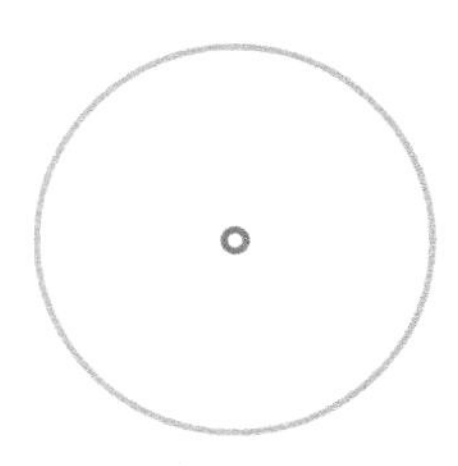

center 1

2. [EUCLIDEAN GEOMETRY] The point from which angles or distances are measured.
3. [NETWORK ANALYSIS] In network allocation, a location from which resources are distributed or to which they are brought.

centerline

1. [DATA CAPTURE] A line digitized along the center of a linear geographic feature, such as a street or a river, that at a large enough scale would be represented by a polygon.

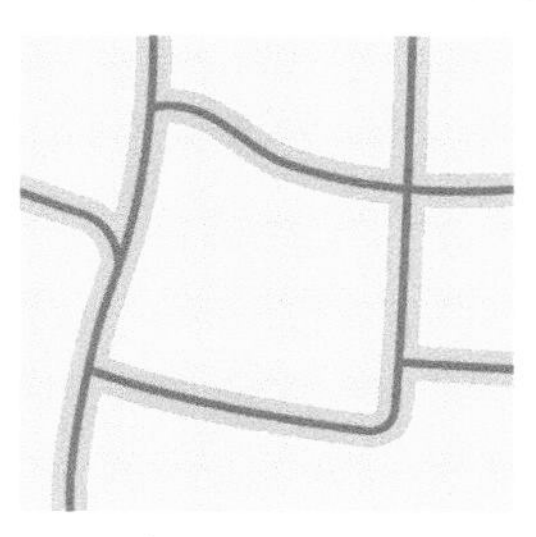

centerline 1

2. [NAVIGATION] The center of the street designated by the painted line down the center of the street or interpolated from the pavement edges.

centerline vectorization [DATA CAPTURE] The generation of vector features along the center of connected cells. Typically used for vectorizing scanned parcel and survey maps. *See also* outline vectorization.

centerpoint [AERIAL PHOTOGRAPHY] In aerial photography, the point at the exact center of an aerial photograph.

central business district [BUSINESS] The commercial and often geographic heart of a city; the downtown or center of a city.

central conic projection [MAP PROJECTIONS] A conformal conic map projection that preserves local shapes and angles—such as coastlines or

C

boundaries—accurately. Central conic projections are commonly used to map regions that have significant east–west latitude variations. *See also* conformality, conic projection, generating globe.

C

central cylindrical projection [MAP PROJECTIONS] A conformal cylindrical map projection that preserves straight lines and angles along the meridians while sacrificing accuracy in other areas. Central cylindrical projections are commonly used to map regions that have significant north–south longitude variations. *See also* cylindrical projection, generating globe.

central meridian [COORDINATE SYSTEMS]

1. The line of longitude that defines the center and often the x-origin of a projected coordinate system. In planar rectangular coordinate systems of limited extent, such as state plane, grid north coincides with true north at the central meridian.

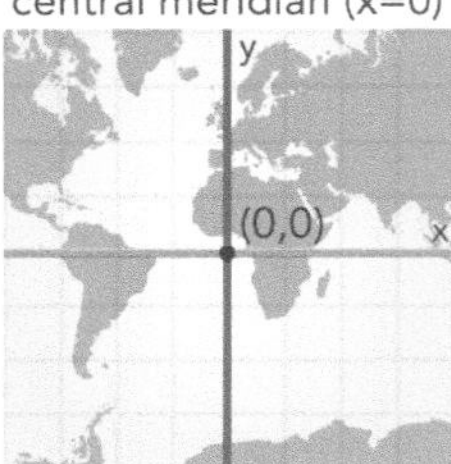

central meridian 1

2. The geographic origin of the longitudinal x-coordinates for a map grid. Also called the longitude of origin or, less commonly, the longitude of center.

See also central parallel, longitude, meridian.

central parallel [COORDINATE SYSTEMS] The geographic origin of the y-coordinates for a map grid. Also called the latitude of origin or, less commonly, the latitude of center. *See also* central meridian, latitude of center, latitude of origin.

centroid

1. [SPATIAL ANALYSIS] The geometric center or average location of a spatial feature. For line, polygon, or three-dimensional features, it is the center of mass (or center of gravity) and may fall inside the feature, as shown in the example image for a triangle, or outside the feature, as shown in the example image for a complex line. For multipoints, polylines, or polygons with multiple parts, it is computed using the weighted mean center of all feature parts.

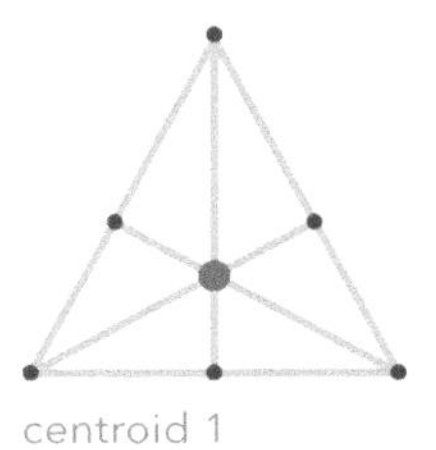

centroid 1

2. [CARTOGRAPHY] The center of area; the balance point or center of gravity.

See also center, geometry, multipoint.

centuria [CARTOGRAPHY] Subdivisions of land under the Centuriation system divided into squares approximately 132 acres (53.5 hectares) in area. *See also* Centuriation system.

Centuriation system [CARTOGRAPHY] A system of land surveying that originated during the Roman Empire in which the land was subdivided into a grid of square parcels. *See also* cardo maximus.

CGI [GRAPHICS (COMPUTING)] Acronym for *computer-generated imagery*. An image or video created or improved with technology.

chain [SURVEYING] A unit of length equal to 66 feet, used specifically in U.S. public land surveys. Ten square chains equal 1 acre. *See also* PLSS, surveying.

chain code [DATA CAPTURE] A method of drawing a polygon as a series of straight line segments defined as a set of directional codes, with each code following the last like links in a chain.

change detection [REMOTE SENSING] A process that measures how the attributes of a particular area have changed between two or more time periods. Change detection often involves comparing aerial photographs or satellite imagery of the area taken at different times. The process is most frequently associated with environmental monitoring, natural resource management, or measuring urban development. *See also* satellite imagery.

change map [CARTOGRAPHY] A map used to show areas in which attributes have changed over a certain period of time.

C

charge-coupled device *See* CCD.

chart

1. [CARTOGRAPHY] A map used for navigational purposes, especially air or marine navigation. *See also* atlas, histogram, map, navigation.
2. [MATHEMATICS] A graphic representation of tabular data; a diagram showing the relationship between two or more variable quantities, usually measured along two perpendicular axes. A chart may also be referred to as a graph. *See also* atlas, histogram, map, navigation.
3. [ESRI SOFTWARE] A visual presentation of the attribute data contained in a table. *See also* atlas, histogram, map, navigation.

chi-square statistic [STATISTICS] A statistic used to assess how well a model fits the data. It compares categorized data with a multinomial model that predicts the relative frequency of outcomes in each category to see to what extent they agree. *See also* binomial distribution, normal distribution.

chord [MATHEMATICS] A straight line that joins two points on a curve. *See also* geometry.

C

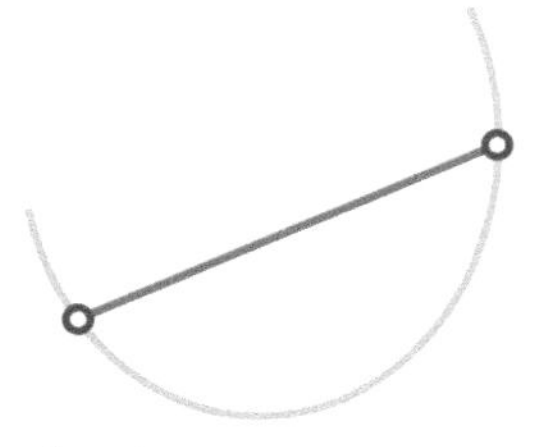

chord

choropleth map [CARTOGRAPHY] A thematic map in which areas are styled to represent the variation of values for a single variable, such as Average Household Income. From the Greek terms *choro* meaning place or land, and *pleth* meaning full.

A bivariate choropleth map—also called a relationship style—is a thematic map that displays two or more variables on a single map by combining different sets of symbols.

See also bivariate map, style, thematic map.

chroma [GRAPHICS (COMPUTING)] The saturation, purity, or intensity of a color. *See also* saturation, value, intensity, hue.

ChromaDepth map [MAP DESIGN] A map that uses the colors in the rainbow (red, orange, yellow, green, blue, indigo, and violet) to create the impression of different heights within a display when viewed with Chroma-Depth 3D glasses.

chronometer [PHYSICS] An extremely accurate portable clock that remains accurate through all conditions of temperature and pressure. The chronometer was developed in the eighteenth century for determining longitude at sea, but its scientific and navigational use has been made obsolete by the invention of quartz and atomic clocks. *See also* atomic clock.

CHUM [NAVIGATION] An abbreviation for *chart updating manual.* A document containing updates to aeronautical information, used by the U.S. military to update published products with the latest information. *See also* chart.

CIR film [PHOTOGRAMMETRY] CIR is an abbreviation of *color infrared*; more commonly known as CIR imagery. A false color image that renders its photographic subjects using reflected electromagnetic waves, such as red for vegetation.

CIR imagery *See* CIR film.

circle [EUCLIDEAN GEOMETRY] A two-dimensional geometric shape for which the distance from the center to any point on the edge is equal; the closed curve defining such a shape. *See also* circular arc, ellipse, sphere.

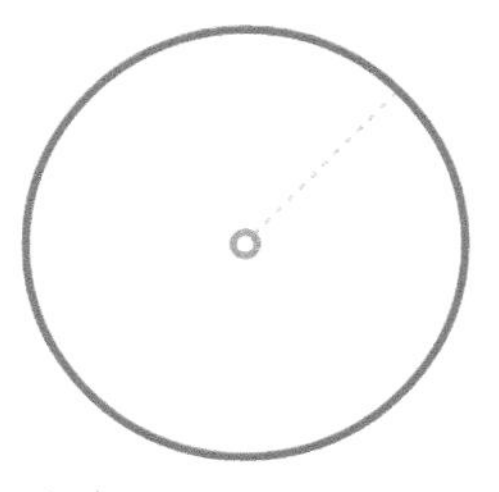

circle

circle sector [CARTOGRAPHY] A portion of a circle, like a pie piece, enclosed by two radii and an arc.

circular arc [EUCLIDEAN GEOMETRY] A curved line that is a section of a circle, with two vertices, one situated at each endpoint. *See also* arc, circle.

circular arc

circular variance [SPATIAL STATISTICS (USE FOR GEOSTATISTICS)] A measure of directional variation, on a scale from zero to one, among a set of line vectors. Circular variance approaches zero when all vectors point in roughly the same direction and approaches one when the vectors point in markedly different directions. *See also* vector.

circumscribing circle [CARTOGRAPHY] The smallest circle that contains a given polygon.

cirque [GEOGRAPHY] A bowl-shaped, steep-walled mountain basin carved by glaciation, often containing a small, round lake.

civilian code [GPS] The standard PRN code used by most civilian GPS receivers. *See also* P-code, PRN code.

Clarke Belt [ASTRONOMY] An orbit 22,245 miles (35,800 kilometers) above the equator in which a satellite travels at the same speed that the earth rotates. Proposed by the writer and scientist Arthur C. Clarke in 1945. It is also referred to as a geostationary orbit. *See also* geostationary.

Clarke ellipsoid of 1866 [GEODESY] A reference ellipsoid used as the basis for the North American Datum of 1927 (NAD27) and other datums, as well as for latitude and longitude on topographic and other maps produced in Canada, Mexico, and the United States from the late 1800s to the late 1970s. Largely replaced by the Geodetic Reference System 1980 (GRS80) or the World Geodetic System 1984 (WGS84). The Clarke ellipsoid of 1866 is also known as the Clarke spheroid of 1866. *See also* datum, ellipsoid, Geodetic Reference System of 1980, WGS84.

Clarke spheroid of 1866 *See* Clarke ellipsoid of 1866.

class

1. [DATA ANALYSIS] A set of entities grouped together based on shared attribute values.
2. [DATA MODELS] Pixels in a raster file that represent the same condition.

class intervals [DATA MANAGEMENT]

1. A set of categories for classification that divide the range of all values so that each piece of data is contained within a nonoverlapping category.
2. The numeric range for a data class.

See also classification, equal-interval classification, Jenks' optimization, natural breaks classification, quantile.

C

classed data [DATA EDITING] Numeric data grouped into classes with ranges of values. *See also* class intervals, classification.

classification

1. [DATA MANAGEMENT] The ordering, scaling, or grouping of data into classes to simplify features and their attributes.
2. [CARTOGRAPHY] The process of sorting or arranging entities into groups or categories; on a map, the process of representing members of a group by the same symbol, usually defined in a legend.
3. [PHOTOGRAMMETRY] In imagery, the computational process of assigning pixels or objects to a set of categories, or classes, which share common spectral, shape, elevation, or other characteristics.

See also equal-area classification, equal-interval classification, Jenks' optimization, natural breaks classification, quantile classification, standard deviation classification.

classification method [DATA MANAGEMENT] The procedure used to assign class intervals to numeric distributions. Also called a classing method or a data classification method. *See also* class intervals, Jenks' optimization, manual classification, natural breaks classification.

classification schema [DIGITAL IMAGE PROCESSING] A set of classes or categories to which image pixels or objects are assigned. Appropriate classification schemes depend on the type of sensors used, their resolution, feature types of interest, and the biome or land cover/land use imaged. *See also* classification, classification method.

cleaning [DATA CONVERSION] Improving the appearance of scanned or digitized data by correcting overshoots and undershoots, closing polygons, performing coordinate editing, and so on. *See also* dirty areas, overshoot, undershoot.

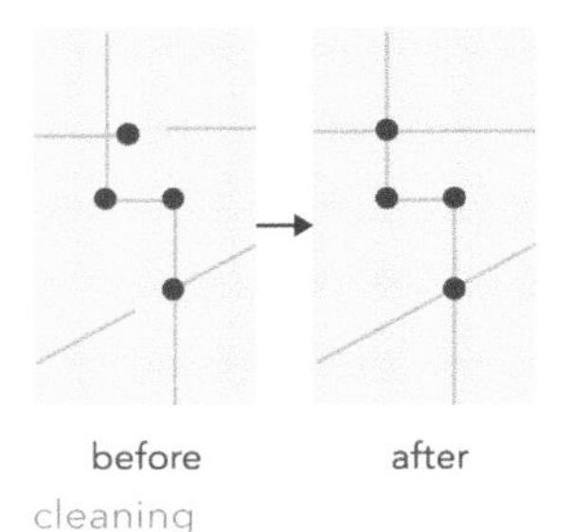

cleaning

clearinghouse [DATA SHARING] A repository structure, physical or virtual, that collects, stores, and disseminates information, metadata, and data. A clearinghouse provides widespread access to information and is generally thought of as reaching or existing outside organizational boundaries. *See also* data sharing, NSDI Clearinghouse Network, NSDI Clearinghouse Node.

client [COMPUTING] A system (app, program, computer, or device) in a client/server model that makes requests to a server. *See also* client-side address locator.

client-side address locator [ESRI SOFTWARE] A locator that is created and used on a client system. *See also* locator.

climate [CLIMATOLOGY] The meteorological conditions, including temperature, precipitation, and wind, that characteristically prevail in a particular region.

climate map [CARTOGRAPHY] A map that shows long-term average monthly or yearly atmospheric conditions. *See also* climate.

climate types [CLIMATOLOGY] Types of climate classified according to the average and the typical ranges of different variables, most commonly temperature, precipitation, and evapotranspiration. *See also* climate.

clinometric map [CARTOGRAPHY] A map that represents slope with colors or shading. *See also* slope.

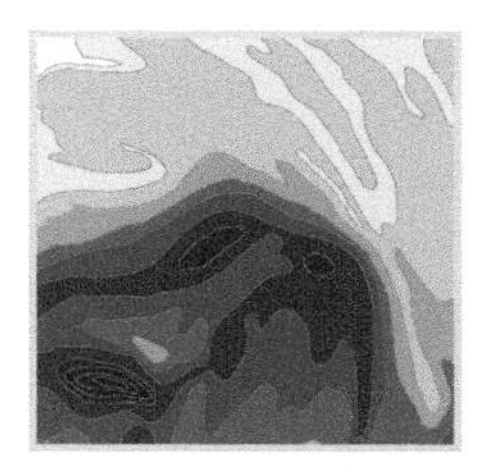

clinometric map

close coupling *See* tight coupling.

closed loop traverse [SURVEYING] A survey of a ground path that ends at the point of beginning (POB) or at a previously surveyed position. This type of traverse provides a check against errors in the distances and bearings because they must ultimately lead back to the original starting point. *See also* point of beginning, traverse.

closest facility analysis [NETWORK ANALYSIS] The process of finding the closest locations (facilities) from sites (incidents), based on the impedance chosen—for example, finding hospitals near a car accident. When finding the closest facilities, users can specify how many to find and whether the direction of travel is toward or away from the site (incident). Users can also specify a cutoff threshold beyond which network analysis will not search for a facility—for example, finding hospitals within 6 miles of a car accident. *See also* facility, incident.

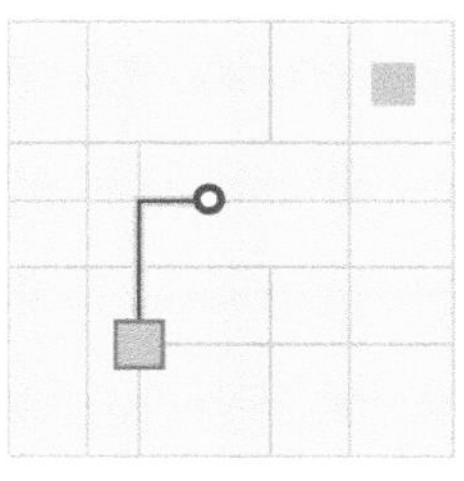

closest facility analysis

closure error [SURVEYING] A discrepancy between existing coordinates and computed coordinates that occurs when the final point of a closed traverse has known coordinates and the final course of a traverse computes different coordinates for the same survey point. *See also* Crandall rule.

C

cluster analysis [STATISTICS] A statistical classification technique for dividing a population into relatively homogeneous groups. The similarities between members belonging to a class, or cluster, are high, whereas similarities between members belonging to different clusters are low. Cluster analysis is frequently used in market analysis for consumer segmentation and locating customers, but it is also applied to other fields. *See also* analysis, classification.

cluster tolerance [ESRI SOFTWARE] The minimum tolerated distance between vertices in a topology. Vertices that fall within the set cluster tolerance are snapped together during the topology validation process. *See also* snapping.

clustered spatial arrangement [SPATIAL ANALYSIS] A grouping of features into one or a few small areas. *See also* cluster analysis.

clustering [ESRI SOFTWARE] A process of grouping or aggregating spatial data based on proximity or similarity, typically to create classes or categories, reduce noise, or improve clarity. *See also* snapping, cluster tolerance.

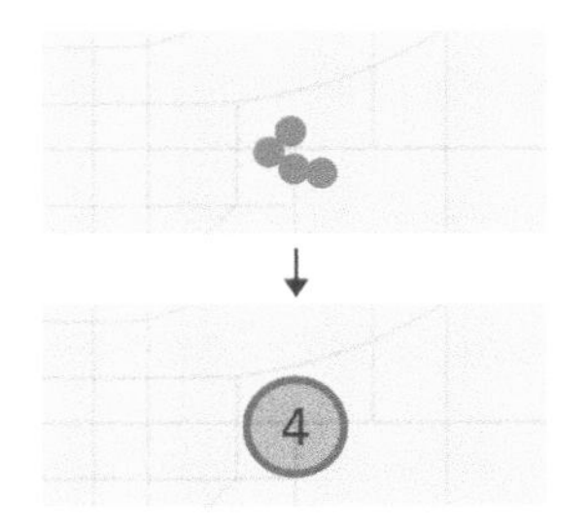

clustering

CMYK [PRINTING] An acronym for *cyan, magenta, yellow, and key* (black). A color model used in commercial printing that combines the printing inks cyan, magenta, yellow, and black to create a range of colors. *See also* color model, HSV, RGB.

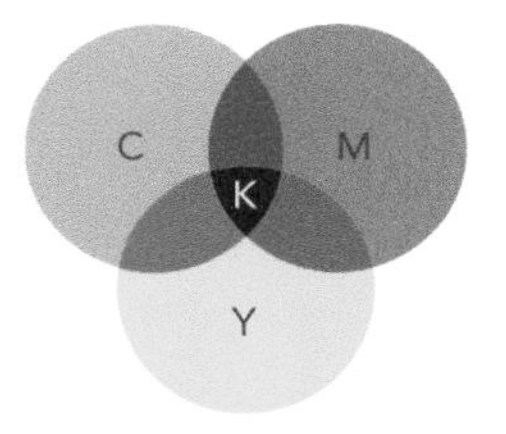

CMYK

CNT *See* condition table.

Coarse/Acquisition code *See* civilian code.

coded value domain [ESRI SOFTWARE, DATABASE STRUCTURES] A type of attribute domain that defines a set of permissible values for an attribute in a geodatabase. A coded value domain consists of a code and its equivalent value. For example, for a road feature class, the numbers 1, 2, and 3 might correspond to three types of road surface: gravel, asphalt, and concrete. Codes are stored in a geodatabase, and corresponding values appear in an attribute table. *See also* domain.

code-phase GPS [GPS] GPS measurements calculated using the PRN code transmitted by a GPS satellite. *See also* carrier-phase GPS, PRN code.

coefficient of areal correspondence [STATISTICS] A commonly used measure of the amount of overlap between two categories of area features.

coefficient of determination [STATISTICS] The square of the correlation coefficient in a regression that provides a measure of the strength of the correlation. *See also* regression.

cognitive map *See* mental map.

COGO [COORDINATE GEOMETRY (COGO)] An abbreviation of *coordinate geometry*. Automated mapping software used in land surveying that calculates locations using distances and bearings from known reference points. *See also* coordinate geometry, distance, simple measurement, surveying.

COGO-enabled [ESRI SOFTWARE] A property assigned to a line feature class that enables it to store coordinate geometry (COGO) dimensions. COGO-enabled lines have additional attribute fields that store COGO dimensions and are drawn with labeled dimensions and COGO symbology. When creating two-point lines, the direction, distance, and radius dimensions provided during data entry are captured and stored in COGO attribute fields *See also* attribute, COGO, feature class, line.

C

coincident [EUCLIDEAN GEOMETRY] Occupying the same space. Coincident features or parts of features occupy the same space in the same plane.

coincident geometry [ESRI SOFTWARE] In a geodatabase, the method of storing coincident feature coordinates. For example, if two lines are coincident, they will be drawn with one line lying precisely on top of the other. For two adjacent polygons, the coordinates for the shared boundary will be stored with each polygon, and the boundary will be drawn twice. *See also* topology.

cokriging [SPATIAL STATISTICS (USE FOR GEOSTATISTICS)] A form of kriging in which the distribution of a second, highly correlated variable (covariate) is used along with the primary variable to provide interpolation estimates. Cokriging can improve estimates if the primary variable is difficult, impossible, or expensive to measure, and the second variable is sampled more intensely than the primary variable. *See also* kriging.

col [GEOGRAPHY] A saddle in the middle of an arête.

colatitude [CARTOGRAPHY] The angular distance from the pole to a boundary point along a meridian.

C

cold front [METEOROLOGY] The edge of a relatively cold, dry air mass.

collection characteristics [DIGITAL IMAGE PROCESSING] The attributes describing how and under what conditions imagery was collected, including spectral, radiometric, spatial, and temporal resolutions, viewing angle, extent, and the physical and environmental conditions at the time of image acquisition.

color [GRAPHICS (COMPUTING)] A property of an object that is derived from the visible portion of the electromagnetic spectrum interacting with human vision. Color categories and specifications are based on physical properties such as light absorption, scattering, and reflection. *See also* image element.

color balancing [MAP DISPLAY] A technique used to adjust the color rendition between images in a mosaicked image to make transitions from one image to an adjoining image appear seamless.

color composite [REMOTE SENSING] A color image made by assigning red, green, and blue colors to three bands of a multispectral image. *See also* color separation, multispectral image.

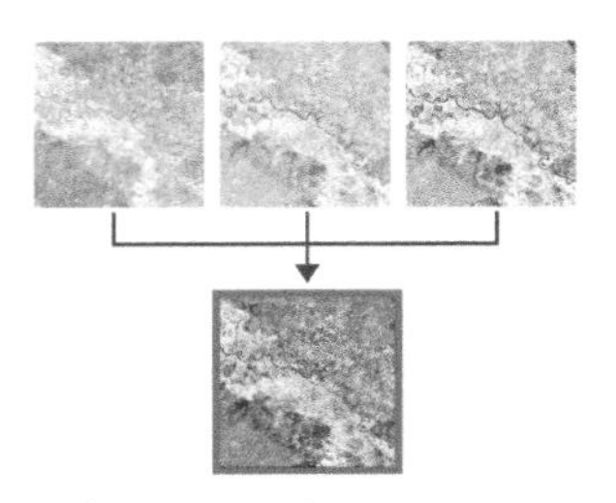

color composite

color contrast [PHOTOGRAMMETRY] The ability to distinguish a foreground color from its background color.

color depth [PHOTOGRAMMETRY] The number of bits used for the red, green, or blue components of a pixel color. *See also* RGB.

color gamut [GRAPHICS (COMPUTING)] The number of colors that can be sensed by a color camera or reproduced on a display or in a print process.

color management [DIGITAL IMAGE PROCESSING] In digital imaging, the controlled conversion between the color representations of various devices, such as sensors (cameras), displays, and printers. The goal of this process is to ensure consistency across all devices and resolutions.

color map [GRAPHICS (COMPUTING)] A set of alphanumeric definitions associated with specific colors. Color maps are most commonly used to display a raster dataset consistently on many different platforms. Not to be confused with a color picker. *See also* raster.

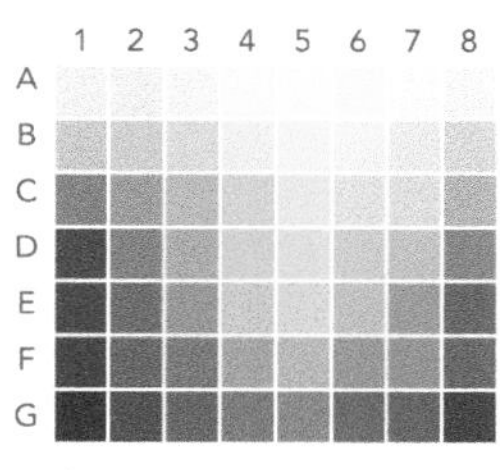

color map

color model [GRAPHICS (COMPUTING)] Any system that organizes colors according to their properties for printing or display. Examples include RGB (red, green, blue), CMYK (cyan, magenta, yellow, black), HSB (hue, saturation, brightness), HSV (hue, saturation, value), HLS (hue, lightness, saturation), and CIE-L*a*b (Commission Internationale de l'Eclairage-luminance, a, b). *See also* CMYK, RGB.

color ramp [SYMBOLOGY] A thematic set of colors used to represent variation, order, sequential values, or different categories on a map, such as topographic elevation or a heat map. *See also* continuous hypsometric tinting, graduated color map, heat map, topographic map.

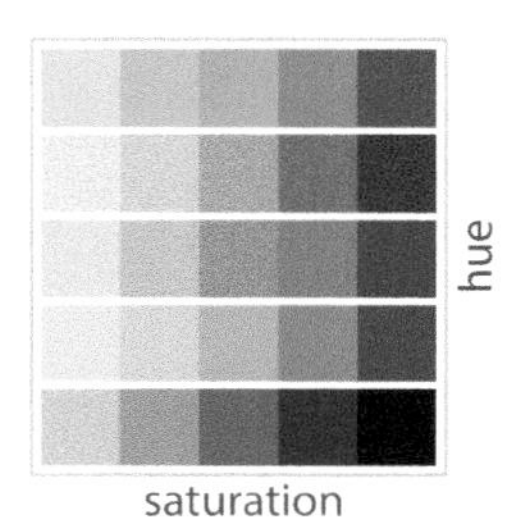

color ramp

color saturation *See* saturation.

color scheme [SYMBOLOGY] An arrangement or combination of colors used for map data. Also called a color progression.

color separation [PRINTING]

1. In printing, the use of a separate printing plate for each ink color used.
2. The process of scanning with color filters to separate the original image into single-color negatives.

See also color composite.

color system [SYMBOLOGY] A system for specifying colors numerically according to their individual components.

colorimetry [USABILITY] The science of quantifying and defining human color perception.

column

1. [COMPUTING] The vertical dimension of a table. Each column stores the values of one type of attribute for all the records, or rows, in the table. All the values in a given column are of the same data type; for example, number, string, BLOB, or date.
2. [DATA MODELS] A vertical group of cells in a raster, or pixels in an image.

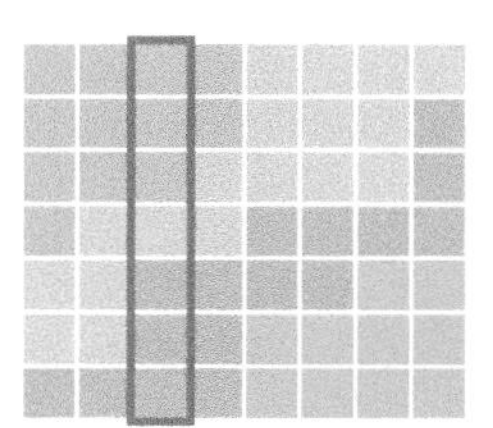
column 2

See also cell, pixel, raster, table.

combinatorial operator [MATHEMATICS] A type of mathematical operator that interprets input with Boolean values. Combinatorial operators assign a different number to each unique combination of input values. *See also* Boolean operator.

C

combined mapping method [CARTOGRAPHY] A way to show multiple quantitative themes on maps by combining two or more mapping methods. *See also* mapping methods.

community sensor model standard [REMOTE SENSING] A group of sensor models that are established, standardized, maintained as open source code, and are overseen by the U.S. National Geospatial-Intelligence Agency. *See also* sensor model.

compact shape [CARTOGRAPHY] A shape in which all points on the boundary are as close as possible to the center. The circle is the most compact two-dimensional shape because its boundary is everywhere equidistant from its center point, and the length of its boundary (perimeter) relative to its area is also minimal.

compactness [CARTOGRAPHY] The relationship of a feature's perimeter to its area.

comparison threshold [COMPUTING] The degree of uncertainty that can be tolerated in the spelling of a keyword used in a search, including phonetic errors and the random insertion, deletion, replacement, or transposition of characters.

compass [NAVIGATION] An instrument used for indicating geographic directions from one's current location. It consists of a case with compass points marked around its edge and a floating magnetic needle or disk that is aligned to the earth's magnetic field and pivots to point to magnetic north. *See also* cardinal point, compass point, magnetic north.

compass

compass balancing zones [CARTOGRAPHY] Zones that divide the earth based on magnetic dip; invented by the compass industry. In a compass balancing zone, a compass needle "balances" perfectly rather than drags or sticks on the compass card or housing. Standard compasses are balanced for the zone in which they are sold.

compass method [WAYFINDING] Use of a compass and a map to show the direction of magnetic north to orient a map. *See also* inspection method, magnetic north.

compass north *See* magnetic north.

compass point [CARTOGRAPHY] An indication of direction. One of the 32 divisions into which the circle around the needle of a compass is divided, each equal to 11.25 degrees. *See also* compass.

compass rose
1. [SYMBOLOGY] A diagram of compass points drawn on a map or chart, subdivided clockwise from 0 to 360 degrees with 0 indicating true north. On older maps and charts, a compass rose was a decorated diagram of cardinal points divided into 16 or 32 points.
2. [CARTOGRAPHY] A circular direction indicator printed in one or more places on charts. The outer circle of the compass rose is oriented to true north and is subdivided clockwise from 0 to 360 degrees with 0 indicating true north, whereas the inner ring is oriented to magnetic north.

See also chart.

compass rule [SURVEYING] A widely used rule for adjusting a traverse that assumes the precision in angles or directions is equivalent to the precision in distances. This rule distributes the closure error over the whole traverse by changing the northings and eastings of each traverse point in proportion to the distance from the beginning of the traverse. More specifically, a correction factor is computed for each point as the sum of the distances along the traverse from the first point to the point in question, divided by the total length of the traverse. The correction factor at each point is multiplied by the overall closure error to get the amount of error correction distributed to the point's coordinates. The compass rule is also known as the Bowditch rule, named for the American mathematician and navigator Nathaniel Bowditch (1773–1838). *See also* closure error, Crandall rule.

compass variation [CARTOGRAPHY] The magnetic declination shown on a navigational chart. *See also* chart, magnetic declination.

complementary metal-oxide-semiconductor array [PHOTOGRAMMETRY] A class of focal plane array that is an active-pixel sensor, most commonly found in commercial grade digital video cameras. *See also* focal plane array.

complementary-formats display [VISUALIZATION] A mapping method in which maps are combined with graphs, plots, tables, text, images, photographs, and information in other formats for the display of multivariate data. *See also* multiple display map.

complex edge feature [ESRI SOFTWARE] In a geodatabase, a linear network feature that corresponds to one or more network elements in the logical network. *See also* edge.

complex feature class [DATA MODELS] A point, line, polygon, or

C

multipoint feature class that participates in a geodatabase topology or is part of a relationship class or a controller dataset (such as a network dataset, terrain dataset, trace network, utility network, or a parcel fabric). This includes dimensions, annotation, and 3D object feature classes. *See also* feature class, geodatabase, simple feature, simple feature class, topology.

complex market area [BUSINESS] An area calculated by finding the outermost customers of a store along several vectors and connecting them. Complex market areas are more accurate than simple market areas because they respond to physical and cultural barriers. They are sometimes called amoebas because of their irregular shapes. *See also* simple market area.

composite index [CARTOGRAPHY] A single numeric index that represents the combination of several attribute values. *See also* composite relationship, composite variable map.

composite relationship [DATA MANAGEMENT] A link or association between objects in which the lifetime of one object controls the lifetime of its related objects, often used in geodatabase tables. For example, the association between a building (origin) and its mailbox (destination) is a composite relationship; while the building can exist on its own, the mailbox cannot. *See also* simple relationship.

composite variable map
[CARTOGRAPHY] A map that contains a composite index. Composite variable maps use GIS software multiple criteria evaluation (MCE) capabilities. Also called a cartographic modeling map or composite index map. *See also* composite index.

composition

1. [DATA MODELS] An association in which the lifetime of the whole controls the lifetime of the parts. In a composition, the instances of two classes depend on each other; the whole controls the location and lifetime of its parts. For example, a digital map is composed of layers. When you move the map, the layers also move, and if you delete the map, its layers get deleted; therefore, the lifetime of these objects depends on one another.
2. [CARTOGRAPHY] The variety and quantity of patch types without consideration of the spatial arrangement of patches within a landscape mosaic.

compound element [STANDARDS] Within metadata, a group of data elements (including other compound elements) that together describe a characteristic of a spatial dataset in more detail than can be described by an individual data element. *See also* metadata.

compression

1. [COMPUTING] The process of reducing the size of a file or database. Typically clarified as lossy (high compression/some data loss) or lossless (lower compression/larger file size). Compression improves data handling, storage, and database performance. Examples of compression methods include run-length encoding and wavelets.
2. [ESRI SOFTWARE] A process that removes unreferenced rows from geodatabase system tables and user delta tables. Compression helps maintain versioned geodatabase performance.

See also JPEG, lossless compression, lossy compression, PNG.

compromise projection [MAP PROJECTIONS] A map projection that is neither equal area, conformal, nor equidistant, but rather a balance between these geometric properties; often used in thematic mapping. *See also* projection.

computational geometry [MATHEMATICS] A branch of mathematics that uses algorithms to solve geometry problems. Computational geometry is used in many GIS operations, including proximity analysis, feature generalization, and automated text placement. *See also* algorithm, geometry.

computer-aided design [GRAPHICS (COMPUTING)] Also known by the acronym *CAD*. The use of software to manipulate, create, organize, or optimize a design. Sometimes called computer-aided drafting. CAD systems are commonly used to support engineering, planning, and illustrating activities. *See also* drafting, computational geometry.

computer-assisted learning [EDUCATION] Instruction or training that uses a computer-based interface, such as a digital classroom. Not to be confused with machine learning or AI.

concatenate [COMPUTING] To join two or more character strings together, end to end; for example, to combine the two strings "spatial" and "analysis" into the single string "spatial analysis." *See also* joining.

concatenate events [ESRI SOFTWARE] In linear referencing, combining event records on the same route for specified values. Events where the to-measure of one event matches the from-measure of the next event are combined. *See also* concatenate.

concatenated key [COMPUTING] In a relational database table, a primary key made by combining two or more keys that together form a unique identifier.

concave hillside [GEOGRAPHY] An inward curving hillside that has a progressively decreasing angle of incline.

concave slope [MATHEMATICS] An inwardly curving slope.

concentric rings *See* overlapping rings.

concentric zone model [URBAN PLANNING] A model of a city with the central business district in the middle, working-class residential areas surrounding that as a ring, and a more affluent residential area as the outermost ring.

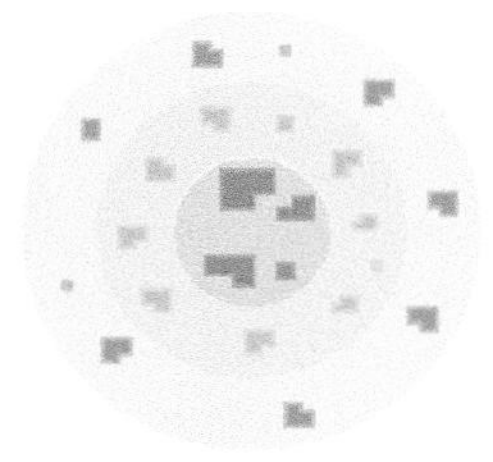

concentric zone model

conceptual accuracy [ACCURACY] Accuracy that is determined by the amount of information used and how it is classified into appropriate categories through appropriate mapping methods.

condition table [DATA QUALITY] A component of attribute rules that contains SQL statements and custom code for feature validation extended beyond standard geodatabase domains. Attribute rules use condition tables for enhanced validation during both database production and quality control. *See also* knowledge base.

conditional operator [COMPUTING] A symbol or keyword that specifies the relationship between two values and is used to construct queries to a database. Examples include = (equal to), < (less than), and > (greater than).

conditional statement [COMPUTING] A logical statement that performs one option if a statement is true, and another if it is false. An *if-then-else* statement is an example of a conditional statement.

confidence level [STATISTICS] In a statistical test, the risk, expressed as a percentage, that the null hypothesis will be incorrectly rejected because of sampling error when the null hypothesis is true. For example, a confidence level of 95 percent means that if the same test were performed 100 times on 100 different samples, the null hypothesis would be incorrectly rejected 5 times. *See also* null hypothesis, significance level.

conflation [DATA EDITING] A set of procedures that aligns the features of two geographic data layers and then transfers the attributes of one to the other. *See also* rubber sheeting.

conformal projection [MAP PROJECTIONS] A map projection that preserves the relative proportions and angles of small areas. In a conformal projection, graticule lines intersect at 90-degree angles, and at any point on the map the scale is the same in all directions. A conformal projection maintains all angles at each point, including those between the intersections of arcs; therefore, the size of areas enclosed by many arcs may be distorted. Conformal projections are particularly useful for mapping coastlines, boundaries, or other detailed features. Examples of conformal projections include the Mercator projection, the Lambert conformal conic projection, and the transverse Mercator projection. *See also* Lambert conformal conic projection, Mercator projection, projection, transverse Mercator projection.

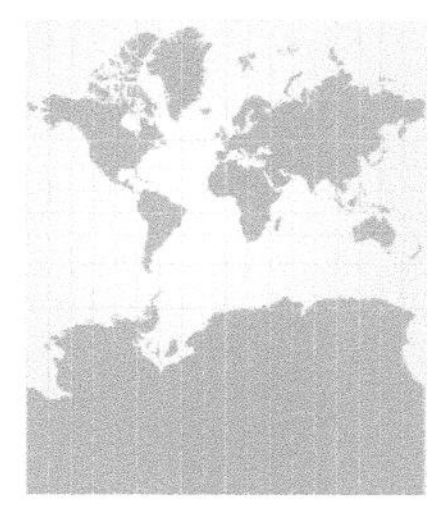

conformal projection

conformality [MAP PROJECTIONS] The characteristic of a map projection that preserves the shape of any small geographic area. *See also* projection.

congressional district [GOVERNMENT] A geographical and political division used by Japan, the Philippines, and the United States. *See also* redistricting.

conic projection [MAP PROJECTIONS] A map projection that transforms points from a spheroid or sphere onto a tangent or secant cone that is wrapped around the sphere. The cone is then sliced from the apex (top) to the bottom and flattened into a plane. Typically used for mapping the earth. *See also* projection.

conic projection family [MAP PROJECTIONS] A map projection family based on the use of a cone as the developable surface. *See also* conic projection, developable surface, map projection families.

conjoint boundary *See* shared boundary.

connection line [CADASTRAL AND LAND RECORDS] A parcel fabric line with bearing and distance COGO dimensions, commonly used to tie parcels across roads, tie in control points, or tie the point of survey commencement to the point of beginning for a particular parcel. Connection lines do not necessarily indicate parcel boundaries. *See also* parcel.

connectivity

1. [DATA MODELS] The way in which features in GIS data are attached to one another functionally or spatially.
2. [ESRI SOFTWARE, NETWORK ANALYSIS] In a geodatabase, the state of association between edges and junctions in a network system for network data models. Connectivity helps define and control flow, tracing, and pathfinding in a network.
3. [NETWORK ANALYSIS] In a coverage, topological identification of connected arcs by recording the from-node and to-node for each arc. Arcs that share a common node are connected.

See also arc-node topology, connectivity group, connectivity policy.

connectivity analysis *See* network analysis.

C

connectivity group [NETWORK ANALYSIS] In a network dataset, a logical grouping of point features, line features, or both, that controls how network elements are connected. Connectivity groups are defined when a network dataset is built. A network dataset may have multiple connectivity groups. *See also* connectivity.

connectivity measure [NETWORK ANALYSIS] An objective comparison of multiple networks to determine a network's connectivity relative to its flow. A connectivity measure can interpret the degree to which line features are interconnected. *See also* connectivity, line feature, network.

connectivity policy [NETWORK ANALYSIS] In a network dataset, a property of network sources that defines how network elements connect to each other within a connectivity group. There are two types of edge-edge connectivity policies (end-point connectivity and any-vertex connectivity) and two types of edge-junction connectivity policies (honor and override). *See also* any-vertex connectivity, connectivity group, endpoint connectivity, override.

connector [DATA MODELS] A visual representation of the relationship between elements in a model. Connectors join elements together to create processes. Typical processes connect an input data element, a tool element, and a derived data element. *See also* intersection connector, model.

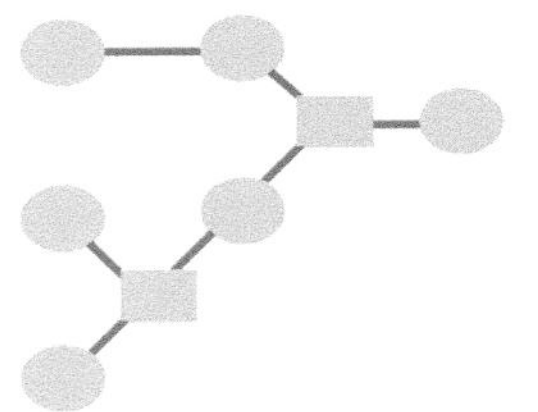

connector

constant azimuth *See* rhumb line.

constant isoline interval [CARTOGRAPHY] An isoline interval that is the same for the entire map. *See also* isoline, isoline interval.

constant-slope path [NAVIGATION] A route—whether physical, such as a road or trail, or predefined, such as an aviation flight path—that maintains a persistent incline.

constraint [DATA MODELS] A limit imposed on a model to maintain data integrity. For example, in a water network model, an 8-inch pipe cannot connect to a 4-inch pipe. *See also* data integrity, restriction.

construction line [CARTOGRAPHY] A line drawn for the purpose of creating a profile. *See also* profile.

containment [DATA EDITING] A spatial relationship in which a point, line, or polygon feature or set of features is enclosed completely within a polygon.

content [INTERNET] Refers collectively to the maps, layers, services, and tools used with ArcGIS software and associated with an ArcGIS account. *See also* ArcGIS account, item.

Content Standard for Digital Geospatial Metadata [STANDARDS] A publication authored by the FGDC that specifies the information content of metadata for digital geospatial datasets. The purpose of the standard is to provide a common set of terminology and definitions for concepts related to the metadata. All U.S. government agencies (federal, state, and local) that receive federal funds to create metadata must follow this standard. *See also* Federal Geographic Data Committee.

conterminous [EUCLIDEAN GEOMETRY] Having the same or coincident boundaries. *See also* adjacency, boundary, shared boundary.

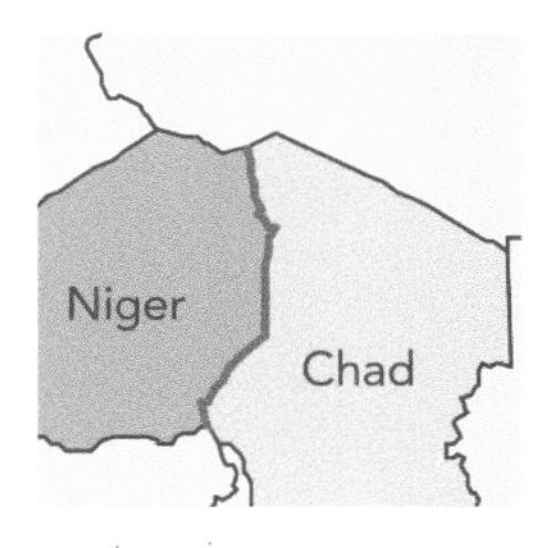

conterminous

context

1. [DIGITAL IMAGE PROCESSING] An image element that provides information in space and time related to an image's pixels that affects the meaning associated with those pixels.
2. [DATA MANAGEMENT] Information associated with the neighbors of an object of interest, used to help identify the object.

See also geographic context, image element.

contiguity

1. [DATA ANALYSIS] In a coverage, the topological identification of adjacent polygons by recording the left and right polygon for each arc.
2. [CARTOGRAPHY] A numeric description of boundary connectedness.

See also king's case, polygon-arc topology.

contiguous

1. [EUCLIDEAN GEOMETRY] Next to or close to one another.

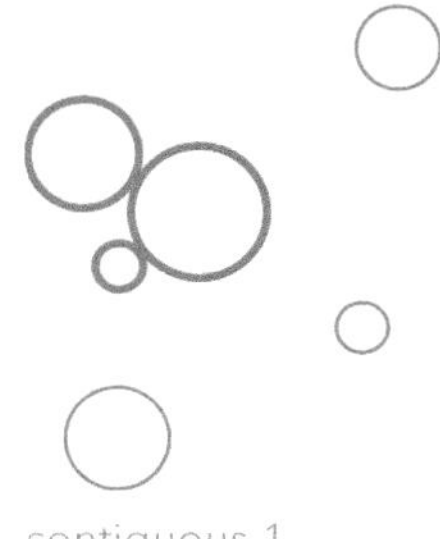

contiguous 1

2. [DATA STRUCTURES] Of polygons: adjacent; having a common boundary; sharing an edge.
3. [DATA STRUCTURES] Of raster cells: connected orthogonally or diagonally; or, sometimes, connected strictly orthogonally.
4. [DATA STRUCTURES] Of TIN edges: having no gaps or overlaps.

See also adjacency, boundary.

C

contiguous cartogram [CARTOGRAPHY] A cartogram on which the proximity and contiguity of neighboring areas are maintained, although sometimes at the expense of shape distortion. *See also* cartogram, contiguity.

continental slope [GEOGRAPHY] The precipitous drop-off between the outer edge of the gradual, gentle change in elevation of the continental shelf and the deep ocean floor.

continuous data [DATA MODELS] Data such as elevation or temperature that varies without discrete steps. Since computers store data discretely, continuous data is usually represented by TINs, rasters, or contour lines, so that any location has either a specified value or one that can be derived. *See also* discrete data.

continuous feature [DATA MODELS] A feature that is not spatially discrete. The transition between possible values on a continuous surface is without abrupt or well-defined breaks. *See also* discrete feature.

continuous hypsometric tinting [MAP DESIGN] A hypsometric tinting method in which the abrupt change between hypsometric tints is minimized by gradually merging one tint into the next, giving a smooth appearance to the tonal gradation. *See also* hypsometric tinting.

continuous phenomena [CARTOGRAPHY] Geographic phenomena that change gradually over space. *See also* continuous surface, continuous-surface map, geographic phenomena.

continuous raster [DATA MODELS] A raster in which cell values vary continuously to form a surface. In a continuous raster, the phenomena represented have no clear boundaries. Values exist on a scale relative to each other. It is assumed that the value assigned to each cell is what is found at the center of the cell. Rasters representing elevation, precipitation, chemical concentrations, suitability models, or distance from a road are examples of continuous rasters. *See also* discrete raster.

continuous raster

continuous surface [CARTOGRAPHY] A surface that represents a geographic phenomenon with continuously changing values—that is, a continuous phenomenon; a surface that changes smoothly in numeric value across the landscape. *See also* continuous phenomena, continuous-surface map, geographic phenomena.

continuous tone image [GRAPHICS (MAP DISPLAY)] A photograph that has not been screened and so displays all the varying tones from dark to light. *See also* dot screen, halftone image.

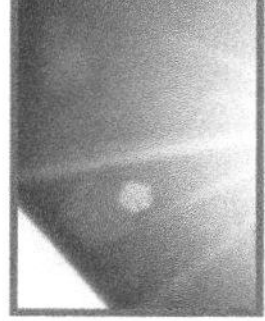

continuous tone image

Continuously Operating Reference Stations *See* CORS.

continuous-surface map [CARTOGRAPHY] A type of map that represents a single geographic phenomenon, such as temperature, precipitation, or elevation, as a smooth and uninterrupted surface. It is created by interpolating or extrapolating data points to estimate values for locations where data is not available. *See also* continuous phenomena, continuous surface, geographic phenomena.

contour [CARTOGRAPHY] A line of equal elevation above or below a datum; a line that shows the shape of the land surface. *See also* form lines, contour interval, supplemental contour.

contour interval [CARTOGRAPHY] The difference in elevation between adjacent contour lines. *See also* contour line.

contour line [CARTOGRAPHY] A line on a map that connects points of equal elevation based on a vertical datum, usually sea level. *See also* form lines, contour interval, supplemental contour.

contour line

contour tagging [DATA CAPTURE] Assigning elevation values to contour lines. *See also* contour line.

contrast [REMOTE SENSING] In remote sensing and photogrammetry, the ratio between the energy emitted or reflected by an object and that emitted or reflected by its immediate surroundings.

contrast ratio [GRAPHICS (COMPUTING)] The ratio between the maximum and minimum brightness values in an image.

contrast stretch [GRAPHICS (COMPUTING)] Increasing the contrast in an image by expanding its grayscale range to the range of the display device.

control measure [DEFENSE, ESRI SOFTWARE] A type of military solutions graphic related to the regulation of unit movement, such as lane boundaries and obstacles.

control point

1. [SURVEYING] An accurately surveyed coordinate location for a physical feature that can be identified on the ground. Control points are used in least-squares analysis and adjustment

C

as the basis for improving the spatial accuracy of all other points to which they are connected. Control points have known integrity, such as a Public Land Survey System (PLSS) section corner.

2. [ACCURACY] A point of known accuracy used when georeferencing map data, digitizing paper map data, or performing spatial adjustment operations, such as rubber sheeting.

See also least-squares analysis and adjustment, precision, rubber sheeting.

conventional alternative hypothesis [SPATIAL STATISTICS (USE FOR GEOSTATISTICS)] In statistical testing, a set of assumptions that will be accepted by test data if the null hypothesis is rejected. In surveying, the alternative hypothesis assumes that there is an outlier present in a single measurement in a measurement network. The test associated with this hypothesis is the W-test. *See also* null hypothesis.

convergence angle [CARTOGRAPHY] The angle between a vertical line (grid north) and true north on a map. *See also* grid north, true north.

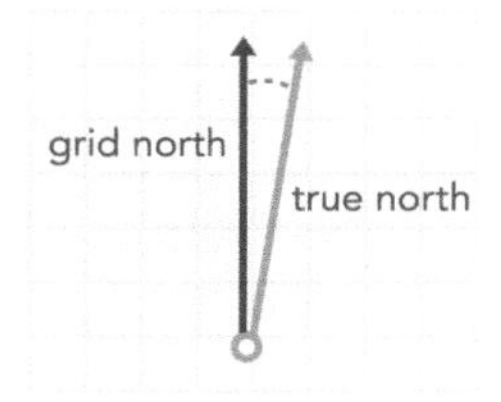

convergence angle

converging meridians [CARTOGRAPHY] The progressively decreasing east–west ground distance between two meridians extending from the equator to the north or south pole. *See also* meridian.

conversion [DATA CONVERSION] The process of changing input data from one representation or format to another, such as from raster to vector, or from one file format to another, such as from x,y coordinate table to point shapefile.

convex hillside [GEOGRAPHY] An outward curving hillside that has a progressively increasing angle of incline.

convex hull [MATHEMATICS] The smallest convex polygon that can enclose a group of objects, such as a group of points. *See also* convex polygon.

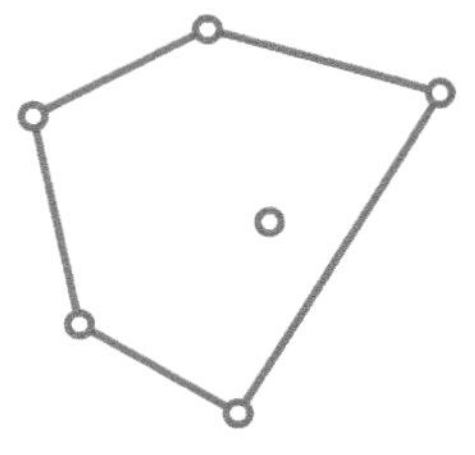

convex hull

convex polygon [MATHEMATICS] A polygon in which a straight line drawn between any two points inside the polygon is completely contained within the polygon. Visually, the boundary of a convex polygon is the shape a rubber band would take around a group of objects. *See also* convex hull.

convex slope [EUCLIDEAN GEOMETRY] An outwardly curving slope.

convolution filter [DIGITAL IMAGE PROCESSING] A kernel or matrix of values that is applied to an image's pixel values. It is used to sharpen, blur, or detect the edges of objects in imagery or to provide other kernel-based image enhancements.

coordinate geometry [COORDINATE GEOMETRY (COGO)] Also known by the abbreviation *COGO*. A method for calculating coordinate points from surveyed bearings, distances, and angles. *See also* COGO.

coordinate reference system *See* coordinate system.

coordinate system [COORDINATE SYSTEMS] A framework used to define, represent, and measure the spatial location of features on the earth's surface. A coordinate system provides a standardized method for referencing and integrating spatial data from various sources and ensures accurate spatial relationships between features. It includes a reference point (origin), coordinate axes (such as latitude and longitude or easting and northing), and units of measurement (such as degrees, meters, or feet). Common coordinate systems used in GIS include geographic, projected, and vertical coordinate systems. *See also* Cartesian coordinate system, coordinates, geographic coordinate system, projected coordinate system, vertical coordinate system.

coordinate transformation [COORDINATE SYSTEMS, SPATIAL ANALYSIS] The process of converting the coordinates in a map or image from one coordinate system to another, typically through rotation and rescaling. Coordinate transformation typically involves mathematical calculations and algorithms to convert the coordinates accurately and may include projection transformation, geographic transformation, and georeferencing. *See also* coordinate system, geographic transformation, georeferencing, projection transformation, rescale, rotate.

coordinated universal time [ASTRONOMY] The official timekeeping system of the world's nations since 1972. It refers local time throughout the world to time at the prime meridian. It is based on atomic clocks but is periodically artificially adjusted to always remain within 0.9 seconds of universal time. The adjustment is made by the addition of leap seconds to the course of atomic time. Coordinated universal time is also known as UTC. The acronym UTC does not represent the word order of *coordinated universal time* in either English or French—the official languages of the International Telecommunication Union (ITU). It is an extension of the *UT** pattern established for variants of universal time, including UT0, UT1, UT2, UT1R,

UT1D, and others. *See also* atomic clock, prime meridian, universal time.

coordinates [COORDINATE SYSTEMS] A set of values represented by the letters *x*, *y*, and optionally *z* (elevation, density, or quantity) or *m* (measure), that define an object's position within a coordinate system. *See also* geographic coordinates, m-value, x,y coordinates, z-value.

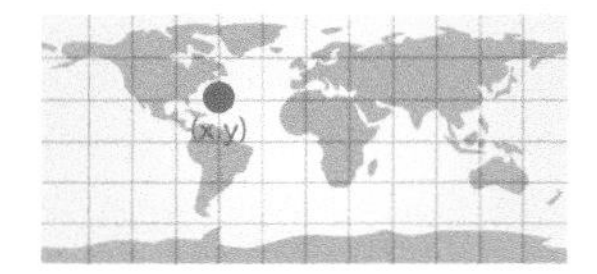

coordinates

core-based statistical area [GOVERNMENT] Also known by the acronym *CBSA*. A geographic region containing at least one urban area with a population of at least 10,000, defined by the U.S. Office of Management and Budget for use by federal statistical agencies, including the U.S. Census Bureau. A core-based statistical area can be a metropolitan statistical area or a micropolitan statistical area. *See also* metropolitan statistical area, micropolitan statistical area.

correction line [SURVEYING] A township line established every 24 miles north and south of the baseline as part of the U.S. Public Land Survey System and Canada's Dominion Land Survey. Surveyors created correction lines where they readjusted range lines to compensate for the convergence of meridians. *See also* meridian, PLSS.

correlation [STATISTICS, ESRI SOFTWARE] An association between data or variables that change or occur together. For example, a positive correlation exists between housing costs and distance from the beach; generally, the closer a home is to the beach, the more it costs. Correlation does not imply causation. For example, there may be a statistical correlation between ice cream sales and crime rates, but neither causes the other. The correlation coefficient is an index number between –1 and 1 indicating the strength of the association between two variables. *See also* multivariate analysis, probability.

correspondence [MAP PROJECTIONS] The agreement between points on the earth and points on the projected surface, such as a paper map or computer screen.

corridor [DATA MODELS] A buffer drawn around a line. *See also* buffer.

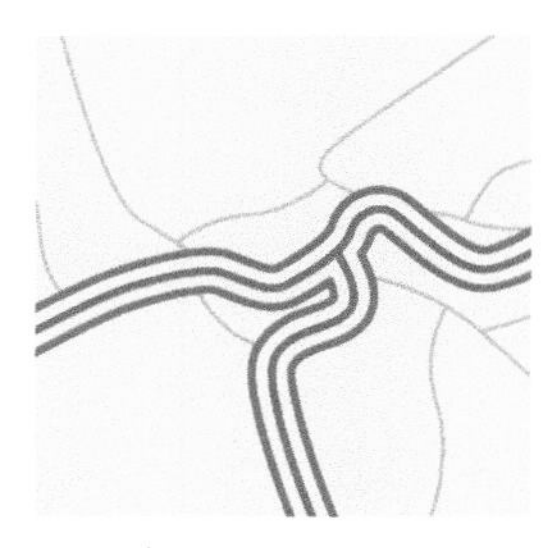

corridor

corridor analysis [SPATIAL ANALYSIS] A form of spatial analysis usually applied to environmental and land-use data to find the best locations for building roads, pipelines, and other linear transportation features. *See also* buffer.

CORS [GEODESY] Acronym for *Continuously Operating Reference Station.* A permanent GNSS tracking station that provides continuous real-time or near real-time positioning information. CORS stations have applications in surveying, meteorology, space, weather, and geophysics. In the United States, the national CORS network is managed by the U.S. National Geodetic Survey (NGS) and is also known as the *NOAA CORS Network (NCN). See also* base station, GPS, NOAA, U.S. National Geodetic Survey.

cost

1. [DATA ANALYSIS] A function of time, distance, or any other factor that incurs difficulty or an outlay of resources.
2. [NETWORK ANALYSIS] An attribute of a network element used to model impedance and demand in network datasets. Cost is an attribute that is accumulated during traversal of a network.

See also cost-distance analysis, descriptor, hierarchy, impedance, least-cost path, restriction, shortest path, weight.

cost grid *See* cost raster.

cost raster [SPATIAL ANALYSIS] A raster dataset that identifies the cost of traveling through each cell in the raster. A cost raster can be used to calculate the cumulative cost of traveling from every cell in the raster to a source or a set of sources. *See also* cost.

cost-benefit analysis [ORGANIZATIONAL ISSUES] Also known by the acronym *CBA*. An appraisal that attempts to compare the benefits (including social benefits) expected from a project with the costs (sometimes including social costs) incurred by the project over its lifetime. Generally, cost-benefit analyses are used to compare alternative proposals or to make a case for the implementation of a particular plan or system.

cost-distance analysis [ESRI SOFTWARE] The calculation of the least cumulative cost from each cell to specified source locations over a cost raster. *See also* cost raster.

cost-weighted allocation [SPATIAL ANALYSIS] An analysis function that identifies the nearest source from each cell in a cost-weighted distance grid. Each cell is assigned to its nearest source cell, in terms of accumulated travel cost. *See also* allocation, cost, cost raster.

cost-weighted direction [SPATIAL ANALYSIS] An analysis function that provides a road map from the cost-weighted distance grid, identifying the route to take from any cell, along the least-cost path, back to the nearest source. *See also* least-cost path, route.

cost-weighted distance [SPATIAL ANALYSIS] An analysis function that uses a cost grid to assign a value—the least accumulative cost of getting back to the source—to each cell of an output grid. *See also* function, source, spatial analysis.

C

counting dial [AERIAL PHOTOGRAPHY] A dial on a polar planimeter that records the number of whole and partial revolutions the measuring wheel makes. Also called a recording dial. *See also* polar planimeter, vernier scale.

county [FEDERAL GOVERNMENT] The primary legal subdivision of all U.S. states except Alaska and Louisiana. The U.S. Census Bureau uses counties or equivalent entities (boroughs in Alaska, parishes in Louisiana, the District of Columbia in its entirety, and municipios in Puerto Rico) as statistical subdivisions.

county subdivision [FEDERAL GOVERNMENT] A statistical division of a county recognized by the U.S. Census Bureau for data presentation. County subdivisions can include census county divisions, census subareas, minor civil divisions, and unorganized territories. *See also* census geography.

course [NAVIGATION] A navigational route.

course heading [NAVIGATION] Navigational direction. *See also* bearing.

covariance [STATISTICS] A statistical measure of the linear relationship between two variables. Covariance measures the degree to which two variables move together relative to their individual mean returns. *See also* variogram, correlation.

coverage [ESRI SOFTWARE] A data model for storing geographic features. A coverage stores a set of thematically associated data considered to be a unit. It usually represents a single layer, such as soils, streams, roads, or land use. In a coverage, features are stored as both primary features (points, arcs, polygons) and secondary features (tics, links, annotation). *See also* attribute table.

coverage units [COORDINATE SYSTEMS] The units of the coordinate system in which a coverage is stored (for example, feet, meters, inches). *See also* coverage.

cracking [DATA EDITING, ESRI SOFTWARE] A part of the topology validation process in which vertices are created at the intersection of feature edges. *See also* edge, topology, vertex.

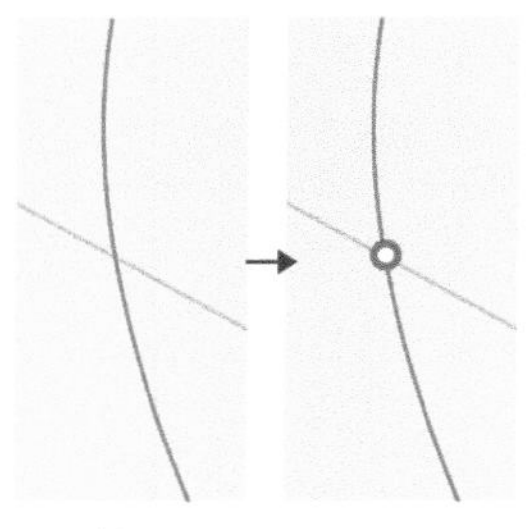

cracking

Crandall rule [SURVEYING] A special-case, least-squares-based method for adjusting the closure error in a traverse. The Crandall rule is most frequently used in a closed traverse that represents a parcel from a subdivision plan to ensure that tangency between courses remains intact as, for example, when applied to a tangent curve. It assumes that course directions and angles have no errors and, therefore, all error corrections are applied only to the distances. This method uses a least-squares analysis and adjustment to distribute the closure error and applies infinite weight to the angles or direction measurements to ensure that they are not adjusted. In some circumstances, the results of this adjustment method may be unexpected, or the adjustment may not be possible, and an alternative method is required. The Crandall rule was developed by C. L. Crandall around 1901. *See also* closure error, least-squares analysis and adjustment.

credits [ESRI SOFTWARE] The currency used across ArcGIS connected to a user account. Credits are consumed for specific types of actions and storage, such as performing analytics, storing features, and using premium content. *See also* ArcGIS account.

critical value

1. [SPATIAL STATISTICS (USE FOR GEOSTATISTICS)] The specific cutoff point that determines acceptance or rejection of a hypothesis. Critical values are determined by the choice of a level of significance (α).
2. [MAP DESIGN] Data with special relevance to a map's theme that is used to set class interval limits.

See also F statistic, significance level.

cross correlation [SPATIAL STATISTICS (USE FOR GEOSTATISTICS)] Statistical correlation between spatial random variables of different types, attributes, names, and so on, where the correlation depends on the distance or direction that separates the locations. *See also* autocorrelation, cross covariance.

cross covariance [STATISTICS] The statistical tendency of variables of different types, attributes, names, and so on, to vary in ways that are related to each other. Positive cross covariance occurs when both variables tend to be above their respective means together, and negative cross covariance occurs if one variable tends to be above its mean when the other variable is below its mean. *See also* covariance.

cross fix [PHOTOGRAMMETRY] A location determined by resection methods in which lines cross at the viewer's location. *See also* resection.

cross section [CARTOGRAPHY] A diagram of the vertical section of the ground surface below a profile line. *See also* profile line.

cross tabulation

1. [DATA ANALYSIS] In a GIS, comparing attributes in different coverages or map layers according to location.

2. [STATISTICS] A method for showing the relationship between two or more data characteristics by repeating each of the categories of one variable for each category of the other variables. For example, a cross tabulation of census data might show households by number of occupants by income.

See also coverage, layer.

C

cross validation [STATISTICS] A procedure for testing the quality of a predicted data distribution. In cross validation, a piece of data whose value is known independently is removed from the dataset and the rest of the data is used to predict its value. Full cross validation is done by removing, in turn, each piece of data from the dataset and using the rest of the data to predict its value. *See also* prediction, validation.

cross variogram [SPATIAL STATISTICS (USE FOR GEOSTATISTICS)] A function of the distance and direction separating two locations, used to quantify cross correlation. The cross variogram is defined as the variance of the difference between two variables of different types or attributes at two locations. The cross variogram generally increases with distance and is described by nugget, sill, and range parameters. *See also* cross correlation.

cross-reference database [ESRI SOFTWARE] A database containing tables with information defining the mapping between a data source schema and an output geodatabase schema. *See also* schema.

cross-tile indexing [ESRI SOFTWARE] A technique for indexing features that cross tile boundaries in a map library by storing them as one or more features in each tile, instead of storing them each as a single feature. *See also* index.

cross-track error [NAVIGATION] The difference between the actual position and the intended track. *See also* bearing, heading.

crosswalk [DATA MANAGEMENT] A table illustrating equivalent categories in two or more classification schemes.

CRS *See* coordinate system.

CSDGM *See* Content Standard for Digital Geospatial Metadata.

CSV [DATABASE STRUCTURES] Acronym for *comma-separated values.* A format for structuring text within a file that uses a comma to mark where one value ends and another begins.

cubic convolution [MATHEMATICS] A technique for resampling raster data in which the average of the nearest 16 cells is used to calculate the new cell value. *See also* resampling.

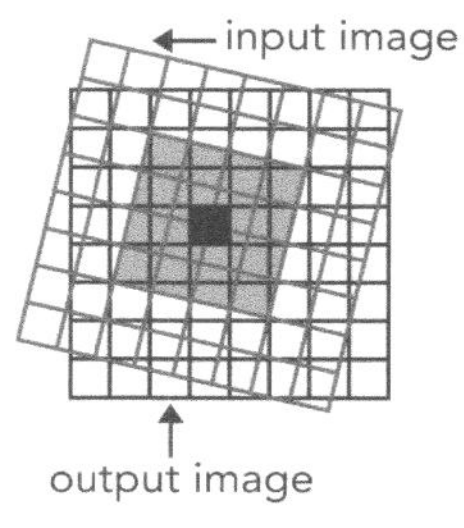

cubic convolution

cultural feature [CARTOGRAPHY] A human-made feature represented on a map, such as a building, road, tower, or bridge. *See also* natural feature.

cultural geography [ENVIRONMENTAL GIS] The field of geography concerning the spatial distribution and patterns created by human cultures and their effects on the earth.

curb approach [NETWORK ANALYSIS] In network analysis, a network location property that models a path for approaching a stop from a specific side based on edge direction. For example, a school bus must approach a school from its door side so that students exiting the bus will not have to cross the street. There are three types of curb approaches: left, right, or both.

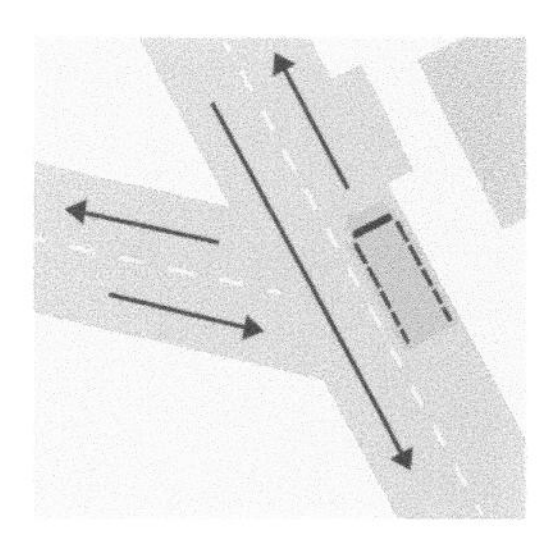

curb approach

Current Atlas [NAVIGATION] A document that shows the currents used for marine navigation, often includes a collection of maps and tide tables.

curvature [EUCLIDEAN GEOMETRY] The amount that a surface deviates from being linear. Mathematically speaking, curvature is a second derivative of a surface, or the slope of a slope. *See also* linear surface, slope, surface.

curve fitting [DATA EDITING] Converting short connected straight lines into smooth curves to represent features such as rivers, shorelines, and contour lines. The curves that result pass through or close to the existing points.

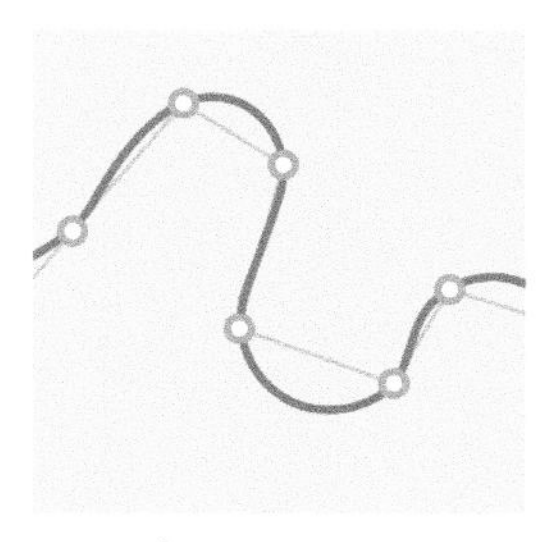

curve fitting

customer market analysis [BUSINESS] A type of market analysis that focuses on data about customers, rather than about a store or stores. An example is desire line analysis. *See also* store market analysis.

customer profiling [BUSINESS] A process that establishes common demographic characteristics for a set of customers within a geographic area. *See also* customer prospecting, demographics.

C

customer prospecting [BUSINESS] A type of market analysis that locates regions with appropriate demographic characteristics for targeting new customers. *See also* customer profiling, demographics.

cut [DEVELOPMENT] A landscape feature that is made when earth is removed from above the desired ground height. *See also* cut/fill.

cut contour [DEVELOPMENT] A type of depression or hachure contour used to identify the place where a roadway has been dug through the landscape, drastically lowering (cutting) the terrain. *See also* hachured contour.

cut/fill [SPATIAL ANALYSIS, ESRI SOFTWARE] A spatial analysis function that summarizes areas and volumes of change between two surfaces.

cutoff meander [GEOGRAPHY] A wandering channel that closes back on itself.

cycle

1. [DATA MODELS] A set of lines forming a closed polygon.
2. [NETWORK ANALYSIS] In network analysis, a path or tour beginning and ending at the same location.
3. [PHYSICS] One oscillation of a wave.

cylindrical equal-area projection [MAP PROJECTIONS] A true-perspective cylindrical map projection that preserves the relative sizes of areas but can introduce distortions of shape and distance, particularly as one moves farther from the equator. *See also* cylindrical projection, equal-area projection.

cylindrical projection [MAP PROJECTIONS] A map projection that transforms points from a spheroid or sphere onto a tangent or secant cylinder; similar to wrapping a piece of paper around a globe. Meridians (lines of longitude) and parallels (lines of latitude) are projected as straight lines that intersect at right angles on the map. Distance, shapes, and area become increasingly distorted as one moves away from the origin. *See also* meridian, origin, parallel, projection.

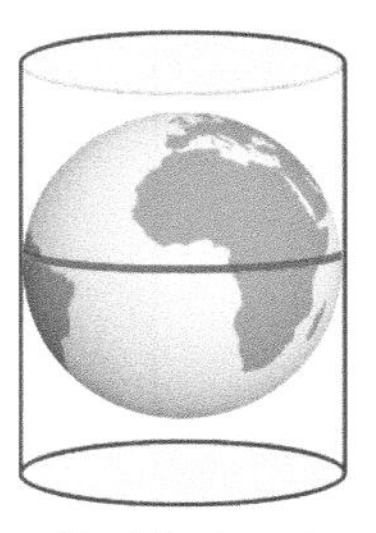

cylindrical projection

cylindrical projection family [MAP PROJECTIONS] A map projection family based on the use of a cylinder as the developable surface. *See also* map projection families, projection.

D

dangle [DATA CAPTURE] The endpoint of a dangling arc. *See also* dangling arc.

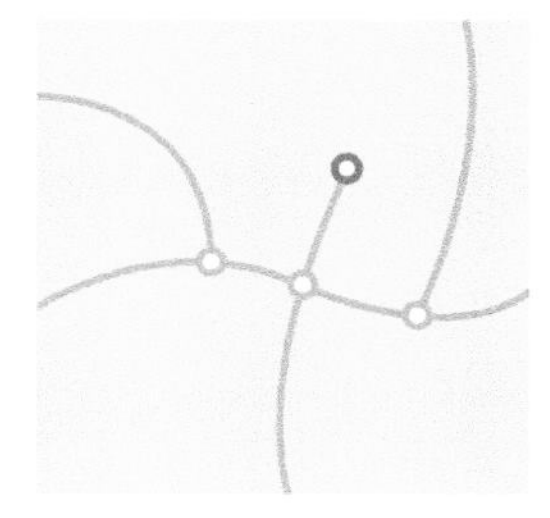

dangle

dangling arc [DATA CAPTURE] An arc with the same polygon on both its left and right sides and at least one node that does not connect to any other arc. A dangling arc often occurs where a polygon does not close properly, where arcs do not connect properly (an undershoot), or where an arc was digitized past its intersection with another arc (an overshoot). It is not always an error; for example, it can represent a cul-de-sac in a street network. *See also* undershoot, overshoot.

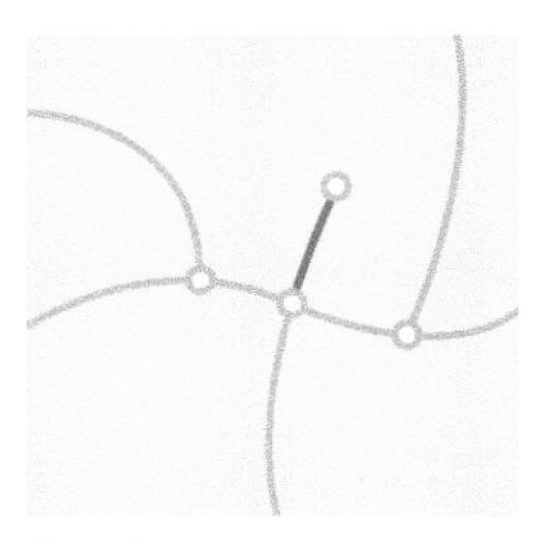

dangling arc

dangling node *See* dangle.

dashboard [VISUALIZATION] A dynamic view of geographic information and supporting data used to monitor events, make decisions, and see trends.

dasymetric mapping [DATA ANALYSIS] A method of reorganizing map data gathered from one data collection unit into inherently more precise areas. The boundaries of the original data collection units are modified using related supporting data, and the attribute values are redistributed within the newly drawn units. For example, a population attribute organized by census tract might be more visually meaningful when areas within which it is reasonable to infer that people do not live (water bodies, vacant land) have been erased. *See also* data collection unit.

data [DATA MANAGEMENT] A collection of facts, statistics, or measurements represented in a structured or unstructured form. In the context of GIS, data typically refers to spatial or geographic information. *See also* data element, spatial data.

data binning [DATA MODELS] A method used to place point data into regular (usually square or hexagonal) cells so that the value of each cell represents the count of points that fall within it. *See also* hexmap.

data capture [DATA CAPTURE] Any operation that converts GIS data into computer-readable form. Geographic data can be captured by being downloaded directly into a GIS from sources such as remote sensing or GPS data, or it can be digitized, scanned, or keyed in manually from paper maps or photographs. *See also* data.

data classification *See* classification.

data collection unit [DATA CAPTURE] A natural or human-defined unit within which information is gathered for inventory or analysis. *See also* dasymetric mapping.

data conversion [DATA CONVERSION] The process of translating data from one format to another.

data dictionary [DATA MANAGEMENT] A catalog or table containing information about the datasets stored in a database. In a GIS, a data dictionary might contain the full names of attributes, meanings of codes, scale of source data, accuracy of locations, and map projections used. *See also* metadata.

data element [DATA TRANSFER] The smallest unit of information used to describe a particular characteristic of a spatial dataset. A data element is a logically primitive description that cannot be further subdivided. *See also* compound element.

data integrity [DATA QUALITY] The degree to which the data in a database is accurate and consistent according to data model and data type. *See also* constraint.

data logger *See* data recorder.

data message [GPS] Information in a satellite's GPS signal that reports its orbital position, operating health, and clock corrections. *See also* data, GPS.

data model

1. [DATA MODELS] In GIS, a mathematical construct for representing geographic objects or surfaces as data. For example, the vector data model represents geography as collections of points, lines, and polygons; the raster data model represents geography as cell matrixes that store numeric values; and the TIN data model represents geography as sets of contiguous, nonoverlapping triangles.
2. [ESRI SOFTWARE] A set of database design specifications for objects in a GIS application. A data model describes the thematic layers used in the application (for example, hamburger stands, roads, and counties); their spatial representation (for example, point, line, or polygon); their attributes; their integrity rules and relationships (for example, counties must nest within states); their cartographic portrayal; and their metadata requirements.

3. [DATA MODELS] In information theory, a description of the rules by which data is defined, organized, queried, and updated within an information system (usually a database management system).

data quality report [DATA MANAGEMENT] Part of the metadata files for a digital dataset that include information about characteristics such as lineage, positional accuracy, and attribute accuracy.

data recorder [GPS] A lightweight, handheld field computer used to store data collected by a GPS receiver.

data repository *See* clearinghouse.

data sharing [DATA SHARING] Making data available and accessible to organizations or individuals other than the creator of the data.

data source [DATA MANAGEMENT] The origin of any data. Data sources may include shapefiles, services, rasters, or feature classes. *See also* data, feature class, source.

data source error [DATA MANAGEMENT] A type of error introduced by using data that is out of date, does not cover the entire area, or does not include all the required features or attributes. *See also* error.

data transfer [DATA TRANSFER] The process of moving data from one system to another or from one point on a network to another.

data type [DATA STORAGE] The attribute of a variable, field, or column in a table that determines the kind of data it can store. Common data types include character, integer, decimal, single, double, and string.

database [DATA STORAGE] One or more structured sets of persistent data, managed and stored as a unit and generally associated with software to update and query the data. A simple database might be a single file with many records, each of which references the same set of fields. A GIS database includes data about the spatial locations and shapes of geographic features recorded as points, lines, areas, pixels, grid cells, or TINs, as well as their attributes. *See also* geodatabase.

database generalization [DATABASE STRUCTURES] The abstraction, reduction, and simplification of features and feature classes for deriving a simpler model of reality or decreasing stored data volumes. *See also* generalization.

data-driven ring analysis [DATA ANALYSIS] A type of market analysis primarily used to look at competing sites or to select potential new locations. *See also* ring study.

dataset [DATA MANAGEMENT] Any collection of related data, usually grouped or stored together. *See also* feature dataset.

dataset precision [DATA QUALITY] The mathematical exactness or detail with which a value is stored within a dataset, based on the number of significant digits that can be stored for each coordinate.

date [DIGITAL IMAGE PROCESSING] An image element that provides information about the date and time an image was acquired; used to help identify an object of interest by providing metadata context. *See also* image element.

datum

1. [GEODESY] The reference specifications of a measurement system, usually a system of coordinate positions on a surface (a horizontal datum) or heights above or below a surface (a vertical datum).
2. [COORDINATE SYSTEMS, PHOTOGRAMMETRY] In imagery, the origin and orientation of lines on a surface. Datums are used as part of a coordinate reference system to provide a frame of reference when measuring location on the earth's surface.
3. [GEOREFERENCING] A collection of highly accurate control points used for georeferencing map data so that it aligns with the geodetic latitude and longitude coordinate system; a standard position of level from which measurements are taken.

See also geocentric datum, geodetic datum, geographic transformation, local datum.

datum contour *See* zero contour.

datum level [GEODESY] A surface to which heights, elevations, or depths are referenced. *See also* vertical geodetic datum.

datum plane *See* datum level.

datum shift *See* geographic transformation.

datum transformation *See* geographic transformation.

DD *See* decimal degrees.

dead reckoning [NAVIGATION] A navigation method of last resort that uses the most recently recorded position of a ship or aircraft, along with its speed and drift, to calculate a new position. *See also* navigation, position, wayfinding.

decimal degrees [MAP PROJECTIONS] Values of latitude and longitude expressed in decimal format rather than in degrees, minutes, and seconds. The decimal degree calculation is dd + mm / 60 + ss / 3600, where dd is the number of whole degrees, mm is the number of minutes, and ss is the number of seconds.

declination

1. [COORDINATE SYSTEMS] In a spherical coordinate system, the angle between the equatorial plane and a line to a point somewhere on the sphere.

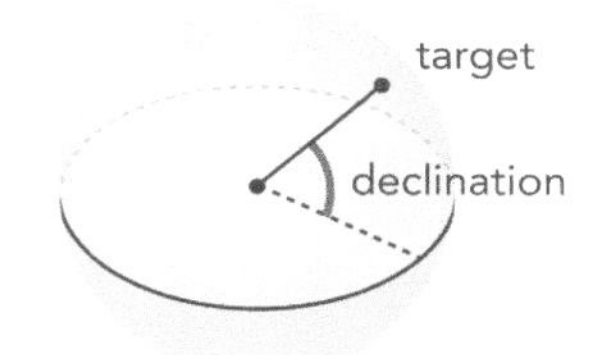

declination 1

2. [COORDINATE SYSTEMS] The arc between the equator and a point on a great circle perpendicular to the equator.
3. [ASTRONOMY] The angular distance between a star or planet and the celestial equator.

See also magnetic declination, celestial sphere.

declination diagram [GEOGRAPHY] A diagram that shows the angular relationship among grid north, magnetic north, and true north. *See also* grid north, magnetic north, true north.

deep learning [PROGRAMMING] In image analysis, a machine learning technique that teaches a computer to filter inputs through layers to learn how to predict and classify information in imagery. Deep learning relies on a large number of training samples representing features and objects of interest, then employs complex algorithms to predict outcomes that are compared to training data iteratively to develop a training model. The training model is then run against new imagery with characteristics similar to the training samples to predict and identify features and objects defined by the training samples. *See also* AI, inferencing, machine learning, neural network, training, training samples.

Defence Geospatial Information Working Group [ORGANIZATIONAL ISSUES] Also known by the acronym *DGIWG*. A group established in 1983 to develop standards for spatial data exchange among nations participating in the North Atlantic Treaty Organization (NATO). The goals of the group are interoperability and burden sharing among nations, and its membership has recently expanded beyond NATO nations. Although DGIWG is not an official NATO body, its work on standards has been recognized by the NATO Geographic Conference (NGC). *See also* data sharing, Digital Geographic Information Exchange Standard, interoperability, Open Geospatial Consortium.

defined study area [ESRI SOFTWARE] A study area with a defined boundary, such as a city.

DEFLEC2022 *See* NAPGD2022.

deflection [DATA EDITING] The creation of a segment at an angle relative to an existing segment.

degree

1. [GEODESY] A unit of angular measure represented by the symbol °. The earth is divided into 360 degrees of longitude and 180 degrees of latitude.

2. [MATHEMATICS] The angle equal to 1/360th of the circumference of a circle. A degree can be divided into 60 minutes of arc or 3600 seconds of arc.

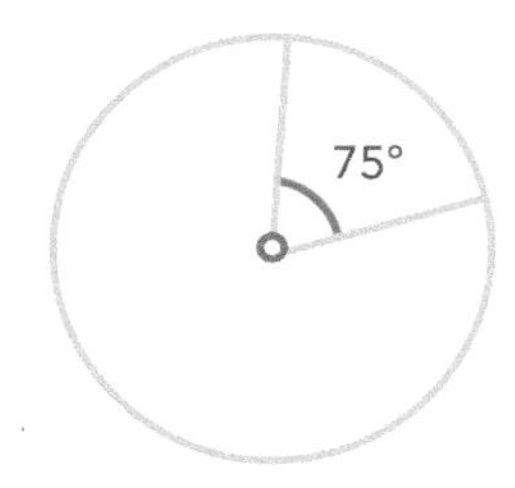

degree 2

See also minute, second.

degree slope [3D ANALYSIS] One method for representing the measurement of an inclined surface. The steepness of a slope may be measured from 0 to 90 degrees. *See also* percent slope.

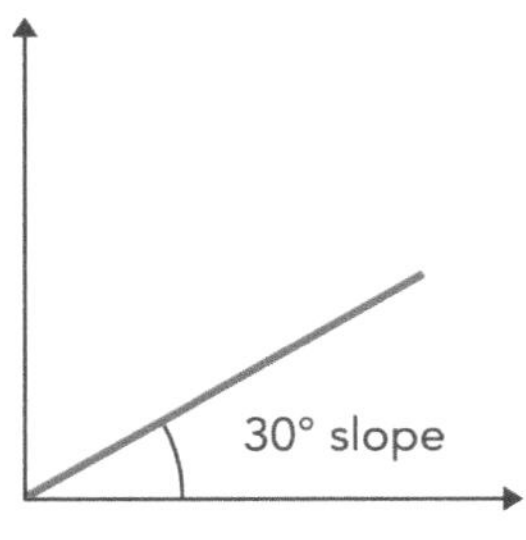

degree slope

degrees-minutes-seconds [COORDINATE SYSTEMS] The unit of measure for describing latitude and longitude. A degree is 1/360th of a circle. A degree is further divided into 60 minutes, and a minute is divided into 60 seconds.

Delaunay triangles [3D ANALYSIS] The components of Delaunay triangulation. Delaunay triangles cannot exist alone; they must exist as part of a set or collection that is typically referred to as a triangulated irregular network (TIN). A circle circumscribed through the three nodes of a Delaunay triangle will not contain any other points from the collection in its interior. *See also* Delaunay triangulation, Voronoi diagram, Thiessen polygon.

Delaunay triangulation [3D ANALYSIS] A technique for creating a mesh of contiguous, nonoverlapping triangles from a dataset of points. Each triangle's circumscribing circle contains no points from the dataset in its interior. Delaunay triangulation is named for the Russian mathematician Boris Nikolaevich Delaunay. *See also* distance weighting.

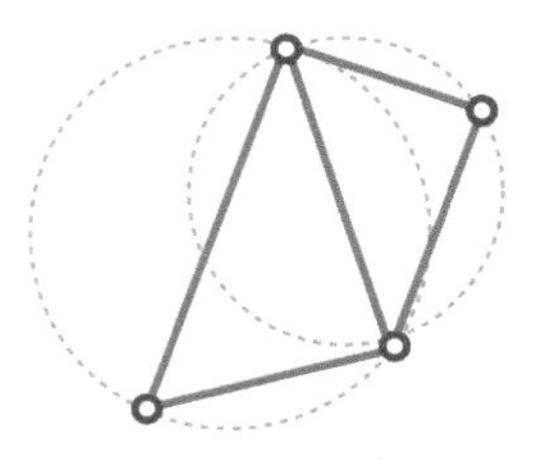

Delaunay triangulation

delimiter [COMPUTING] A character, such as a space or comma, that separates words or values. *See also* CSV.

DEM *See* digital elevation model.

demographic map [GEOGRAPHY] A map that shows human population details, such as socioeconomics, age, and income. *See also* demographics.

demographics [GEOGRAPHY] The statistical characteristics (such as age, birth rate, and income) of a human population. *See also* customer profiling, customer prospecting, ecological fallacy.

demography [GEOGRAPHY] The statistical study of human populations, especially their locations, distribution, economic statistics, and vital statistics. *See also* demographics.

dendogram [DATA MODELS, LINEAR REFERENCING] A diagram with a treelike structure that represents hierarchical clustering or relationships; a decision tree. *See also* dendritic, dendritic drainage pattern.

dendritic [DATA MODELS, LINEAR REFERENCING] Branching.

dendritic drainage pattern [DATA MODELS, LINEAR REFERENCING] A drainage pattern in which the main river has tributaries that have their own tributaries. From the air, these can look like the veins in a leaf or the branching roots of a tree. *See also* dendritic.

densify [DATA EDITING] To add vertices to a line at specified distances without altering the line's shape. *See also* spline.

densitometer [GRAPHICS (MAP DISPLAY)] An instrument for measuring the opacity of translucent materials such as photographic negatives and optical filters. *See also* microdensitometer.

density

1. [SPATIAL STATISTICS (USE FOR GEOSTATISTICS)] The measurement of the intensity or concentration of a particular phenomenon within a given area. Density is a spatial analysis technique used to understand the distribution and pattern of features or events across a geographic area.
2. [PHYSICS] In a substance such as a gas, solid, or liquid, a measurement of the ratio of mass to volume.
3. [SPATIAL ANALYSIS] A function that distributes the quantity or magnitude of point or line observations over a unit of area to create a continuous raster—for example, population per square kilometer.

See also data collection unit, population density.

density measure [SPATIAL ANALYSIS] A measure that describes the congestion or sparseness of the spatial pattern.

density slicing [REMOTE SENSING] A technique normally applied to a single-band monochrome image for highlighting areas that appear to be uniform in tone but are not. Grayscale values (0–255) are converted into a series of intervals, or slices, and distinct colors are assigned to each slice. Density slicing is often used to highlight

D

variations in vegetation. *See also* digital image processing, monochromatic, remote sensing.

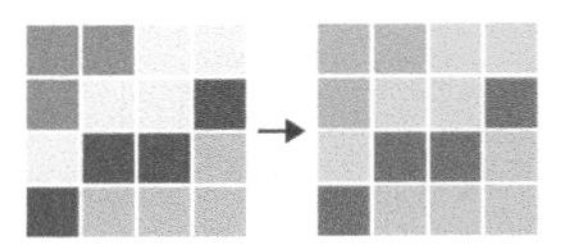

density slicing

density value [SPATIAL ANALYSIS] A value obtained by dividing a count value by the area of its data collection unit. *See also* data collection unit.

dependent variable [STATISTICS] The variable representing the process being predicted or modeled, such as crime, foreclosure, or rainfall. The dependent variable is a function of the independent variables. Regression can be used to predict the dependent variable, using known (observed) values to build (calibrate) the regression model. In the regression equation, the dependent variable appears on the left side of the equal sign. *See also* independent variable, regression equation.

depot [NETWORK ANALYSIS] A network location used to represent a starting, stopping, or renewal location for routes in vehicle routing problem (VRP) analysis. Users can specify multiple depots. Depots are used as locations for loading/unloading vehicles within the fleet. *See also* route renewal, vehicle routing problem.

depot visit [NETWORK ANALYSIS] An object used to represent a single visit to a specific depot in vehicle routing problem (VRP) analysis. A depot visit may occur at the start of a route, the end of the route, or as a renewal midway along a route. *See also* vehicle routing problem.

depression [CARTOGRAPHY] An area surrounded on all sides by higher ground.

depression contour *See* hachured contour.

depth contour [SYMBOLOGY] A contour line on a map connecting points of equal depth below a hydrographic datum. *See also* bathymetric map, bathymetry, contour line, hydrographic datum, submerged contour.

depth curve *See* depth contour.

derivative bands [REMOTE SENSING] The result of processing imagery to create transformed bands containing information or characteristics that differ from the original bands. For example, the result might be a vegetation index.

derived data [MODELING] Data created by running a geoprocessing operation on existing data. Derived data from one process can serve as input data for another process.

derived value *See* derived data.

descending node [ASTRONOMY] The point at which a satellite traveling north to south crosses the equator. *See also* ascending node.

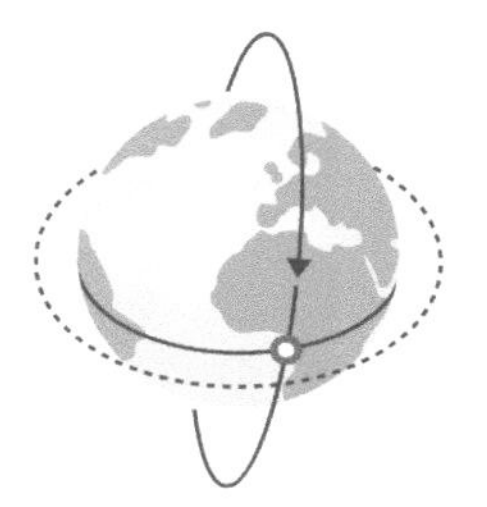

descending node

descriptive data *See* attribute data.

descriptive statistics [DATA ANALYSIS] The analysis of data that helps describe, show, or summarize data in a meaningful way. There are two general types of descriptive statistics: measures of central tendency and measures of spread. *See also* measures of central tendency, measures of spread.

descriptor [NETWORK ANALYSIS] A type of attribute for network elements that cannot be apportioned. The value of a descriptor stays the same through the length of an edge element in a network dataset. Descriptors describe characteristics of the element; for example, the number of lanes for a particular road in a road network. *See also* cost, hierarchy, network attribute.

designated market area [MEDIA, RETAIL] Also known by the acronym *DMA*. A United States region of television viewers as defined by Nielsen Media Research. Most DMAs correspond to whole counties, but there are a few exceptions where counties are split into different DMAs. *See also* market area.

desire-line analysis [DATA ANALYSIS] A type of market analysis that draws lines from a set of geocoded points (usually customers) to a single, central point (usually a store). Desire lines can be weighted. *See also* weight.

destination

1. [COMPUTING] The secondary object in a relationship class, such as a table containing attributes associated with features in a related table.
2. [NETWORK ANALYSIS] A location that defines the ending point of a route or network.

See also least-cost path, origin, origin-destination cost matrix, path, route, stop.

determinate flow direction [NETWORK ANALYSIS] A conclusively definitive line or course in which something is issuing or moving in a stream. For an edge feature, this occurs when the flow direction can be ascertained from the connectivity of a network, the locations of sources and sinks, and the enabled or disabled states of features. *See also* indeterminate flow direction.

Détermination d'Orbite et Radiopositionnement Intégré par Satellite *See* DORIS.

D

deterministic model [SPATIAL STATISTICS (USE FOR GEOSTATISTICS)] In spatial modeling, a type of model or a part of a model in which the outcome is completely and exactly known based on known input; the fixed or nonrandom components of a spatial model. The spline and inverse distance weighted interpolation methods are deterministic since they have no random components. The kriging and cokriging interpolation methods may have a deterministic component, often called the trend. *See also* kriging, model, stochastic model, trend.

detrending [STATISTICS] The process of removing the trend from a spatial model by subtracting the trend surface (usually polynomial functions of the spatial x- and y-coordinates) from the original data values. The resulting detrended values are called residuals. *See also* trend.

developable surface [MAP PROJECTIONS] A surface that can be flattened on a plane without geometric distortion; in the case of map projections, the flat surface on which the earth's features are projected. There are three types of developable surfaces used in map projections: planes, cones, and cylinders. *See also* projection.

device coordinates [GRAPHICS (COMPUTING)] The coordinates shown on a digitizer or display, as opposed to those of a recognized datum or coordinate system. *See also* coordinates, digitizer.

DGIWG *See* Defence Geospatial Information Working Group.

DGPS *See* differential correction.

diazo process [OUTPUT] A way of quickly and inexpensively copying maps using a diazo compound, ultraviolet light, and ammonia.

difference image [DIGITAL IMAGE PROCESSING] In image processing, an image made by subtracting the pixel values of one image from those in another. *See also* change detection.

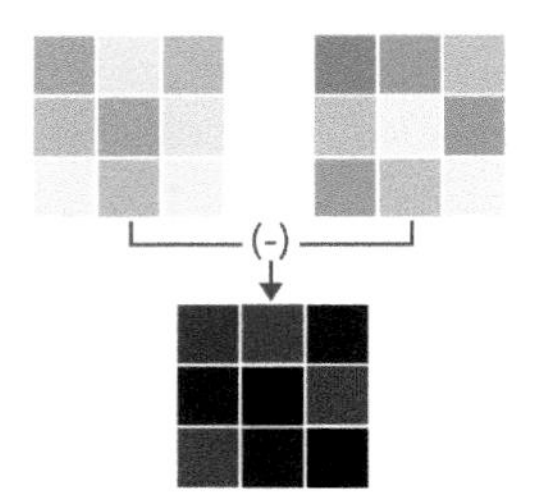

difference image

differential correction [GPS]

1. A technique for increasing the accuracy of GPS/GNSS measurements by comparing observed with known ranges or coordinates at a fixed base station and transmitting corrections to roving receivers.

differential correction 1

2. A method of improving the accuracy of a GPS/GNSS location using corrections broadcast from a GPS/GNSS base station.

See also base station, GNSS, GPS.

differential Global Positioning System *See* differential correction.

diffusion

1. [DIFFUSION] The spread of an innovation or technology use among a group of people or organizations.
2. [GEOGRAPHY] The spread or scatter of a geographic phenomenon over space.

See also geographic phenomena.

DIGEST *See* Digital Geographic Information Exchange Standard.

digital elevation model [DATA MODELS]

1. Also known by the acronym *DEM*. A general term referencing modeled elevation values over a topographic surface, whether a gridded raster or an irregular TIN. DEMs can refer to bare earth or a top surface that includes trees and buildings.

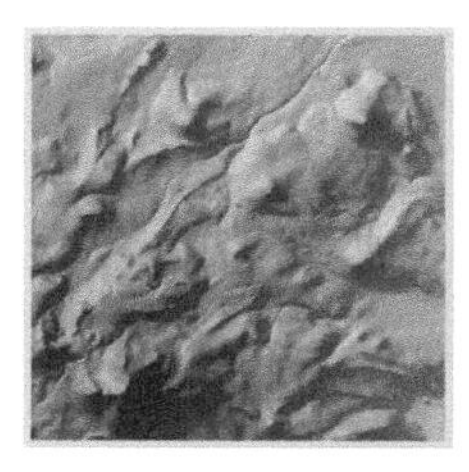

digital elevation model 1

2. A raster dataset produced from a sample of elevations or depths taken on a regular grid, typically used to represent the bare-earth terrain, void of vegetation and human-made features; sometimes confused with digital terrain model or digital surface model.
3. A format for elevation data, tiled by map sheet, produced by the National Mapping Division of the USGS.

See also digital surface model, digital terrain model, raster, TIN dataset, z-value.

Digital Geographic Information Exchange Standard [STANDARDS] Also known by the acronym *DIGEST*. A standard for spatial data transfer among nations, data producers, and data users. The Defence Geospatial Information Working Group (DGIWG) developed the standard to support interoperability within and between nations and share the burden of digital data production. The standard addresses the exchange of raster, matrix, and vector data (and associated text) and a range of levels of topological structures. *See also* Defence Geospatial Information Working Group, Open Geospatial Consortium.

Digital Geographic Information Working Group *See* Defence Geospatial Information Working Group.

digital image

1. [DIGITAL IMAGE PROCESSING] An image composed of pixels with numbers representing grayscale or color shades.

D

2. [GRAPHICS (COMPUTING)] An image stored in binary form and divided into a matrix of pixels. Each pixel consists of a digital value of one or more bits, defined by the bit depth. The digital value may represent, but is not limited to, energy, brightness, color, intensity, sound, elevation, or a classified value derived through image processing. A digital image is stored as a raster and may contain one or more bands.

See also image, analog image, raster.

digital image processing [REMOTE SENSING] Any technique that changes the digital values of an image for the purpose of analysis or enhanced display, such as density slicing or low- and high-pass filtering. *See also* density slicing.

digital line graph [DATA MODELS] Also known by the acronym *DLG*. Data files containing vector representations of cartographic information derived from USGS maps and related sources. DLGs include information from the USGS planimetric map base categories such as transportation, hydrography, contours, and public land survey boundaries. *See also* planimetric map, U.S. Geological Survey, vector.

digital nautical chart [NAVIGATION] Also known by the acronym *DNC*. A nautical database developed from existing hard-copy charts, digital data, bathymetric survey information, imagery, and various raster data. DNCs are used by the U.S. military and its allies for marine navigation. *See also* electronic navigational chart.

digital number [REMOTE SENSING] In a digital image, a value assigned to a pixel. *See also* digital image, pixel.

digital orthophoto quadrangle [AERIAL PHOTOGRAPHY] Also known by the acronym *DOQ*. A 1-meter resolution digital orthorectified image basemap, comprising an area measuring 7.5-minutes longitude by 7.5-minutes latitude, covering the United States at a scale of 1:24,000, created from 1:40,000-scale aerial photography. Produced for, and distributed by, the United States Geological Survey (USGS). Digital orthophoto quadrangles are true photographic maps in which the effects of tilt and relief are removed by transformation or rectification. The uniform scale of a DOQ allows accurate measurement of distances. *See also* orthophoto, transformation.

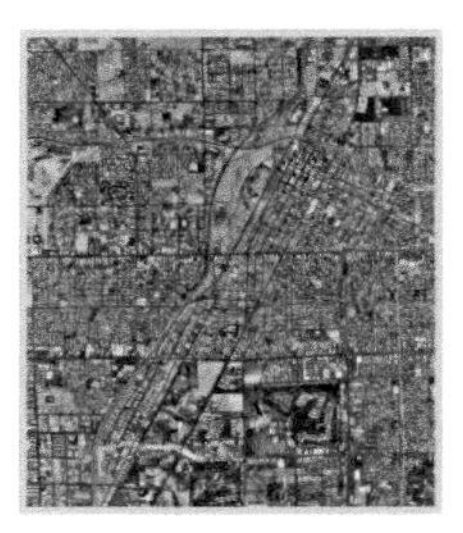

digital orthophoto quadrangle

digital orthophoto quarter quadrangle [AERIAL PHOTOGRAPHY] Also known by the acronym *DOQQ*. A digital orthophoto quadrangle (DOQ) divided into four quadrants. A 1-meter resolution digital orthorectified image basemap comprising an area measuring 3.75-minutes longitude by 3.75-minutes latitude, covering the United States at a scale of 1:12,000, in the UTM projection, referenced to the NAD27 or NAD83 datum. May be black and white (B/W), natural color, or color infrared (CIR). Produced for, and distributed by, the United States Geological Survey (USGS). *See also* digital orthophoto quadrangle, orthophoto.

digital raster graphic [DATA MODELS] Also known by the acronym *DRG*. A raster image of a scanned USGS standard series topographic map, usually including the original border information, referred to as the map collar, map surround, or marginalia. Source maps are georeferenced to the surface of the earth, fit to the universal transverse Mercator (UTM) projection, and scanned at a minimum resolution of 250 dpi. The accuracy and datum of a DRG match the accuracy and datum of the source map. *See also* map surround, UTM.

digital surface model [DATA MODELS] Also known by the acronym *DSM*. A gridded raster representing the highest visible surface, including vegetation and human-made features, at every pixel. *See also* digital elevation model, digital terrain model.

digital terrain elevation data *See* DTED.

digital terrain model [DATA MODELS] Also known by the acronym *DTM*. The representation of continuous elevation values over a topographic surface by an array of z-values that may be irregularly spaced to incorporate the elevation of significant topographic features—including mass points and breaklines—to better characterize the true shape of the bare-earth terrain and referenced to a common datum. With a DTM, the distinctive terrain features are more clearly defined and precisely located, and contours generated from DTMs more closely approximate the real shape of the terrain. *See also* digital elevation model, digital surface model.

digital twin [DATA MODELS] A virtual simulation of a physical, real-world object, concept, or process. The first practical digital twin was developed by NASA in 2010 to simulate spacecraft prior to construction. Digital twins are specifically useful in risk analysis by modeling potentially high-cost or high-risk scenarios. *See also* AI, machine learning.

digitize [DATA CONVERSION] To convert maps or data to digital format. *See also* data capture, device coordinates, digitizing.

digitizer [DATA CAPTURE]

1. A field device used to convert real-world objects to digital x,y

coordinates, yielding vector data consisting of points, lines, and polygons.
2. The title of a person who uses a digitizer.
3. An optical device that translates an analog image into an array of digital pixel values. A video digitizer can be used in place of a manual digitizer, but since it produces a raster image, additional software must be used to convert the data into vector format before topological analysis can be done.

See also data capture.

digitizing [DATA CAPTURE] The process of converting the geographic features on an analog map into digital x,y coordinates, or spatial data. *See also* digitizer.

Dijkstra's algorithm [NETWORK ANALYSIS] An algorithm that examines the connectivity of a network to find the shortest path between two points. Conceived by Dutch computer scientist Edsger Dijkstra (1930–2002). *See also* shortest path.

dilution of precision [GEODESY] Also known by the acronym *DOP*. An indicator of satellite geometry for a constellation of satellites used to determine a position. Positions with a lower DOP value generally constitute better measurement results than those with higher DOP. Factors determining the total GDOP (geometric DOP) for a set of satellites include PDOP (positional DOP), HDOP (horizontal DOP), VDOP (vertical DOP), and TDOP (time DOP). *See also* horizontal dilution of precision, precision, satellite constellation.

DIME *See* Dual Independent Map Encoding.

dimension

1. [PHYSICS] A length of a certain distance and bearing.
2. [PHYSICS] The area over which an entity extends.
3. [SPATIAL ANALYSIS] A measure of spatial extent, especially width, height, or length. A point has zero dimensions, a line has one dimension, a plane or polygon has two dimensions, and a feature with volume has three dimensions.
4. [PHYSICS] The number of axes that are essential to the existence of an entity in space. For example, the identity of a location on a plane requires two axes; therefore, a plane exists in the second dimension, and an entity with two axes, or dimensions, may be uniquely identified as a plane.

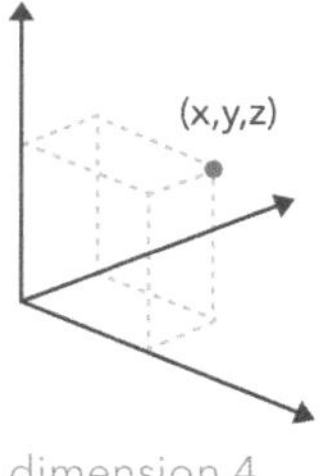

dimension 4

See also axis, bearing, distance, extent.

dimension construction method
[ESRI SOFTWARE] One of several procedures that dictate what type of dimension feature is created and the number of points required to complete the feature's geometry. Construction methods include simple aligned, aligned, linear, rotated linear, free aligned, and free linear. *See also* annotation construction method, dimension.

dimension feature class [ESRI SOFTWARE] A geodatabase feature class that stores dimension features. *See also* feature class.

DIP *See* digital image processing.

direct georeferencing [REMOTE SENSING] A methodology that provides accurate exterior orientation for remotely sensed imagery without input from ground survey data. Direct georeferencing connects aerial images to geographic positioning on the earth's surface through integrated GPS (GNSS), inertial measurement units (IMU), and other aiding sensors.

directed network flow [NETWORK ANALYSIS] A network state in which edges have an associated direction of flow. In a directed network flow, the resource that traverses a network's components cannot choose a direction to take, as in hydrologic and utility systems. *See also* downstream, undirected network flow, upstream.

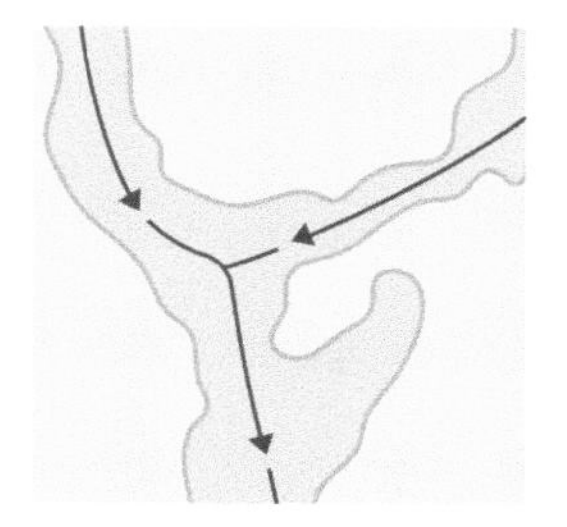

directed network flow

direction

1. [ESRI SOFTWARE] In a vertical coordinate system, an indicator that z-values are either positive up or positive down. Heights or elevations are typically positive up, or against the force of gravity; depths are typically positive down, or with the force of gravity.
2. [COORDINATE SYSTEMS] The difference between the reference line and the direction line given in angular units.

See also cost-weighted direction, direction line, directional influences.

direction line [COORDINATE SYSTEMS] The line from an observer to a distant object. *See also* direction, grid azimuth, grid north, reference line.

directional filter [DIGITAL IMAGE PROCESSING] In image processing, an edge-detection filter that enhances those linear features in an image that are oriented in a particular direction. *See also* digital image processing, edge detection.

D

directional influences [SPATIAL STATISTICS (USE FOR GEOSTATISTICS)] Natural or physical processes that affect a measured trait or attribute so that the magnitude of the effects on the attribute vary in different directions.

Dirichlet tessellation *See* Voronoi diagram.

dirty areas [DATA QUALITY] Regions surrounding features that have been altered after the initial topology validation process and require additional topology validation. *See also* cleaning, error, validation.

disconnected editing [ESRI SOFTWARE] The process of copying data to another geodatabase, editing that data, and merging the changes with the source geodatabase.

discrete cosine transform [DIGITAL IMAGE PROCESSING] A method of lossy image compression; used by JPEG format, for example.

discrete data [DATA MODELS] Data that represents phenomena with distinct boundaries. Property lines and streets are examples of discrete data. *See also* continuous data.

discrete digitizing [DATA CAPTURE] A method of digitizing in which points are placed individually to define a feature's shape. *See also* digitizing, stream mode digitizing.

discrete feature [ESRI SOFTWARE] A feature that has definite feature boundaries.

discrete raster [DATA MODELS, ESRI SOFTWARE] A raster that typically represents phenomena that have clear boundaries with attributes that are descriptions, classes, or categories. It is assumed that the phenomenon represented by each value fills the entire area of its cell. Rasters representing land use, political boundaries, or ownership are examples of discrete rasters. *See also* continuous raster.

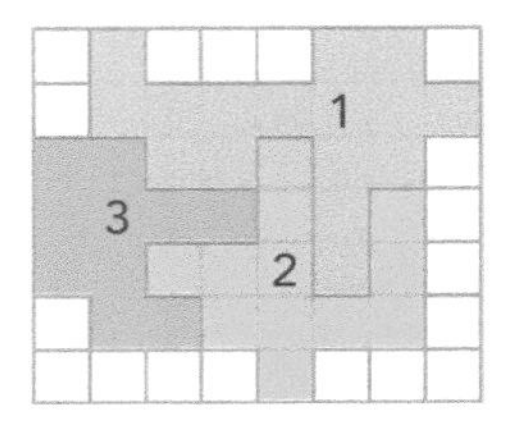

discrete raster

displacement [MAP DESIGN] The adding of space between features on a map so that the features can be visually distinguished and a correct interpretation of their ground relationships can be portrayed.

display projection

1. [MAP PROJECTIONS] The coordinate system used for displaying geographic data on an output device.
2. [MAP PROJECTIONS, ESRI SOFTWARE] A pseudo plate carrée projection used to display data that is in a geographic coordinate system. The angular values of the geographic coordinate system are directly mapped to the display, just as values from a projected coordinate system are mapped.

See also equirectangular projection.

display scale [MAP DISPLAY] The scale at which data is rendered on a computer screen or on a printed map. *See also* automation scale.

display unit [MAP DISPLAY] The unit of measure used to render dimensions of shapes, distance tolerances, and offsets on a computer screen or on a printed map. Although they are stored with consistent units in the dataset, users can choose the units in which coordinates and measurements are displayed—for example, feet, miles, meters, or kilometers. *See also* unit of measure.

disproportionate symbol error [CARTOGRAPHY] An error that occurs when the size of the symbol used to map a feature is not proportional to its ground size. *See also* error.

dissemination *See* diffusion.

dissolve

1. [ESRI SOFTWARE] A geoprocessing command that removes boundaries between adjacent polygons that have the same value for a specified attribute.

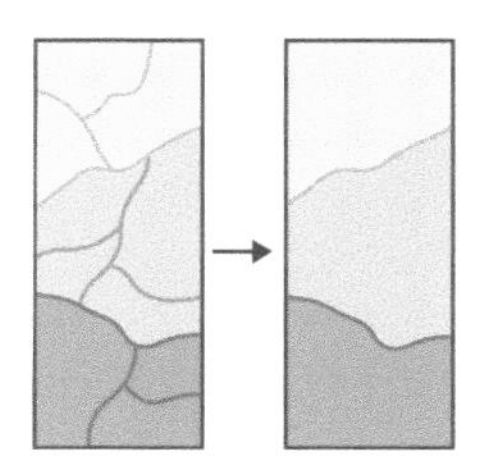

dissolve 1

2. [DATA EDITING] The process of removing unnecessary boundaries between features, such as the edges of adjacent map sheets, after data has been captured.

dissolve join *See* dissolve.

distance [PHYSICS] The measure of separation between two entities or locations that may or may not be connected, such as two points. Distance is differentiated from length, which implies a physical connection between entities or locations. *See also* distance unit, spatial statistics.

distance band [SPATIAL ANALYSIS] The radius, or circle of influence assigned to points in fixed-distance weighting.

distance decay [SPATIAL ANALYSIS] A mathematical representation of the effect of distance on the accessibility of locations and the number of interactions between them, reflecting the notion that demand drops as distance increases. Distance decay can be expressed as a power function or as an exponential function.

distance inset [CARTOGRAPHY] A map notation, usually a double-edged arrow, that indicates the distance between significant features, such as cities.

distance segment [CARTOGRAPHY] A ground distance annotation often added between neighboring places, such as highway intersections, or towns on highway maps.

distance table [CARTOGRAPHY] A table that identifies measured ground distances between a select set of locations.

distance unit [PHYSICS] The unit of measurement for distance, such as feet, miles, meters, and kilometers. *See also* unit of measure.

distance weight [CARTOGRAPHY] A numeric value associated with a feature; the higher the value, the greater the weight for that feature. A feature with a larger weight has more influence on the pattern than something with a smaller weight.

distance weighting [SPATIAL ANALYSIS] A method for factoring the numeric value associated with a feature based on distance so that nearer features have larger distance weights and farther features have smaller distance weights.

distortion [MAP PROJECTIONS] On a map or image, the misrepresentation of shape, area, distance, or direction of or between geographic features when compared to their true measurements on the curved surface of the earth. *See also* equal-area projection, orthophoto, radiometric correction, Tissot's indicatrix.

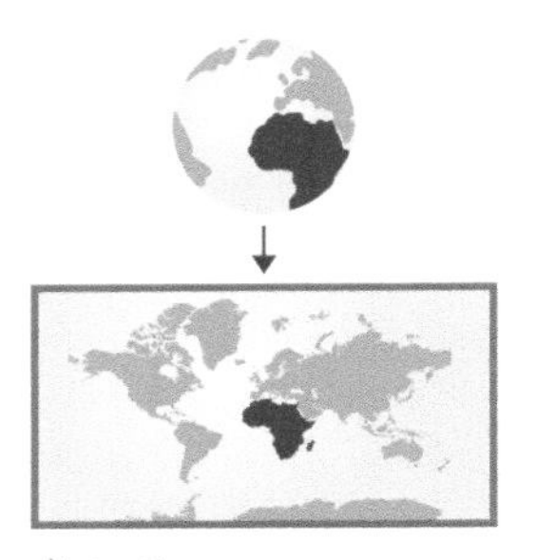

distortion

distortion circle *See* Tissot's indicatrix.

distributed data [ESRI SOFTWARE] Data spread over multiple platforms or a network by a process referred to as replication.

distributed database [DATABASE STRUCTURES] A database with records that are dispersed between two or more physical locations. Data distribution allows two or more people to be working on the same data in separate locations. *See also* database.

distribution [STATISTICS]

1. The frequency or amount at which a thing or things occur within a given area.
2. The set of probabilities that a variable will have a particular value.

See also estimation, histogram.

distributive flow map [CARTOGRAPHY] A map that uses line symbols of variable thickness to show the distribution of commodities or some other flow that diffuses from one or a few origins to multiple destinations.

The flow line's width is typically made proportional to some magnitude. *See also* flow map.

dithering [GRAPHICS (COMPUTING)] The approximation of shades of gray or colors in a computer image made by arranging pixels of black and white or other colors in alternate layers. The technique gives the appearance of a wider range of color or shades than is actually present in the image. It is widely used to improve the appearance of images displayed on devices with limited color palettes.

dithering

diurnal [ASTRONOMY] Daily, as in the revolution of the earth. *See also* diurnal arc.

diurnal arc [ASTRONOMY] The apparent path from rise to set made by a celestial body across the sky. *See also* arc, celestial sphere, diurnal, sphere.

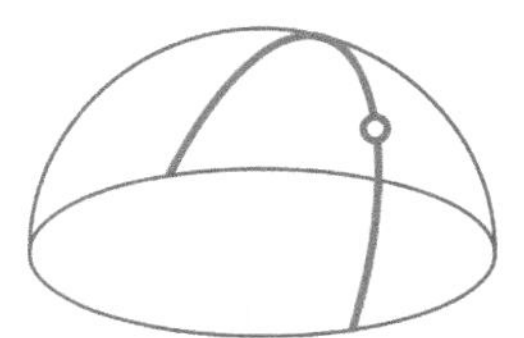

diurnal arc

divergence analysis [PHOTOGRAMMETRY] A statistical technique used for classifying pixels or objects according to how far they diverge from a reference spectra.

diverging color scheme [SYMBOLOGY] A color scheme with variations in two hues that range from light to dark, with a light color such as white or gray in the middle; used to show data that has a critical value in the midrange of the distribution from which other values differ progressively.

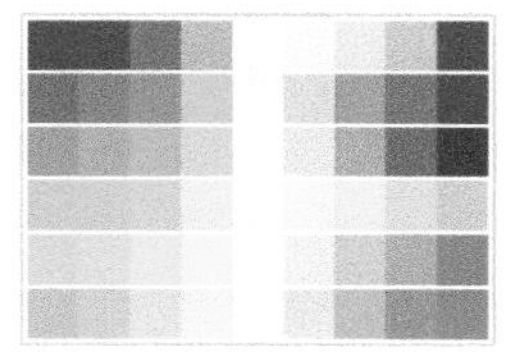

diverging color scheme

diversity [CARTOGRAPHY] A measure that describes spatial variation in various patch types (such as land-cover classes or ecosystems) within the landscape.

diversity index

1. [MODELING] A measure of pattern arrangement that considers the relative abundance of each category for the kernel at each cell location.
2. [ESRI SOFTWARE] Summarizes racial and ethnic diversity, indicating the likelihood that two individuals, chosen at random from the same area, belong to the same race or ethnic group. The index ranges from 0 (no diversity) to 100 (highest diversity).

An area's Diversity Index increases when the population includes more race/ethnic groups.

See also kernel.

diversity index range [DATA ANALYSIS] The range between the minimum and maximum diversity index values possible for the collection of categories on a map. *See also* diversity index.

diversity measure [ECOLOGY] A measure used in ecological analysis to describe the spatial arrangement of habitat units.

DLG *See* digital line graph.

DLS [SURVEYING] Acronym for *Dominion Land Survey.* The public land survey system used to divide most of Canada's western provinces into townships and sections for agricultural and other purposes. *See also* PLSS.

DMA *See* designated market area.

DMS *See* degrees-minutes-seconds.

DNC *See* digital nautical chart.

domain [DATA TRANSFER] The range of valid values for a particular metadata element.

donut rings [SPATIAL ANALYSIS] A spatial configuration where one polygon is nested within another polygon, creating a ringlike or annular shape. Donut rings are commonly used to represent complex spatial relationships, such as multipart features with exclusions or voids, where the inner polygon represents an area that does not belong to the outer polygon, such as lakes within a park or islands within a body of water. *See also* concentric rings, ring study.

DOP *See* dilution of precision.

Doppler effect *See* Doppler shift.

Doppler Orbitography and Radiopositioning Integrated by Satellite *See* DORIS.

Doppler radar map [METEOROLOGY] A map made from modern weather radar that uses the Doppler technique to examine the motion of precipitation. *See also* Doppler shift.

Doppler shift [PHYSICS] An apparent change in wave frequency caused by distance between the source and an observer. As they approach one another, the frequency increases; as they draw apart, the frequency decreases. The Doppler shift is also known as the Doppler effect and is named for Austrian physicist and mathematician Christian Andreas Doppler (1803–1853). *See also* azimuth, Doppler radar map, Doppler-aided GPS, frequency.

Doppler-aided GPS [GPS] Signal processing that uses a measured Doppler shift to help the receiver track the GPS signal. *See also* Doppler shift.

DOQ *See* digital orthophoto quadrangle.

DOQQ *See* digital orthophoto quarter quadrangle.

DORIS [GPS] Acronym for *Détermination d'Orbite et Radiopositionnement Intégré par Satellite* or *Doppler Orbitography and Radiopositioning Integrated by Satellite.* A satellite constellation used for the determination of satellite orbits and for positioning. DORIS was initiated and is maintained by the French space agency, Centre National d'Études Spatiales (CNES). *See also* Argos, GNSS, satellite constellation, SPOT.

dot density map [CARTOGRAPHY] A quantitative, thematic map on which point symbols of the same size are randomly placed in proportion to a numeric attribute associated with an area. Dot density maps convey the intensity or distribution of an attribute.

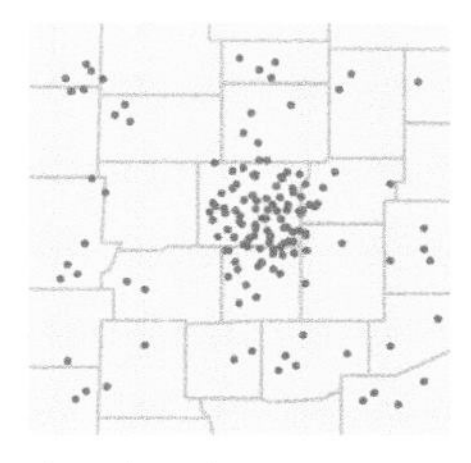

dot density map

dot screen [PRINTING] A photographic film covered with uniformly sized, evenly spaced dots used to break up a solid color, producing an apparently lighter color. *See also* halftone image, continuous tone image.

dot size [CARTOGRAPHY] The diameter of the dots on a dot density map. *See also* dot density map.

dot unit value [CARTOGRAPHY] The amount that each dot represents on a dot density map; the value must be greater than 1. *See also* dot density map.

dots per inch [GRAPHICS (COMPUTING)] Also known by the acronym *dpi.* A measure of the resolution of scanners, printers, and graphic displays. The more dots per inch, the more detail can be displayed in an image. *See also* image, measurement, resolution.

double precision [COMPUTING] The level of coordinate exactness based on the possible number of significant digits that can be stored for each coordinate. Datasets can be stored in either single or double precision. Double-precision geometries store up to 15 significant digits per coordinate (typically 13 to 14 significant digits), retaining the accuracy of much less than 1 meter at a global extent. *See also* single precision.

double-coordinate precision *See* double precision.

Douglas-Peucker algorithm *See* Ramer-Douglas-Peucker algorithm.

downstream [NETWORK ANALYSIS] In network tracing, the direction along a line or edge that is the same as the direction of flow. *See also* upstream, directed network flow.

dpi *See* dots per inch.

drafting [CARTOGRAPHY] A method of drawing with pencil or pen and ink, used in cartographic reproduction.

D

drainage [DATA ANALYSIS] All map features associated with the movement and flow of water, such as rivers, streams, and lakes. *See also* bifurcation ratio.

draping [MAP DISPLAY] A perspective or panoramic rendering of a two-dimensional image superimposed onto a surface. In the most common example, an aerial photograph is draped over a digital terrain model (DTM) or triangulated irregular network (TIN) to create a more realistic terrain visualization. *See also* digital terrain model, triangulated irregular network.

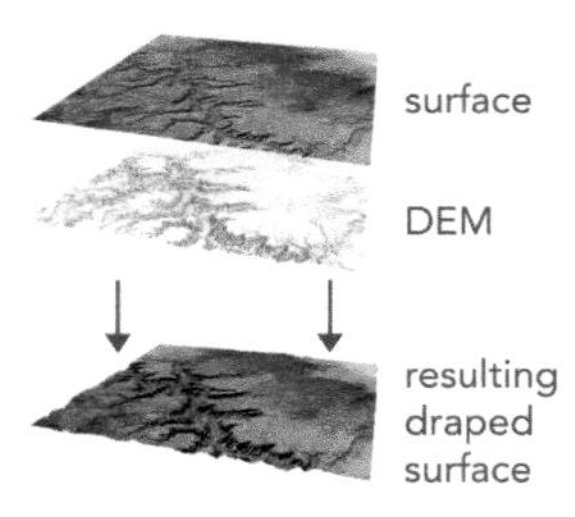

draping

drawing priority [MAP DISPLAY] In 3D analysis, the order in which layers that occupy the same x,y,z positions are drawn in a scene. For example, if a road feature layer and an orthophotograph are draped over the same terrain model, the roads and raster may appear patchy or broken up where they coincide. The drawing priority for the raster can be reduced so it will appear below the features. The drawing priority can only be changed for polygon features and surfaces. *See also* orthophoto.

DRG *See* digital raster graphic.

drift [SPATIAL STATISTICS (USE FOR GEOSTATISTICS)] The general pattern of z-values throughout a kriging model. The drift, or structure, forms the model's basic shape. *See also* kriging, z-value.

drive-time area [SPATIAL ANALYSIS] A zone around a map feature measured in units of time needed for travel by car. For example, a store's 10-minute drive-time area defines the area in which drivers can reach the store in 10 minutes or less. *See also* buffer.

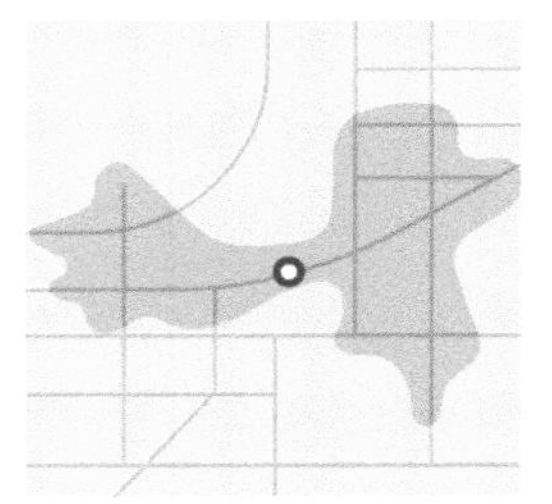

drive-time area

drone imagery [AERIAL PHOTOGRAPHY] Still images and video gathered from sensors mounted on remotely piloted vehicles. Drones can be uncrewed aerial vehicles (UAV) or uncrewed aerial systems (UAS). *See also* UAS.

DSM *See* digital surface model.

DTED [DATA MODELS] Acronym for *digital terrain elevation data.* A format for elevation data, usually tiled in 1-degree cells, produced by the National

Geospatial-Intelligence Agency and U.S. allies for military applications. *See also* digital elevation model, Earth Gravitational Model.

DTM *See* digital terrain model.

Dual Independent Map Encoding [DATA MODELS] Also known by the acronym *DIME*. A data storage format for geographic data developed by the U.S. Census Bureau in the 1960s. DIME-encoded data was stored in Geographic Base Files (GBF). The Census Bureau replaced the DIME format with Topologically Integrated Geocoding and Referencing (TIGER) in 1990. *See also* GBF/DIME, TIGER.

dynamic feature class [ESRI SOFTWARE] A feature class consisting of points associated with address elements in an address data table that change based on changes made to the address data table. *See also* feature class, point, route event source.

dynamic image map [MAP DISPLAY] A display that includes animated sequences of maps shown in a predefined order. The display often allows the map reader to interact with map symbols, so that symbols are linked to text, pictures, or video clips.

dynamic map [MAP DISPLAY] A type of map that allows for interactive and real-time updates or changes in response to user input or changing data.

dynamic qualitative thematic map [MAP DESIGN] A map that focuses on changes in feature locations and qualitative attributes over time. *See also* thematic map.

dynamic segmentation [DATA ANALYSIS] A technique used in GIS to locate, manage, and analyze linear features, such as roads, pipelines, or rivers, as continuous and dynamic entities. It involves dividing linear features into a series of connected segments, each with its own set of attributes, without segmenting the underlying feature. Attributes can include information such as road names, speed limits, elevation profiles, or any other relevant data associated with the linear feature. *See also* linear referencing.

Earth Gravitational Model [DATA MODELS] Geopotential models of the earth published by the National Geospatial-Intelligence Agency (NGA) used as the geoid reference in the World Geodetic System (WGS). *See also* DTED, geoid, National Geospatial-Intelligence Agency, WGS72.

earth-centered datum *See* geocentric datum.

earth-centered earth-fixed [COORDINATE SYSTEMS] A Cartesian coordinate system that is valid for the entire earth, with its origin at the center of mass for the earth and its axes fixed relative to Earth's rotation. The x-axis is in the plane of the equator, passing through the origin and extending from 180 degrees longitude (negative) to the prime meridian (positive); the y-axis is also in the plane of the equator, passing through the origin and extending from 90 degrees west longitude (negative) to 90 degrees east longitude (positive); the z-axis is the line between the north and south poles, with positive values increasing northward, and negative values increasing southward. *See also* Cartesian coordinate system, x-axis, y-axis, z-axis.

easement [CADASTRAL AND LAND RECORDS] A right held by one person to make limited use of another person's land (property), typically for the purpose of access to adjoining land.

easting [COORDINATE SYSTEMS]

1. The distance east of the origin that a point in a Cartesian coordinate system lies, measured in that system's units.

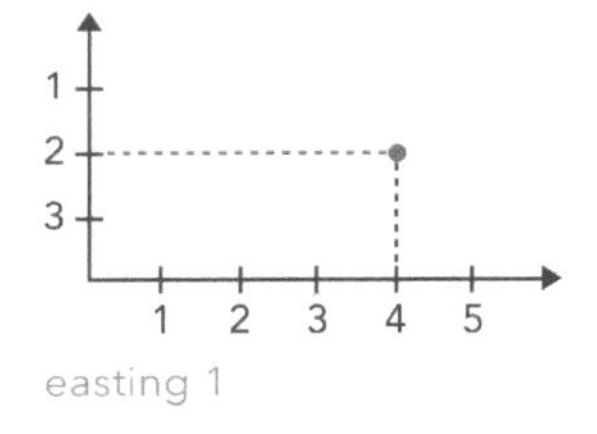

easting 1

2. The x-coordinate distance measured east from the origin in a grid coordinate system.

See also grid coordinate system, northing.

eccentricity [EUCLIDEAN GEOMETRY] A measure of how much an ellipse deviates from a circle, expressed as the ratio of the distance between the center and one focus of an ellipsoid to the length of its semimajor axis. The square of the eccentricity (e^2) is commonly used with the semimajor axis *a* to define a spheroid.

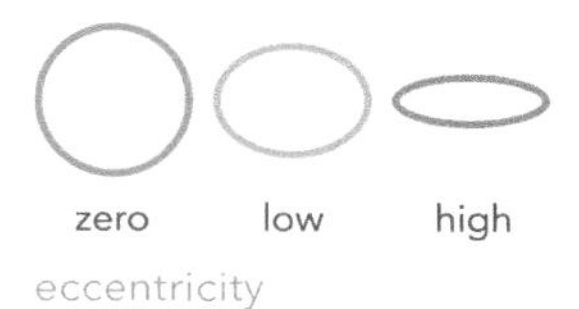

eccentricity

ECEF *See* earth-centered earth-fixed.

ecliptic [ASTRONOMY]

1. The great circle formed by the intersection of the plane of the earth's orbit around the sun (or apparent orbit of the sun around the earth) and the celestial sphere.

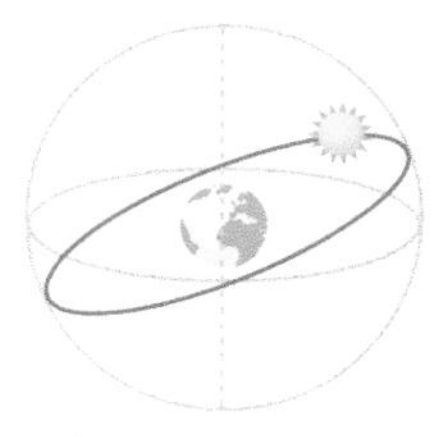
ecliptic 1

2. The mean plane of the earth's orbit around the sun.

See also celestial sphere.

ecological fallacy [STATISTICS] The assumption that an individual from a specific group or area will exhibit a trait that is predominant in the group as a whole.

economic geography [GEOGRAPHY] The field of geography concerning the distribution and variation of economic factors by location, including how economic factors interact with geographic factors such as climate, land use, and geology. *See also* geography.

ecoregion [ECOLOGY] An area defined by its environmental conditions, especially climate, landforms, and soil characteristics.

ED50 [COORDINATE SYSTEMS] Acronym for *European Datum of 1950.* A datum created after World War II to be a consistent reference datum for most of Western Europe; however, Belgium, France, Great Britain, Ireland, Sweden, Switzerland, and the Netherlands continue to retain and use their own national datums. Latitude and longitude coordinates in this datum are based on the International Ellipsoid of 1924. *See also* datum, International Ellipsoid of 1924.

edge

1. [DATA MODELS] A line between two points that forms a boundary. In a geometric shape, an edge forms the boundary between two faces. In an image, edges separate areas of different tones or colors. In topology, an edge defines lines or polygon boundaries.

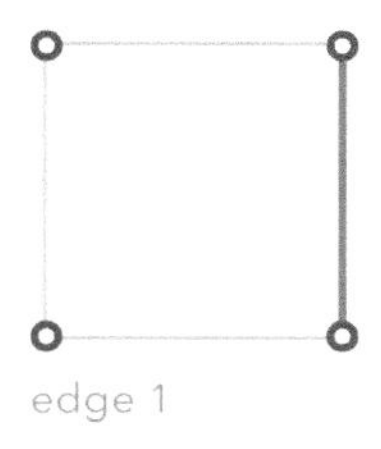
edge 1

2. [NETWORK ANALYSIS] In a network system, a line feature through which a substance, resource, or traffic flows. Examples include a street

in a transportation network and a pipeline in a sewer system.

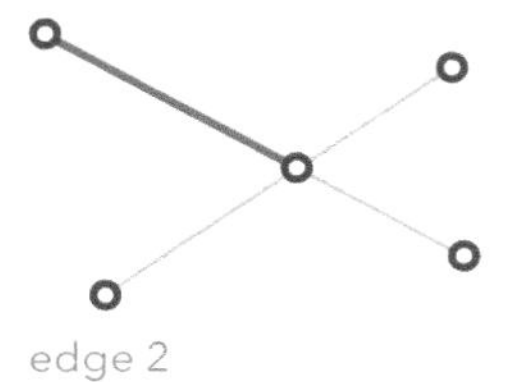

edge 2

E

3. [3D ANALYSIS] In a TIN data model, a line segment between nodes (sample data points). Edges store topologic information about the faces that they border.

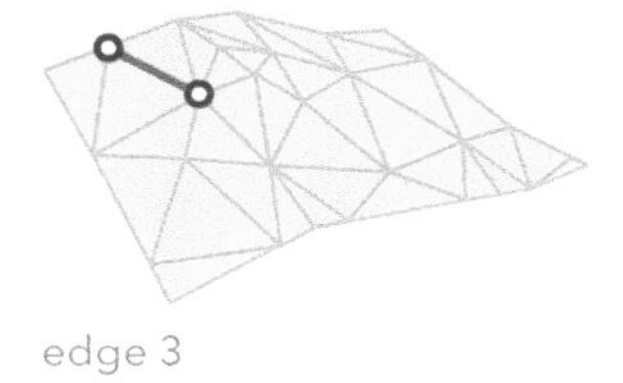

edge 3

See also boundary, endpoint connectivity, junction.

edge connectivity policy [NETWORK ANALYSIS] In a network dataset, a connectivity policy that defines how one edge may connect to another edge mid-span. There are two edge-edge connectivity policies: endpoint connectivity and any-vertex connectivity. *See also* any-vertex connectivity, connectivity policy, endpoint connectivity, junction connectivity policy.

edge detection [DIGITAL IMAGE PROCESSING] A digital image processing technique for isolating edges in a digital image by examining it for abrupt changes in pixel value.

edge element *See* edge.

edge enhancement [DIGITAL IMAGE PROCESSING] A digital image processing technique for emphasizing the appearance of edges and lines in an image. *See also* high-pass filter.

edge-junction cardinality [ESRI SOFTWARE] In a network connectivity relationship, the number of edges of one type that may be associated with junctions of another type. Edge-junction cardinality defines a range of permissible connections that may occur in a one-to-many relationship between a single junction and many edges. *See also* cardinality.

edgematching [DATA EDITING] A spatial adjustment process that aligns features along the edge of an extent to the corresponding features in an adjacent extent. *See also* rubber sheeting.

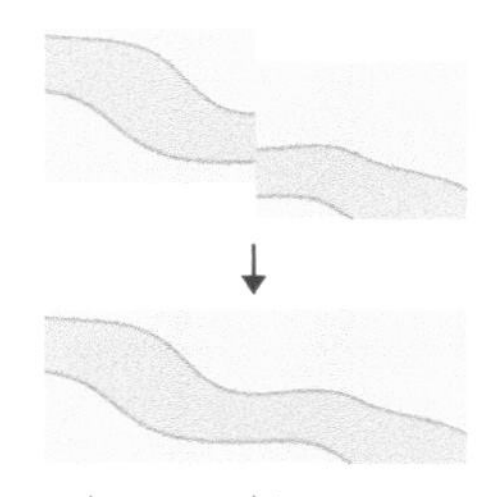

edgematching

edit mask [DATA EDITING] The portion of a coverage where the geometry (or geographic features) has been altered, but where topology has not yet been restored. *See also* mask.

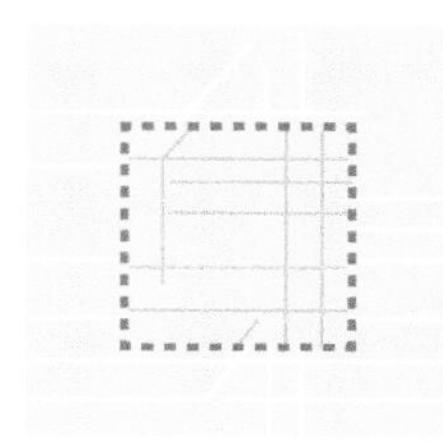

edit mask

EGNOS [GPS] Acronym for *European Geostationary Navigation Overlay Service.* A regional satellite-based augmentation system (SBAS) that supplements Galileo and GPS positioning by providing reliability and accuracy assessments and corrections over European airspace. EGNOS was developed by the European Space Agency and EUROCONTROL on behalf of the European Commission. *See also* Galileo, GPS.

elapsed time [CARTOGRAPHY] The length of time between a map's date of completion and its date of use.

elastic transformation *See* rubber sheeting.

electromagnetic energy [PHYSICS] Radiant energy that exhibits both particle and wave phenomena. It can be characterized by wavelength or frequency. It ranges from gamma rays and X-rays through ultraviolet, visible, and infrared light and out to microwave and radio frequencies. The wave energy model is based on opposing, oscillating electric and magnetic waves that are at 90 degrees to each other and are orthogonal to the direction of propagation. At propagation, in a vacuum, energy waves move at 3 × 108 m/s, the speed of light. The wave energy model was defined by Scottish physicist James C. Maxwell in the nineteenth century. The particle-based model, introduced in the twentieth century by German-born theoretical physicist Albert Einstein, is used to describe how energy interacts with matter, where it is described by behavior of the discrete or quantum nature of the photon. *See also* electromagnetic radiation, electromagnetic spectrum.

electromagnetic radiation [PHYSICS] Energy that moves through space at the speed of light as different wavelengths of time-varying electric and magnetic fields. Types of electromagnetic radiation include gamma, X, ultraviolet, visible, infrared, microwave, and radio. *See also* electromagnetic energy, electromagnetic spectrum.

electromagnetic spectrum [PHYSICS] The range of frequencies of electromagnetic radiation and their respective wavelengths and photon energies. The entire spectrum ranges from frequencies below one hertz to above 1025 hertz. This frequency range is divided into bands characterized by their frequency properties and sensor capabilities, starting with shortwave to longer wavelengths, gamma rays, X-rays, ultraviolet, visible light, infrared, microwaves, and radio waves. The estimated limit of short wavelengths is the Planck length; there is no known limit for long wavelengths. *See also* band, electromagnetic energy, electromagnetic radiation, multispectral.

electronic atlas [MAP DISPLAY] A mapping system that displays but does not allow for the spatial analysis of data.

electronic navigational chart [NAVIGATION] Also known by the acronym *ENC*. A vector data product used for nautical navigation. ENC data is produced by nautical charting agencies throughout the world and uses the IHO (International Hydrographic Organization) S-57 standard for its database structure and attribution. *See also* digital nautical chart.

element [ESRI SOFTWARE] In geoprocessing, a component of a model. Elements can be variables, such as input and derived data, or tools.

elevation [GEODESY] The vertical distance of a point or object above or below a reference surface or datum (generally mean sea level). Elevation generally refers to the vertical height of land. *See also* altitude, vertical geodetic datum.

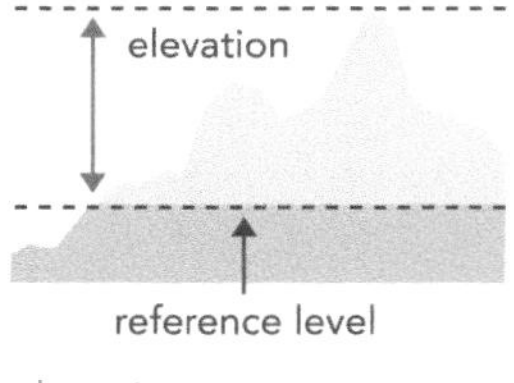

elevation

elevation drawing [ARCHITECTURE] A drawing of a building seen from one side. An architectural term for a profile. *See also* profile.

elevation guide [SYMBOLOGY] A map element that displays a simplified representation of the terrain within a map's extent. Elevation guides are designed to provide a quick overview of topography, including the high and low points. *See also* elevation tints.

elevation tints [SYMBOLOGY] Hypsometric tint bands based on elevation ranges used in an elevation guide. *See also* elevation guide, hypsometric tinting.

ellipse [EUCLIDEAN GEOMETRY] A geometric shape described mathematically as the collection of points whose distances from two given points (the foci) add up to the same sum. An ellipse is shaped like a circle viewed obliquely. *See also* circle, eccentricity, ellipsoid.

ellipsoid

1. [EUCLIDEAN GEOMETRY] A three-dimensional, closed geometric shape, all planar sections of which are ellipses or circles. An ellipsoid has three independent axes and is usually specified by the lengths a,b,c of the three semiaxes. If an ellipsoid is made by rotating an ellipse about one of its axes, two axes of the ellipsoid are the same, and it is called an ellipsoid of revolution, or spheroid. If the lengths of all three of its axes are the same, it is a sphere.

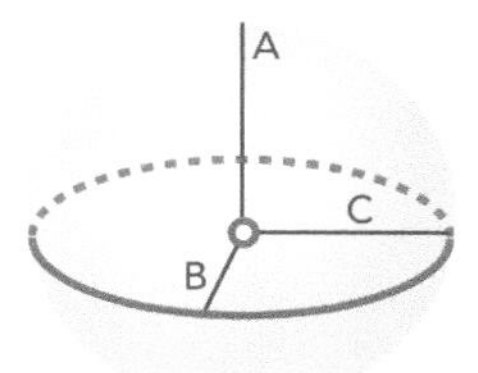

ellipsoid 1

2. [GEODESY] When used to represent the earth, an oblate ellipsoid of revolution made by rotating an ellipse about its minor axis.
3. [CARTOGRAPHY] A 3D figure obtained by rotating an ellipse 180 degrees about its polar (minor) axis.

See also geoid.

ellipsoid height [GEOREFERENCING] In mapping, the elevation above a mathematical model that approximates the shape of the earth, planets, or moons. Ellipsoid height is distinct from height above the geoid, which uses a geopotential surface for mean sea level. *See also* height above ellipsoid.

ellipsoidal geodesics [CARTOGRAPHY] The study of geodesic distances on an ellipsoid. *See also* geodetic distance.

ellipticity *See* eccentricity.

embedded feature class [3D GIS] A multipoint feature class embedded into a terrain dataset. When a feature class is embedded, it is incorporated directly into the terrain pyramid and the terrain becomes the sole container of the data. Embedded feature classes can be used to reduce the amount of disk space required by mass point data such as lidar. *See also* feature class.

empirical [STATISTICS] That property of a quantity that indicates that the quantity depends on data, observations, or experiment only; that is, it is not a model or part of a model. An empirical semivariogram is computed on data only, in contrast to a theoretical semivariogram model.

ENC *See* electronic navigational chart.

encoding [DATA CONVERSION] The recording or reformatting of data into a computer format. Data may be encoded to reduce storage, increase security, or transfer it between systems using different file formats. In GIS, analog graphic data, such as paper maps and images, is encoded into computer formats by scanning or digitizing.

end hatch definition [ESRI SOFTWARE] In linear referencing, a special type of hatch definition that draws hatch marks only at the low and high measure of a linear feature. *See also* hatching, linear referencing.

end offset [ESRI SOFTWARE] An adjustable value that dictates how far away from the end of a line an address location should be placed. Using an end offset prevents the point from being placed directly over the intersection of cross streets if the address happens to fall at the beginning or end of the street. *See also* side offset.

endpoint connectivity [NETWORK ANALYSIS] In network datasets, a type of edge connectivity policy that states that an edge may only connect to another edge at its endpoints. *See also* any-vertex connectivity, connectivity policy, edge connectivity policy.

E

engineering plan [ENGINEERING] An accurate, large-scale reference map often used by city engineers or public works departments.

English unit [MATHEMATICS] A system of measurement, sometimes referred to as "foot-pound-second," after the base units of length, mass, and time, used in Australia, Canada, India, Malaysia, the Republic of Ireland, the United Kingdom, the United States, and other countries. Also known as imperial unit. *See also* metric system, unit of measure.

enhancement [REMOTE SENSING] In remote sensing, applying operations to raster data to improve appearance or usability by making specific features more detectable. Such operations can include contrast stretching, edge enhancement, filtering, smoothing, and sharpening. *See also* contrast stretch, edge enhancement, filter, raster, smoothing.

enroute aeronautical chart [NAVIGATION] A digital chart used for air navigation.

enterprise GIS [ORGANIZATIONAL ISSUES] A geographic information system that is integrated through an entire organization so that a large number of users can manage, share, and use spatial data and related information to address a variety of needs, including data creation, modification, visualization, analysis, and dissemination. *See also* GIS.

entrance pupil [DIGITAL IMAGE PROCESSING] In imagery, the perspective center at the front of the optics in an optical system. It equates to the point where all incoming rays converge.

enumeration district [GOVERNMENT] The geographic area assigned to a census taker, usually representing a specific portion of a city or county in the United States. *See also* block group, census block, census tract, metropolitan statistical area.

envelope [COORDINATE SYSTEMS] The rectangle surrounding one or more geographical features in coordinate space, determined by the minimum and maximum coordinates in the x and y directions, as well as the ranges of any z- or m-values that the features may have. An envelope can be used to filter data for analysis. *See also* bounding rectangle.

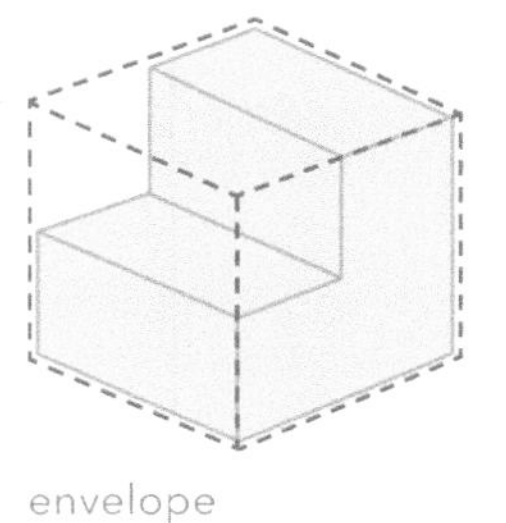

envelope

environmental model [MODELING] An abstract representation of a complex environmental process, emphasizing relationships and patterns in natural systems. Environmental models allow decision-makers to better understand the effects of natural systems or the impact of human activities on natural systems. *See also* model.

ephemeris

1. [GPS] A table of the predicted positions of a satellite within its orbit for each day of the year, or for other regular intervals.
2. [ASTRONOMY] A table that charts the movement of celestial bodies (planets, moons, and stars) and predicts their positions at a given time.

See also GPS, position.

epidemiological map [EPIDEMIOLOGY] A map that shows the distribution of a disease in a region.

epipolar lines [PHOTOGRAMMETRY] The line on stereo images that intersects an epipolar plane with the image focal planes. Epipolar lines are created so that if a ground point appears on the epipolar line of the first stereo image, that point will occur somewhere on the conjugate epipolar line of the other stereo image. When conjugate epipolar lines are aligned, y-parallax is eliminated. *See also* epipolar plane, focal plane, stereopair.

epipolar plane [PHOTOGRAMMETRY] A plane defined by two exposure stations and a ground point. The airbase vector and conjugate epipolar lines are in the epipolar plane. *See also* airbase vector, epipolar lines, exposure station.

epipole

1. [PHYSICS] The geometric point of intersection of two image projections with the image plane, related to stereo vision.
2. [PHOTOGRAMMETRY] In photogrammetry, the point of intersection of an airbase vector with the extended focal plane.

See also airbase vector, focal plane.

EPSG code [COORDINATE SYSTEMS] EPSG is an acronym for *European Petroleum Survey Group*. An identification attached to one of the geodetic parameters established by the EPSG and maintained by the IOGP Geomatics Committee. The EPSG geodetic parameters is a publicly available dataset composed of coordinate reference systems, datums, transformations, and related information, such as ellipsoids and units of measure. Each parameter is assigned a unique EPSG code and a WKT. *See also* spatial reference, WKID, WKT.

equal competition area [BUSINESS] A trade area boundary set halfway between a store or service point and its neighboring stores or service points.

equal-area classification [CARTOGRAPHY] A data classification method that divides polygon features into groups so that the total area of the polygons in each group is approximately the same. *See also* classification, equal-interval classification.

E

equal-area projection [MAP PROJECTIONS] A map projection that preserves the relative area of regions everywhere on the earth. Also called an equivalent projection. An equal-area projection distorts shape, angle, scale, or any combination of these characteristics. *See also* distortion, projection.

equal-area projection

equal-interval classification [DATA MANAGEMENT] A data classification method that divides a set of attribute values into groups, or class intervals, to obtain an equal range of values. This method emphasizes the amount of an attribute value relative to other values. Sometimes called equal-range intervals or equal intervals. *See also* classification, equal-area classification.

equator [GEODESY] The imaginary circle around the earth that is equidistant from the north and south poles; the zero parallel (0°) used as the reference from which latitude north and south are measured. *See also* parallel, latitude.

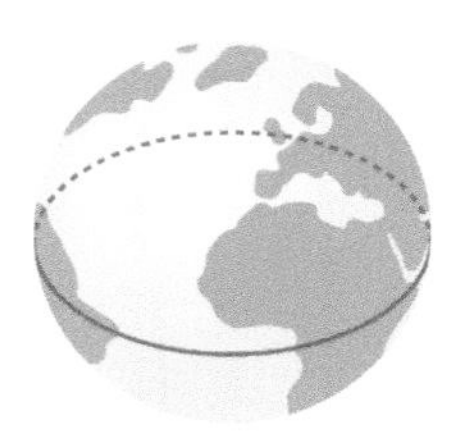

equator

equatorial aspect

1. [MAP PROJECTIONS] A planar (or azimuthal) projection with its central point located at the equator.

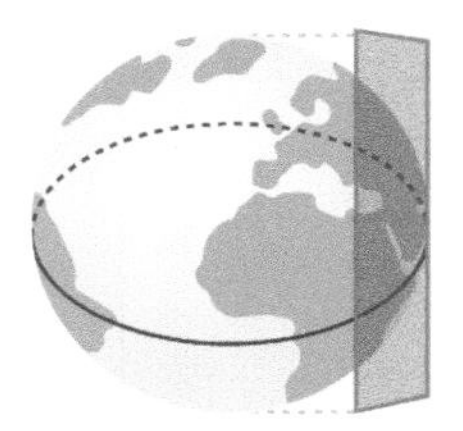

equatorial aspect 1

2. [CARTOGRAPHY] The aspect that occurs when a point or line of tangency is at or along the equator in the tangent case or equidistant from the equator in the secant case.

See also projection.

equatorial axis [GEOGRAPHY] The axis from the center of the earth to the equator. *See also* equator.

equatorial plane [GPS] The plane that the equator would cut through the earth, dividing it into the northern and southern hemispheres. The equatorial plane is perpendicular to the earth's axis of rotation. *See also* axis, hemisphere.

equidistance [MAP PROJECTIONS] The property of preserving the spherical great-circle distance on a map projection. *See also* great circle.

equidistant projection [MAP PROJECTIONS] A projection that maintains scale along one or more lines, or from one or two points to all other points on the map. Lines along which scale (distance) is correct are the same proportional length as the lines they reference on the globe. In the sinusoidal projections, for example, the central meridian and all parallels are their true lengths. An azimuthal equidistant projection centered on Chicago shows the correct distance between Chicago and any other point on the projection but not between any other two points. *See also* projection, scale.

azimuthal equidistant

equidistant projection

equirectangular projection [MAP PROJECTIONS] A cylindrical map projection on which parallels and meridians are mapped as a grid of equally spaced horizontal and vertical lines twice as wide as high. Also known as the equidistant cylindrical projection, geographic projection, or plate carrée projection. *See also* cylindrical projection.

equivalence [MAP PROJECTIONS] The property of preserving the relative size of regions on the earth on a map projection.

equivalent projection *See* equal-area projection.

error

1. [DATA QUALITY] A measured, observed, calculated, or interpreted value that differs from the true value or the value that would be obtained by a perfect observer using perfect equipment and perfect methods under perfect conditions.
2. [ACCURACY] In a GIS database, a spatial or attribute value that differs from the true value. An error may also be understood as the totality of wrong or unreliable information in a database. A spatial error can be positional (incorrect coordinates) or topological (features do not properly connect, intersect, or adjoin). An attribute error can be the result of incorrect quantities or descriptions or due to missing or invalid values.

   There are many processes that can potentially introduce errors to a GIS database, for example, as the result of human errors, flawed instruments, map digitizing errors, data integration errors of scale, data collection or conversion errors, and so on.
3. [ESRI SOFTWARE] In geodatabase topology, violation of a topology rule detected during the validation process.

See also error propagation, uncertainty.

error budget [REMOTE SENSING] In photogrammetry and remote sensing, a set of error estimates that can be propagated collectively in a sensor model. Contains covariance terms to characterize expected geopositioning errors. *See also* sensor model.

error matrix [REMOTE SENSING] In remote sensing, a table used to determine thematic map accuracy by comparing the map classification results with reference data. The accuracy assessment produces various measures of feature identification accuracy such as User's accuracy (errors of omission), Producer's accuracy (errors of commission), and Kappa statistic. *See also* Kappa coefficient.

error propagation [DATA MANAGEMENT] In GIS data processing, the persistence of an error into new datasets calculated or created using datasets that originally contained errors. The study of error propagation is concerned with the effects of combined and accumulated errors throughout a series of data processing operations. *See also* error.

error triangle [WAYFINDING] A small triangle, created using the resection method, in which the observer's position can be assumed to be found. *See also* compass method, inspection method, resection method.

Esri Grid [ESRI SOFTWARE] An Esri data format for storing raster data that defines geographic space as an array of equally sized square cells arranged in rows and columns. Each cell stores a numeric value that represents a geographic attribute (such as elevation) for that unit of space. When the grid is drawn as a map, cells are assigned colors according to their numeric values. Each grid cell is referenced by its x,y coordinate location. *See also* raster.

Esri JSON [PROGRAMMING, ESRI SOFTWARE] A variant of the JavaScript Object Notation (JSON) file and data interchange format used to represent spatial data, such as maps, layers, features, and attributes. Esri JSON includes properties and structures specific to Esri GIS data models and workflows. *See also* GeoJSON, JavaScript Object Notation.

estimation [SPATIAL STATISTICS (USE FOR GEOSTATISTICS)] In spatial modeling, the process of forming a statistic from observed data to assign optimal parameters in a model or distribution. *See also* prediction.

ETRS89 [COORDINATE SYSTEMS] Acronym for *European Terrestrial Reference System 1989*. The reference coordinate system for Europe. When its datum was first defined in 1989, it was coincident with the World Geodetic System of 1984 (WGS84) datum. Because the Eurasian continental tectonic plate is largely static, ETRS89 continues to be Europe's primary reference system. In areas of the world in which the tectonic plates are moving, there can be a difference of more than half a meter between WGS84 and ETRS89. *See also* datum, WGS84.

Euclidean distance [EUCLIDEAN GEOMETRY] The straight-line (shortest) distance between two points on a plane. Euclidean distance can be calculated using the Pythagorean theorem.

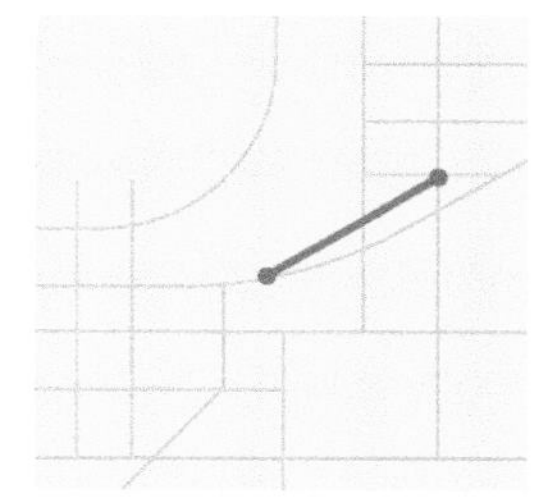

Euclidean distance

Euclidean distance analysis [SPATIAL ANALYSIS] A description of each raster cell's relationship to the closest source. *See also* Euclidean distance, source.

Euclidean geometry [MATHEMATICS] The geometry of planes and solids based on Euclid's axioms, such as the principles that parallel lines never cross, the shortest distance is a straight line, and space is three-dimensional.

European Geostationary Navigation Overlay Service *See* EGNOS.

European Terrestrial Reference System 1989 *See* ETRS89.

evaluator [NETWORK ANALYSIS] A function that determines attribute values for network elements in a network dataset. If a network source does not have an evaluator, the default evaluator for its element type is used. *See also* function, network element.

event [LINEAR REFERENCING] A geographic location stored in tabular rather than spatial form. Event types include address events, route events, x,y events, and temporal events. *See also* address event, route event, x,y event.

E

event overlay [LINEAR REFERENCING] In linear referencing, an operation that produces a route event table that is the logical intersection or union of two input route event tables. Event overlay is one way to perform line-on-line, line-on-point, and event point-on-point overlays. *See also* event, route event.

event table [LINEAR REFERENCING] A data source containing location information in tabular format (called events) that is used to create a spatial dataset. For example, an event table might contain x,y coordinates or routes. *See also* event.

event theme [VISUALIZATION, ESRI SOFTWARE] A spatial data theme created from an event table. *See also* event table.

event transform *See* transform events.

exception [ESRI SOFTWARE] An error that is an acceptable violation of a topology rule. *See also* error.

explanatory variable *See* independent variable.

explode [ESRI SOFTWARE] An editing process that separates a multipart feature into its component features, which become independent features. *See also* multipart feature.

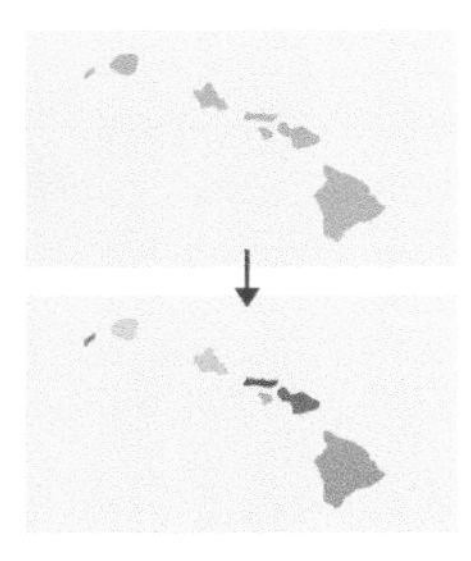

explode

exponent [MATHEMATICS] In mathematics, a number that indicates how many times a base value is multiplied by itself. Exponents are usually indicated with superscripts. *See also* fractal, logarithm.

exposure station [AERIAL PHOTOGRAPHY] In aerial photography, each point in the flight path at which the sensor exposes the film. *See also* aerial photograph, remote sensing.

expression [MATHEMATICS] A sequence of operands and operators constructed according to the syntactic rules of a symbolic language that evaluates to a single number, string, or value. *See also* operator.

extended postal code *See* ZIP+4 code.

extent

1. [VISUALIZATION] The amount of geographic area that can be seen on a map.
2. [DATA ANALYSIS] The minimum bounding rectangle (xmin, ymin and xmax, ymax) defined by coordinate pairs of a data source. All coordinates for the data source fall within this boundary.
3. [MATHEMATICS] The size and scale of an object.

See also map extent, minimum bounding rectangle.

extent rectangle [ESRI SOFTWARE] A rectangle that is displayed in one data frame, showing the size and position of another data frame. *See also* extent.

exterior orientation [REMOTE SENSING] The position and attitude of a sensor at the time of image exposure that establishes the relationship between the ground and the image. Values are provided in the units used by the ground reference system. *See also* sensor.

external polygon *See* universe polygon.

extraction guide [DATA MANAGEMENT] A specification that defines parameters for feature extraction and attribution. Specifications typically include the size of features to be collected, density of feature collection, scale ranges, and attribute assignment. *See also* data dictionary.

extrapolation [STATISTICS] Using known or observed data to infer or calculate values for unobserved times, locations, or other variables outside a sampled area. In the absence of data, extrapolation is a common method for making predictions, but it is not always accurate. For example, based on observed economic indicators, an economist can make predictions about the state of the economy at a future time. These predictions may not be accurate because they cannot consider seemingly random events, such as natural disasters. *See also* prediction.

extrusion [3D ANALYSIS] The process of projecting features in a two-dimensional data source into a three-dimensional representation: points become vertical lines, lines become planes, and polygons become three-dimensional blocks. Uses of extrusion include showing the depth of well point features or the height of building-footprint polygons. *See also* 3D feature, projection.

F

F *See* F statistic.

F statistic [STATISTICS] A ratio of variances, calculated from a sample of data and used to provide information about a whole dataset. For example, statistic F may be used to provide estimates of variance, or differences, in a population, based on observations from two or more random samples. *See also* F test.

F test [STATISTICS] A statistical test for determining the probability that the variances of two different samples are the same. The F test uses a statistic known as statistic F to test statistical hypotheses about the variances of distributions from which samples have been drawn. *See also* F statistic.

FAA [FEDERAL GOVERNMENT] Acronym for *Federal Aviation Administration.* The U.S. government agency responsible for regulating and overseeing all aspects of civil aviation in the United States.

fabric *See* parcel fabric.

face

1. [EUCLIDEAN GEOMETRY] A planar surface of a geometric shape, bounded by edges.

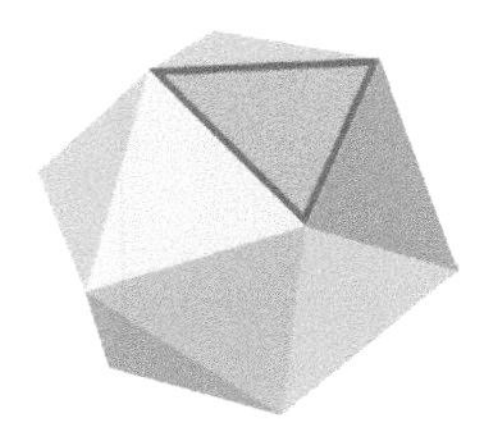

face 1

2. [3D ANALYSIS] In a TIN, the flat surface of a triangle bounded by three edges and three nodes. Faces do not overlap; each face is adjacent to three other faces of the TIN. A face defines a plane with an aspect and slope.

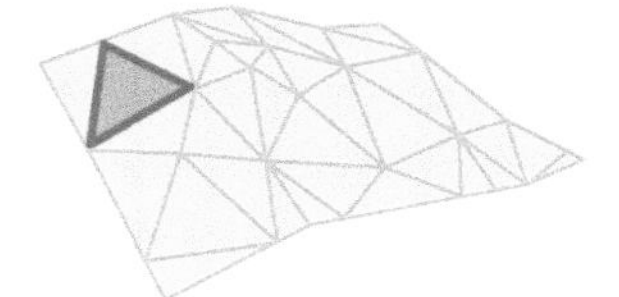

face 2

See also TIN.

facility [NETWORK ANALYSIS] A network location used in closest facility and service area analyses. *See also* closest facility analysis, service area analysis.

factual error [ACCURACY] A type of error resulting from a feature being left off a map by mistake, a feature existing on the map but not in the environment, a symbol being misplaced on the map, or a symbol remaining on the map after the feature has disappeared from the environment. *See also* error.

false easting [MAP PROJECTIONS] A value added to all x-coordinates of a map projection so that there are no negative eastings. *See also* easting, false northing.

false northing [MAP PROJECTIONS] A value added to all y-coordinates of a map projection so that there are no negative northings. *See also* northing, false easting.

false precision [ACCURACY] Deriving information from a map at a higher level of precision than it is legitimately possible to obtain from that map. *See also* precision.

feature [CARTOGRAPHY] Any part of the earth's surface, or anything found on the earth's surface, that can be represented on a map. Features can represent various types of geographic entities, such as points, lines, polygons, or even more complex objects such as networks or surfaces. Each feature typically has associated attributes that provide additional information—such as its name, population, or elevation—and may have additional, spatial characteristics, such as geographic coordinates. *See also* attribute, network, point, polygon, polyline, surface.

feature

feature attribute table *See* attribute table.

feature class [DATA STORAGE, DATA MANAGEMENT] A collection of geographic features with the same geometry type (such as point, line, or polygon), the same attributes, and the same spatial reference. Feature classes can be used to group homogeneous features into a single unit for storage purposes. For example, highways, primary roads, and secondary roads can be grouped into a line feature class named *roads*. In a geodatabase, feature classes can also store annotation and dimensions. *See also* object class.

geodatabase

feature class

feature collection [DATA STRUCTURES, DATA MANAGEMENT] A data structure containing one or more features that share the same set of attributes. *See also* attribute, feature.

feature data [SPATIAL ANALYSIS] Geographic entities (or features) represented as points, lines, and polygons. *See also* data, feature.

feature dataset [DATA MANAGEMENT] A collection of feature classes stored together that share a spatial reference (coordinate system) and a common geographic area. Feature classes with different geometry types may be stored in the same feature dataset. *See also* feature class.

feature displacement [MAP DESIGN] The movement of features that would otherwise overprint or conflict with other features. For example, if a river, a road, and a railway run through a narrow valley, it is necessary, at some scales, to displace at least one of the features that represent them on the map to keep their symbols distinct. *See also* generalization, line simplification.

feature extraction [DIGITAL IMAGE PROCESSING, DATA ANALYSIS] A method of pattern recognition in which patterns within an image are measured and then classified as features based on those measurements. *See also* pattern recognition.

feature layer [DATA ANALYSIS] A layer that references a set of feature data. Feature data represents geographic entities as points, lines, and polygons. *See also* layer.

feature service [ESRI SOFTWARE] A persistent software process that provides access to spatial and nonspatial data. Client applications use a feature service to query and edit data attributes, geometry, related data, and attachments and to access and display features. *See also* feature, service.

feature space analysis [REMOTE SENSING, ESRI SOFTWARE] The process of categorizing ground features present in imagery, in the form of pixels, objects, or segments with distinctive characteristics that set them apart from other entities within a feature space. These are typically designated according to their proximity to similar or dissimilar entities.

feature streaming [ESRI SOFTWARE] The process of defining and delivering vector feature data for a service. On the client side, feature streaming allows users to access a published map and add feature data for overlays, sharing, making notes, and performing analysis. Feature streaming functionality minimizes the need for multiple server requests. *See also* feature, feature data, feature service, streaming.

feature table [DATA MANAGEMENT] A table in a geodatabase that stores geometric shapes for each feature. Feature tables are used in geodatabases that store data as a binary data type.

feature template [DATA MANAGEMENT] A collection of default settings for creating a feature.

feature-linked annotation [DATA MANAGEMENT] Annotation that is stored in the geodatabase with links to features through a geodatabase relationship class. Feature-linked annotation reflects the current state of features in the geodatabase: It is automatically updated when features are moved, edited, or deleted. *See also* annotation.

Federal Geographic Data Committee [SURVEYING, FEDERAL GOVERNMENT] Also known by the acronym *FGDC*. An organization established by the United States Federal Office of Management and Budget responsible for coordinating the development, use, sharing, and dissemination of surveying, mapping, and related spatial data. The committee is composed of representatives from federal and state government agencies, academia, and the private sector. The FGDC defines spatial data metadata standards for the United States in its Content Standard for Digital Geospatial Metadata and manages the development of the National Spatial Data Infrastructure (NSDI). *See also* National Spatial Data Infrastructure.

FGDC *See* Federal Geographic Data Committee.

FGDC Clearinghouse *See* NSDI Clearinghouse Node.

FGDC standard *See* Content Standard for Digital Geospatial Metadata.

FID *See* ObjectID.

fiducial mark [REMOTE SENSING] A notch or etching within a sensor system that appears in a produced image for use as a point of reference. A fiducial mark is rigidly referenced to the principal axis of the optics to locate the precise center of the image (principal point). *See also* principal point.

F

field

1. [DATABASE STRUCTURES] A column in a table that stores the values for a single attribute.

FID	Species	Color
1	oak	brown
2	pine	green
3	fir	green

field 1

2. [DATA MODELS] A synonym for surface.

See also attribute, column, field observation.

field mapping [ANALYSIS/GEOPROCESSING] The process of defining the structure and content of a dataset. *See also* field, geoprocessing.

field observation [ENVIRONMENTAL GIS] A method of collecting data at the place where it is found in the environment. *See also* data capture.

field of regard [REMOTE SENSING] The total area that can potentially be captured by a movable (steerable or agile) sensor. A field of regard is much larger than the actual field of view of the sensor. *See also* field of view, sensor.

field of view [REMOTE SENSING] Also known by the acronym *FOV*. The extent of the observable world that a focal plane, or rendering viewpoint, can contain at a given point in time. Usually measured in degrees. *See also* focal plane, viewpoint.

field precision [DATA MANAGEMENT] The number of digits that can be stored in a field in a table. *See also* dataset precision.

F

field scale [DATA MANAGEMENT] The number of decimal places for float or double-type geodatabase table fields. *See also* floating point.

field sketch [ENVIRONMENTAL GIS] A drawing produced to support data collected within a field study.

field view [COGNITION] A philosophical view of geographic space in which space is completely filled by occurrences of phenomena, and in which phenomena are described by a range of values on a numeric scale. In this view, every spatial location is something, even if it is the zero value of a phenomenon. *See also* geographic phenomena, object view.

figure-ground organization [COGNITION] A perceptual phenomenon in which the mind and eye work together to spontaneously organize what is being viewed into two contrasting impressions—the figure, on which the eye settles, and the ground below or behind it.

fill [SYMBOLOGY] The interior of a polygon; the area inside the perimeter. *See also* fill symbol.

fill contour [CARTOGRAPHY] A contour that shows where the terrain was raised to support a road or railway grade. *See also* contour.

fill symbol [SYMBOLOGY] A color or pattern used to fill polygons on a map. *See also* marker symbol.

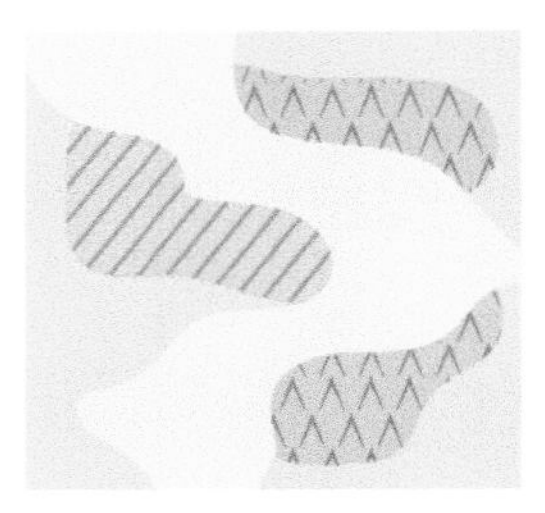

fill symbol

fillet [DATA ANALYSIS] A segment of a circle used to connect two intersecting lines. Fillets are used to create smoothly curving connections between lines, such as pavement edges at street intersections or rounded corners on parcel features. *See also* circle, intersection, joining.

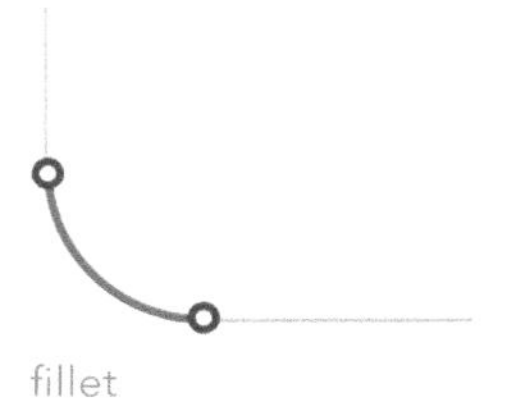

fillet

filter

1. [DATA ANALYSIS] A constraint or set of constraints used to limit data to a subset. For example, in a map layer of global cities, set a filter to show only cities in Nigeria.
2. [SPATIAL ANALYSIS] On a raster, an analysis boundary or processing window within which cell values affect calculations and outside which they do not. Filters are used mainly in cell-based analysis where the value of a center cell is changed to the mean, the sum, or some other function of all cell values inside the filter.

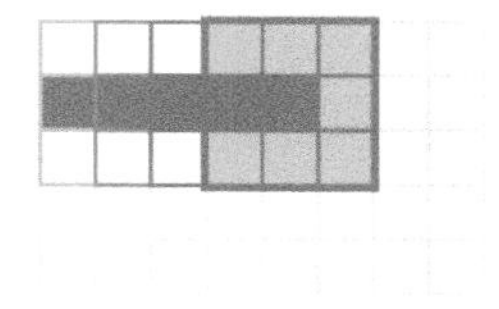

filter 2

3. [DATA MANAGEMENT] A GIS operation used to hide (but not delete) features in a map document or attribute table.

See also constraint, focal analysis, raster.

Find Addresses [ESRI SOFTWARE] A function used to input a street address and return a candidate list of matched locations for display on a map. *See also* Find Places, Find Points of Interest, Find Route.

Find Places [ESRI SOFTWARE] A function used to search for place-names and return a candidate list of places and associated coordinates for display on a map. *See also* Find Addresses, Find Points of Interest, Find Route.

Find Points of Interest [ESRI SOFTWARE] A function used to search for points of interest by name and return a candidate list of points of interest coordinates for display on a map. *See also* Find Addresses, Find Places, Find Route.

Find Route [ESRI SOFTWARE] A function used to generate multipoint driving directions between user-defined locations, determine the shortest or the quickest path between locations, and create a map of the travel route and a list of travel directions. *See also* Find Addresses, Find Places, Find Points of Interest, least-cost path.

Find Similar analysis [DATA ANALYSIS, ESRI SOFTWARE] A process that seeks out new market areas based on the characteristics of an existing market area. *See also* analysis, market area.

first normal form [DATABASE STRUCTURES] The first level of guidelines for designing table and data structures in a relational database. The first normal form guideline recommends creating a unique key for every row in a database table, eliminating duplicate columns from a table, and creating separate tables to contain related data. A database that follows these guidelines is said to be in first normal form. *See also* normal form, second normal form, third normal form.

fishnet map [MAP DESIGN] A map that resembles a net or wire mesh draped over the terrain. It is produced by combining closely spaced, parallel terrain profiles across the landform at right angles to each other. Also called a wireframe map. *See also* topographic profile.

fitness for use [DATA QUALITY] The degree to which a dataset is suitable for a particular application or purpose, encompassing factors such as data quality, scale, interoperability, cost, data format, and so on. *See also* metadata.

F

fix [GEODESY] A single position obtained by surveying, GPS, or astronomical measurements, usually given with altitude, time, date, and latitude-longitude or grid position. *See also* position.

fixed reference point [SURVEYING, ESRI SOFTWARE] A point in parcel fabric that is used as a control in a weighted least-squares analysis and adjustment. The coordinates of the point are constrained and are not updated in the adjustment. *See also* least-squares analysis and adjustment, parcel fabric.

fixed-distance weighting [SPATIAL ANALYSIS] A distance weighting method in which a distance band is imposed on the analysis.

flag [GRAPHICS (MAP DISPLAY)] A marker that identifies or calls attention to something, indicating importance or the need for further attention. *See also* marker symbol.

flattening [GEODESY] A measure of how much an oblate spheroid differs from a sphere. The flattening equals the ratio of the semimajor axis (*a*) minus the semiminor axis (*b*) to the semimajor axis. *See also* oblate ellipsoid, semimajor axis, semiminor axis, smoothing.

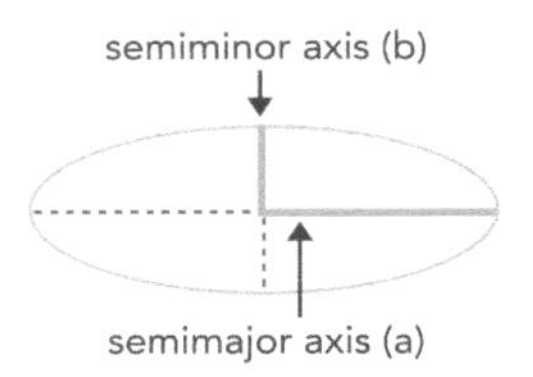

$$\text{flattening} = \frac{a - b}{a}$$

flattening

flight line [REMOTE SENSING] The path that an aircraft or sensor follows, typically north–south or east–west, to acquire aerial photos.

float [MATHEMATICS] A numeric data type that contains a decimal. Represents a real number where an integer is to the left of the decimal and the fraction is to the right, such as 3.5.

floating point [COMPUTING] A type of numeric field for storing real numbers with a decimal point. The decimal point can be in any position in the field and therefore may float from one location to another for different values stored in the field. For example, a floating-point field can store the numbers 23.632, 0.000087, and –96432.15. *See also* float, scale.

flow accumulation [DIGITAL IMAGE PROCESSING] In raster data, a calculation of the number of cells (weighted or not) flowing into each downslope cell in a raster elevation file. When no weight is provided, a weight of 1 is applied to each cell. Typically used to identify stream channels and ridges in topographic raster data.

flow direction [NETWORK ANALYSIS] The route or course followed by commodities proceeding through edge elements in a network. *See also* downstream, network, sink, source.

flow line [UTILITIES] A proportional line symbol used on flow maps. *See also* flow map.

flow map [CARTOGRAPHY] A map that uses line symbols of variable thickness to show the proportion of traffic or flow within a network. *See also* barrier.

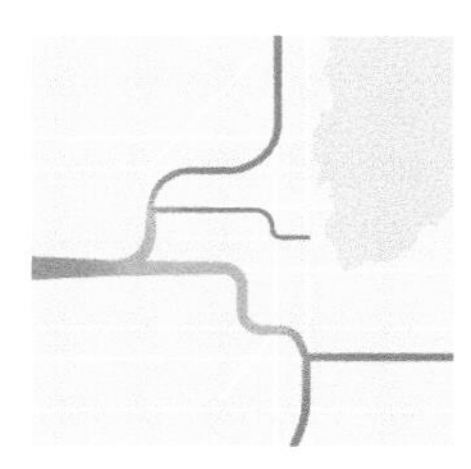

flow map

fly-through [VISUALIZATION] A view of a relief portrayal from an animated sequence of vantage points, giving the impression of flying and that the terrain is passing under and around the viewer. *See also* animated map, animation.

FMV *See* full-motion video.

focal analysis [DATA ANALYSIS] The computation of an output raster where the output value at each cell location is a function of the value at that cell location and the values of the cells within a specified neighborhood around the cell.

focal functions *See* focal analysis.

focal plane [DIGITAL IMAGE PROCESSING] The distance between a sensor lens and the interior point of focus required to capture an image. *See also* reflective optics, refractive optics.

focal plane array [DIGITAL IMAGE PROCESSING] A device that measures the electromagnetic energy received by digital sensors. A focal plane array is a solid-state sensor located at the sensor focal plane and can be either a charge-coupled device (CDD) or a complementary metal-oxide semiconductor (CMOS). *See also* CCD, complementary metal-oxide-semiconductor array, electromagnetic energy.

follow feature mode [ESRI SOFTWARE] A method for placing geodatabase annotation relative to a line or polygon feature. For example, text next to a river may be dragged along it so that the text curves like the river. *See also* annotation, feature.

font [SYMBOLOGY] A single typeface or a set of related patterns representing characters or symbols. Part of a text element. Characteristics include font family and font size. *See also* point size, text element.

footprint [DATA STRUCTURES] A polygon that represents the extent of an image or video frame in ground coordinates. A mosaic dataset or image service contains footprints of all the image items that compose the dataset or service. A footprint can be limited to only valid image data. *See also* extent.

foreground

1. [MAP DISPLAY] In a scene or display, the area that appears to be closest to an observer.
2. [ESRI SOFTWARE] The area in a raster layer where cells are eligible for selection and vectorization.

See also raster, vectorization.

foreshortening [MAP PROJECTIONS] Compression and extension of the projected positions of features equally spaced on the generating globe; this technique gives oblique-perspective projections the appearance of viewing part of the earth obliquely from a point above the surface of the earth. *See also* generating globe, oblique-perspective projection.

form lines [CARTOGRAPHY] Lines on a map that approximate the shape of terrain in lieu of actual contours. Form lines do not refer to a true datum and do not necessarily use regular intervals. *See also* contour line, hachure.

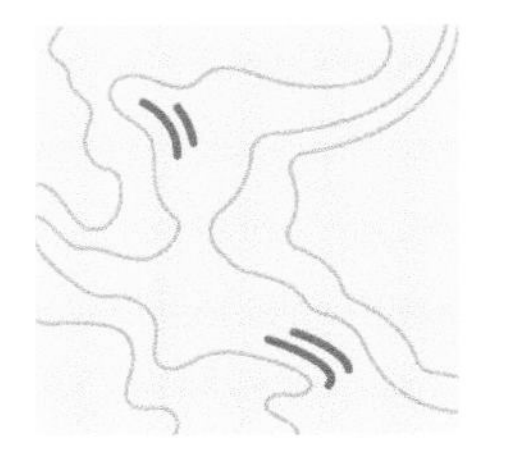

form lines

formal balance [MAP DESIGN] An arbitrary or subjective sense of visual balance that is perceived when the map and the content in the margins are arranged symmetrically to produce an impression of equal distribution. *See also* visual balance.

fractal [MATHEMATICS] A geometric pattern that repeats itself, at least roughly, at ever smaller scales to produce self-similar, irregular shapes and surfaces that cannot be represented using classical geometry. If a fractal curve of infinite length serves as the boundary of a plane region, the region itself will be finite. Fractals can be used to model complex natural shapes such as clouds and coastlines.

fractal

fragmentation map [CARTOGRAPHY] A map that shows the total number of land-cover (or other) categories found near each cell.

frame camera model [PHOTOGRAMMETRY] A physical, mathematical sensor model for a frame camera based on collinearity equations. It consists of an interior and an exterior orientation model.

The camera calibration provides the interior orientation, consisting of the calibrated focal length; the location of the principal point; a 2D transformation of pixel row and column into x, y coordinates in millimeters, referenced to the principal point; and a lens distortion model.

The exterior orientation consists of a specified ground coordinate reference model, a specification for camera station location and attitudes at the moment of exposure, and corrections for nonlinear image rays that are outside the sensor (for example, bending due to the atmosphere).

See also framing camera, sensor.

framing camera [PHOTOGRAMMETRY] In imagery, a sensor that exposes its entire, light-sensitive focal plane array at the same instance. Framing cameras produce 2D images that are very stable geometrically and do not require time-sensitive modeling for exterior orientation. *See also* frame camera model.

free network adjustment [SURVEYING] A method of examining and assessing the overall geometry of a network by processing its measurements only. If reference points exist, they are used for scale and rotation purposes. Emphasis is placed on testing the consistency and quality of the measurements rather than computing coordinates. *See also* least-squares analysis and adjustment, surveying.

frequency [PHYSICS] The number of oscillations per unit of time in a wave of energy, or the number of wavelengths that pass a point in a given amount of time. *See also* electromagnetic energy, wavelength.

from-node [DATA STRUCTURES] One of an arc's two endpoints. From- and to-nodes give an arc a sense of direction. *See also* to-node.

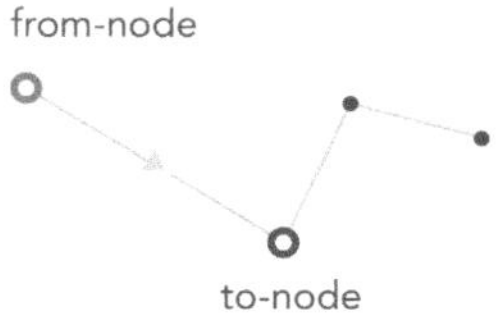

from-node

front

1. [CLIMATOLOGY] The edge of an air mass.
2. [DEFENSE] The foremost line of advance.

full-motion video [PHOTOGRAMMETRY] Also known by the acronym *FMV*. A video format that is a combination of video stream and associated metadata in one video file, making the video geospatially aware. Associated metadata includes camera pointing position, platform position and attitude, and other related data that is encoded into each video frame. Full-motion video is compliant with the Motion Imagery Standards Board (MISB) specifications. FMV typically means FMV-compliant. *See also* metadata.

function [ANALYSIS/GEOPROCESSING]
1. An operation. In GIS, functions typically include data input (editing and management), data query (analysis and visualization), and output operations.
2. A processing operation applied to rasters or mosaic datasets. Functions can be applied in an order defined by a function chain.

See also function chain.

function chain [ANALYSIS/GEOPROCESSING] An ordered list of functions applied to a raster, image, or mosaic dataset as the data is accessed. Displays the results in near real time on the user's screen. *See also* function.

functional surface [SPATIAL ANALYSIS] A surface that stores a single z-value for any given x,y location. *See also* x,y coordinates, z-value.

fuzzy boundary [UNCERTAINTY, DATA MANAGEMENT] A boundary that has a vague or indeterminate location or that is a gradual transition between two zones. *See also* boundary, fuzzy classification, vagueness.

fuzzy classification [UNCERTAINTY, DATA MANAGEMENT] Any method for classifying data that allows attributes to apply to objects by membership values, so that an object may be considered a partial member of a class. Fuzzy classification may be applied to geographic objects, so that an object's boundary is treated as a gradated area rather than an exact line. In GIS, fuzzy classification has been used in the analysis of soil, vegetation, and other phenomena that tend to change gradually in their physical composition and for which attributes are often partly qualitative in nature. *See also* fuzzy set, fuzzy boundary, vagueness.

fuzzy set [MATHEMATICS, UNCERTAINTY] In mathematics, a collection of elements that belong together based on specified criteria, so that elements with partial or uncertain degrees of membership may be included in the collection. *See also* fuzzy classification, vagueness.

fuzzy tolerance [SPATIAL ANALYSIS, ESRI SOFTWARE] The distance within which coordinates of nearby features are adjusted to coincide with each other when topology is being constructed or polygon overlay is performed. Nodes and vertices within the fuzzy tolerance are merged into a single coordinate location. Fuzzy tolerance is a small distance—usually from 1/1,000,000 to 1/10,000 times the width of the coverage extent—and is generally used to correct inexact intersections. *See also* cluster tolerance.

GAGAN [GPS] Acronym for *GPS-aided GEO augmented navigation.* A regional satellite-based augmentation system (SBAS) that improves the accuracy of GNSS receivers over Indian airspace. GAGAN was implemented by the government of India to provide reference signals for communication, navigation, surveillance, and air traffic management. *See also* GNSS, GPS, NavIC.

Galileo [GPS] A global geopositioning satellite constellation used for positioning, navigation, and timing (PNT); operated by the joint European Union Agency for the Space Programme (EUSPA), it is part of the international GNSS constellations spread between several orbital planes. *See also* BeiDou Navigation Satellite System, geopositioning, GLONASS, GNSS, GPS, PNT, satellite constellation, satellite navigation.

Gall-Peters projection [MAP PROJECTIONS] A secant case of the cylindrical equal-area projection that lessens shape distortion in higher latitudes by placing lines of tangency at 45° N and 45° S. *See also* Albers equal-area conic projection, secant projection.

gamma correction [DIGITAL IMAGE PROCESSING] A nonlinear adjustment to every pixel in an image that either compresses or expands the luminance response of gray levels to control the image's overall brightness.

Gantt chart [BUSINESS] A project management graph that displays tasks on a schedule, often used to plan and track projects. The Gantt chart was developed by the American mechanical engineer and management consultant Henry Laurence Gantt.

Gaussian distribution *See* normal distribution.

Gauss-Krüger projection [MAP PROJECTIONS] An ellipsoidal projected coordinate system that uses the transverse Mercator projection to divide the world into standard zones 6 degrees wide. Used mainly in Europe and Asia, the Gauss–Krüger coordinate system is similar to the universal transverse Mercator coordinate system. The Gauss–Krüger projection is named for the German geodesist, mathematician, and physicist Carl Friedrich Gauss and the German mathematician and surveyor/geodesist Johann Heinrich Louis Krüger. *See also* transverse aspect, transverse Mercator projection, UTM.

gazetteer [CARTOGRAPHY] A list of geographic place-names and their coordinates. Entries may include other information as well, such as area, population, or cultural statistics. Atlases often include gazetteers, which are used as indexes to their maps. Well-known digital gazetteers include the U.S. Geological Survey Geographic Names Information System (GNIS) and the Alexandria Digital Library Gazetteer. *See also* atlas.

GBF/DIME [DATA MODELS] Acronym for *Geographic Base Files/Dual Independent Map Encoding.* Vector geographic base files made for the 1970 and 1980 U.S. censuses, containing address ranges, ZIP codes, and the coordinates of street segments and intersections for most metropolitan areas in the United States. TIGER files replaced DIME files for the 1990 and subsequent censuses. *See also* Dual Independent Map Encoding, TIGER.

GDB *See* geodatabase.

generalization

1. [MAP DESIGN] The abstraction, reduction, and simplification of features for change of scale or resolution.
2. [DATA EDITING] The process of reducing the number of points in a line without losing the line's essential shape.
3. [DATA EDITING] The process of enlarging and resampling cells in a raster format.

See also cartographic generalization, database generalization.

generating globe [MAP PROJECTIONS] A globe reduced to the scale of the desired flat map.

genetic algorithm [COMPUTING] A search algorithm inspired by genetics and Darwin's theory of natural selection. The algorithm goes through an iterative process of applying genetic operators, such as reproduction, mutation, and crossover, to a collection of data over several stages. At each stage, the fitness of the results is evaluated, and the best of the results population is retained, until the results present an optimal solution. *See also* algorithm, optimization.

geocentric

1. [GEODESY] Measured from the earth or the earth's center.
2. [ASTRONOMY] Having the earth as a center.

See also geocentric datum.

geocentric coordinate system [COORDINATE SYSTEMS] A three-dimensional, earth-centered reference system in which locations are identified by their x-, y-, and z-values. The x-axis is in the equatorial plane and intersects the prime meridian (usually Greenwich). The y-axis is also in the equatorial plane; it lies at right angles to the x-axis and intersects the 90-degree meridian. The z-axis coincides with the polar axis and is positive toward the north pole. The origin is located at the center of the sphere or spheroid. *See also* coordinate system, geocentric, reference system.

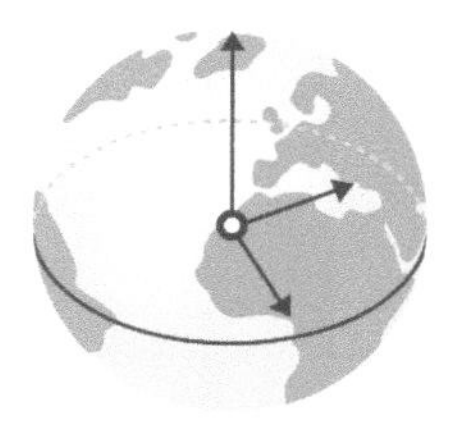

geocentric coordinate system

geocentric coordinates [COORDINATE SYSTEMS] Coordinate-defined location relative to a geocentric coordinate system or geographic direction system. *See also* geocentric coordinate system, geographic direction system.

geocentric datum [GEODESY] A horizontal geodetic datum based on an ellipsoid that has its origin at the earth's center of mass. Examples are the World Geodetic System of 1984, the North American Datum of 1983, and the Geodetic Datum of Australia of 1994. The first uses the WGS84 ellipsoid; the latter two use the GRS80 ellipsoid. Geocentric datums are more compatible with satellite positioning systems, such as GPS, than are local datums. *See also* datum, GPS, horizontal geodetic datum, local datum.

geocentric latitude [COORDINATE SYSTEMS] The angle between the equatorial plane and a line from a point on the surface to the center of the sphere or spheroid. On a sphere, all lines of latitude are geocentric. Latitude generally refers to geodetic latitude. *See also* geodetic latitude.

geocentric longitude [COORDINATE SYSTEMS] The angle between the prime meridian and a line drawn from a point on the surface to the center of a sphere or spheroid. For an ellipsoid of revolution (such as the earth), geocentric longitude is the same as geodetic longitude. *See also* geodetic longitude.

geocode

1. [DATA CAPTURE] To assign a street address or other location coordinates to a place.
2. [COORDINATE GEOMETRY (COGO)] A code that represents the location of an object, such as an address, a census tract, a postal code, latitude-longitude, or x,y coordinates.

See also geocoding.

geocode server [ESRI SOFTWARE] A component that provides web access to a locator and performs interactive geocoding, reverse geocoding, address location suggestions, and batch geocoding. *See also* geocoding.

geocoded feature class [GEOCODING] A feature class created by batch geocoding. *See also* feature class, batch geocoding.

geocoding [DATA CAPTURE] The process of converting street addresses or other location information to spatial data (points on a map). *See also* locator.

geocoding index [GEOCODING] An index of reference data used by a locator that contains indexed address attributes. *See also* locator, geocoding.

G

geocoding reference data [ESRI SOFTWARE] Data that a geocoding service uses to determine the geometric representations for locations. *See also* reference data.

geocoding service [GEOCODING, ESRI SOFTWARE] A persistent software process that translates the nonspatial description of a place, such as a street address, into spatial data that can be displayed as a feature on a map. A geocoding service defines the path to the reference data source and the file of nonspatial data, provides the algorithms and parameters for standardizing, reading, and matching addresses to reference data, and generates output spatial data to client applications. *See also* geocoding, locator.

geocomputation [COMPUTING] The application of computer technology to spatial problems, including problems of collecting, storing, visualizing, and analyzing spatial data, and of modeling spatial system dynamics. *See also* geoinformatics, quantitative geography.

geodata *See* geographic data.

geodatabase [DATA MANAGEMENT] A database or file structure used primarily to store, query, and manipulate spatial data. Geodatabases store geometry, a spatial reference system, attributes, and behavioral rules for data. Various types of geographic datasets can be collected within a geodatabase, including feature classes, attribute tables, raster datasets, network datasets, topologies, and many others. Geodatabases can be stored in relational database management systems or in a system of files, such as a file geodatabase. *See also* database.

geodatabase data model [ESRI SOFTWARE] The schema for the various geographic datasets and tables in an instance of a geodatabase. The schema defines the GIS objects, rules, and relationships used to add GIS behavior and integrity to the datasets in a collection. *See also* database.

geodatabase feature dataset [ESRI SOFTWARE] In a geodatabase, a collection of feature classes stored together so they can participate in topological relationships with one another. All the feature classes in a feature dataset must share the same spatial reference; that is, they must have the same coordinate system and their features must fall within a common geographic area. Feature classes with different geometry types may be stored in a feature dataset.

geodataset [ESRI SOFTWARE] Any organized collection of data in a geodatabase with a common theme. *See also* dataset, geodatabase.

geodesic [EUCLIDEAN GEOMETRY] The shortest distance between two points on the surface of a spheroid. Any two points along a meridian form a geodesic. *See also* geodetic distance, great circle route.

geodesic

geodesy [GEODESY] The science of measuring and representing the shape and size of the earth, and the study of its gravitational and magnetic fields. *See also* geoid, WGS72, WGS84.

geodetic datum [GEODESY] A datum that is the basis for calculating positions on the earth's surface or heights above or below the earth's surface. *See also* datum, horizontal geodetic datum, vertical geodetic datum.

geodetic distance [CARTOGRAPHY] The measurement of the distance of the shortest route between two points on the earth's surface. *See also* geodesic, geodetic measurement.

geodetic latitude [COORDINATE SYSTEMS] Latitude on an ellipsoid; the angle between the horizontal equator line and a line perpendicular to the ellipsoidal surface at the parallel of interest. *See also* geocentric latitude.

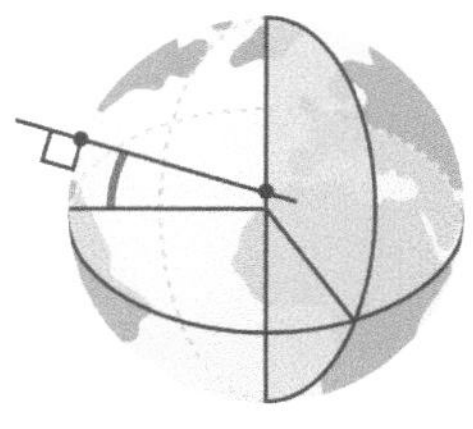

geodetic latitude

geodetic longitude [COORDINATE SYSTEMS] Longitude on an ellipsoid. The angle between the plane of the meridian that passes through a point on the ellipsoidal surface and the plane of a prime meridian, usually the Greenwich meridian. Because of the geometric nature of ellipsoids, the angular distance between the prime meridian and another meridian is typically the same as for its geocentric longitude. *See also* geocentric longitude, longitude, prime meridian.

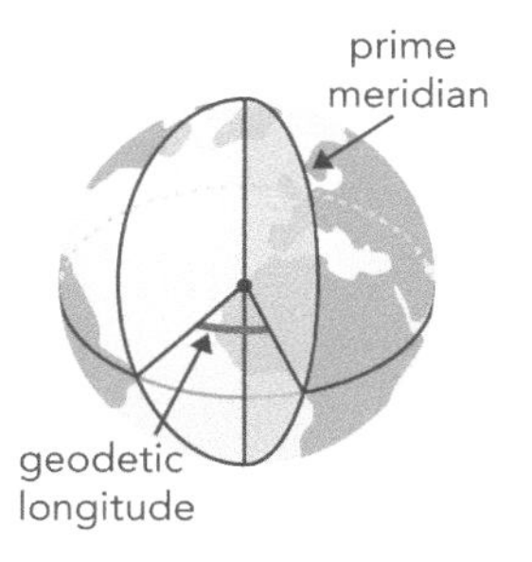

geodetic longitude

geodetic measurement [GEODESY] The quantity associated with length or area on a curved (nonplanar) surface. Geodetic measurement typically refers to the measurement of the earth's surface and features considering the curvatures and irregularities of the planet. The most common geodetic measurement is a great circle route or a geodetic line. *See also* geodetic distance, great circle route, planar measurement.

Geodetic Reference System of 1980 [GEODESY] Also known by the acronym *GRS80*. The standard measurements of the earth's shape and size adopted by the International Union of Geodesy and Geophysics in 1979; an ellipsoid that is essentially identical to the World Geodetic System of 1984 ellipsoid. *See also* North American Datum of 1983, reference system, WGS84.

geodetic survey [GEODESY] A survey that takes the shape and size of the earth into account, used to precisely locate horizontal and vertical positions suitable for controlling other surveys. *See also* plane survey, surveying.

G

geodetic transformation *See* geographic transformation.

geofence [MAP DISPLAY] A designated boundary around a geometry that, if crossed, initiates a notification. Geofences are often used in real-time route web applications. *See also* boundary, route.

geographic [GEOGRAPHY] Of or relating to the earth. *See also* spatial.

geographic constraint *See* extent.

geographic context [GEOGRAPHY] Characteristics of a location that describe how it may be impacted by different events. *See also* context.

geographic coordinate system [COORDINATE SYSTEMS] A reference system that uses latitude and longitude to define the locations of points on the surface of a sphere or spheroid. A geographic coordinate system includes a datum, a prime meridian, and an angular unit of measure. *See also* angular unit, datum, prime meridian.

geographic coordinates [COORDINATE SYSTEMS] A measurement of a location on the earth's surface expressed in degrees of latitude and longitude. *See also* geographic coordinate system.

geographic data [GIS TECHNOLOGY] Information describing the location and attributes of things, including their shapes and representation. Geographic data is the composite of spatial data and attribute data. *See also* spatial data, attribute data.

geographic database *See* geodatabase.

geographic direction system [COORDINATE SYSTEMS] A system in which direction is measured in the angular units of a circle, with north at the top so that the north reference line points to 12 o'clock on the clock face.

geographic information science *See* GIScience.

geographic information system *See* GIS.

geographic north *See* true north.

geographic phenomena [GEOGRAPHY] Measurable or observable events occurring on the earth's surface.

geographic poles [CARTOGRAPHY] The ends of the earth's axis of rotation.

geographic primitive [GRAPHICS (MAP DISPLAY)] A graphic representation of a location; for example, a point to represent the location of a smokestack or a polygon to represent the location of a toxic plume.

geographic projection *See* display projection.

geographic transformation [COORDINATE SYSTEMS, DATA CONVERSION] A systematic conversion of coordinates from one geographic coordinate system to equivalent values in another. Typically, equations are used to model the position and orientation of the source and target systems in three-dimensional coordinate space, adjusting for differences in the size, shape, and orientation of the earth between coordinate systems, as well as the position and orientation of the coordinate axes; transformation parameters may include translation, rotation, and rescaling. *See also* datum, geographic coordinate system, rescale, rotate, translate.

geography [GEOGRAPHY]
1. The study of the earth's surface, encompassing the description and distribution of the various physical, biological, economic, and cultural features found on the earth and the interaction between those features.
2. The arrangement of the geographic features of an area.

See also biogeography, cultural geography, economic geography, human geography, physical geography, quantitative geography, urban geography.

geography level [GOVERNMENT] A division of statistical geographic data, such as country, province, postal code, tract, or block group. *See also* geographic data.

geoid [GEODESY] The equipotential surface of the earth's gravity field that coincides with the mean sea level of the unperturbed oceans. The geoid is used as a reference surface for measuring gravity-based heights (orthometric, normal, and dynamic). *See also* ellipsoid, spheroid.

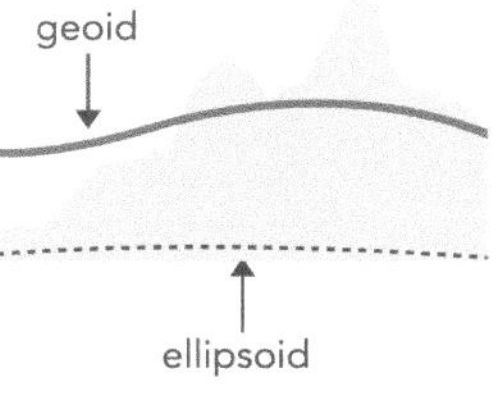

geoid

geoid height [GEODESY] The height of the geoid above the ellipsoid. *See also* geoid-ellipsoid separation.

GEOID2022 [MODELING] The regional, gridded geoid undulation model for the NAPGD2022 component of the National Spatial Reference System (NSRS). *See also* geoid, NAPGD2022, National Spatial Reference System, U.S. National Geodetic Survey.

geoidal undulation [GEODESY] The elevation difference between the geoid and the ellipsoid. *See also* ellipsoid, geoid, geoid-ellipsoid separation, geoid height.

geoid-ellipsoid separation [GEODESY] The distance from the surface of an ellipsoid to the surface of the geoid, measured along a line perpendicular to the ellipsoid. The separation is positive if the geoid lies above the ellipsoid, negative if it lies below. *See also* geoid, geoid height.

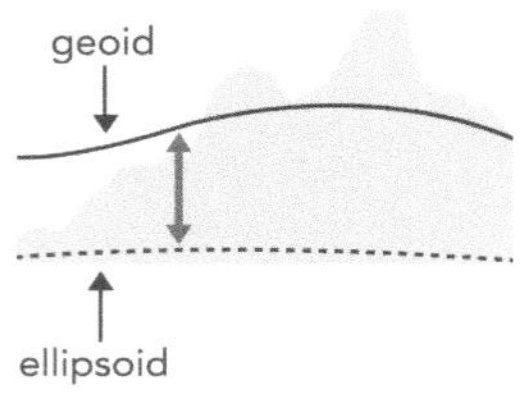

geoid-ellipsoid separation

geoinformatics [COMPUTING] The application and use of data, software, and computer technology to address the concerns of science and engineering, specifically in the fields of cartography, spatial analysis, geodesy, and related disciplines. *See also* geocomputation.

GeoJSON [PROGRAMMING] An open-standard file and data interchange format for representing geospatial data using JavaScript Object Notation (JSON). *See also* Esri JSON, JavaScript Object Notation.

geolocation [GEOLOCATING] The process of creating geographic features from tabular data by matching the tabular data to a spatial location. An example of geolocation is creating point features from a table of x,y coordinates. *See also* location, tabular data.

geologic cross section [GEOLOGY] A view of rock units in the earth's interior as if it were cut open and seen from the side. *See also* geologic map.

geologic map [GEOLOGY] A map that shows the distribution of rock units and geologic structures within an area. *See also* geologic cross section.

geometric correction [REMOTE SENSING] The correction of errors in remotely sensed data, such as those caused by satellites or aircraft not staying at a constant altitude or by sensors deviating from the primary focus plane. Images are often compared to ground control points on accurate basemaps and resampled so that exact locations and appropriate pixel values can be calculated. *See also* error, orthorectification, rectification, resampling.

geometric dilution of precision *See* dilution of precision.

geometric distortion [ACCURACY] The distortion that occurs in map projections that can be detected by comparing the graticule (latitude and longitude lines) on the projected surface with the same lines on a globe.

geometric element [EUCLIDEAN GEOMETRY] One of the most basic parts or components of a geometric figure: that is, a surface, shape, point, line, angle, or solid. *See also* element, multipoint, segment.

geometric interval classification [DATA MODELS, STATISTICS] A data classification method that creates class breaks based on class intervals that have a geometric series. The algorithm creates geometric intervals by minimizing the sum of squares of the number of elements in each class. This method typically provides results that are cartographically comprehensive, highlighting neither the middle nor extreme ranges of values. *See also* classification.

geometric transformation [COORDINATE SYSTEMS, SPATIAL ANALYSIS] The process of altering the spatial location, shape, or orientation of geographic data. A geometric transformation involves applying mathematical functions or algorithms to manipulate the coordinates of points, lines, or polygons in a dataset. Geometric transformations are commonly used in GIS for purposes such as projection, registration, rescaling, rotation, translation, and affine transformation. *See also* affine transformation, Helmert transformation, transformation.

geometry

1. [EUCLIDEAN GEOMETRY] The measures and properties of points, lines, and surfaces. In a GIS, geometry is used to represent the spatial component of geographic features.
2. [MATHEMATICS] The branch of mathematics concerning points, lines, and polygons and their properties and relationships.

See also analysis, centroid.

geomorphology [GEOGRAPHY] The study of the nature and origin of landforms, including relationships to underlying structures and processes of formation. *See also* geography, landform, photogeology.

geopositioning [NAVIGATION] The process of determining an object's location. *See also* georeferencing, linear referencing, satellite navigation.

geoprocessing [ANALYSIS/GEOPROCESSING] A GIS operation that is used to manipulate an input spatial dataset and return the result as an output dataset. Common examples of geoprocessing operations include the creation of a mosaic dataset, image enhancements, geographic feature overlays, topology processing, calculation of variables (such as the extent and strength of a hurricane over time and distance), and data analysis/conversion. *See also* mosaic dataset, overlay, topology.

georectification [DATA EDITING] The digital alignment of a satellite or aerial image with a map of the same area. In georectification, a number of corresponding control points, such as street intersections, are marked on both the image and the map. These locations become reference points in the subsequent processing of the image. *See also* control point, georeferencing, orthorectification.

G

georeferencing [COORDINATE SYSTEMS, GEOREFERENCING] The process of aligning geographic data to a known coordinate system so it can be viewed, queried, and analyzed with other geographic data. Georeferencing may involve shifting, rotating, rescaling, skewing, and in some cases, warping, rubber sheeting, or orthorectifying the data. *See also* coordinate system, coordinate transformation, rubber sheeting, transformation.

georelational data model [DATA MODELS] A geographic data model that represents geographic features as an interrelated set of spatial and attribute data. The georelational model is the fundamental data model used in a coverage. *See also* attribute data, coverage, spatial data.

geosearch [GEOLOCATING] The process of converting a text string into potential corresponding map coordinates; the results can be filtered by proximity. *See also* geocoding, locator.

geospatial data clearinghouse *See* NSDI Clearinghouse Network.

geospatial intelligence [DATA ANALYSIS] Also known by the abbreviation *GEOINT*. Information derived from a thorough analysis of images, data, and calculations about a location, typically used to solve geographic problems. It includes, but is not limited to, the analysis of literal imagery; geospatial data; and information and data from, for example, processed spectral, spatial, and temporally fused products.

geospatial technology [GIS TECHNOLOGY] A set of technological approaches, such as GIS, photogrammetry, and remote sensing, for acquiring and manipulating geographic data. *See also* GIS, photogrammetry, remote sensing.

geospecific model [SYMBOLOGY] A model used to represent a real-world feature. For example, a geospecific model for the White House would look exactly like the White House and be used to represent the White House on a map of Washington, D.C. *See also* geotypical model.

geostationary [ASTRONOMY] An orbital path above the earth's equator with an angular velocity that follows the earth's rotation and an inclination and eccentricity approaching zero. A geostationary satellite has the same orbital period as the earth's rotation (23 hours, 56 minutes, and 4 seconds) traveling at a rate of approximately 3 km per second, causing it to appear effectively stationary. Geostationary satellites are typically positioned at much higher altitudes (typically 35,786 km above the earth's equator). While geostationary satellites are geosynchronous, a geosynchronous satellite is not necessarily geostationary. *See also* Clarke Belt, geosynchronous, satellite constellation.

geostatistics [STATISTICS] A class of statistics used to analyze and predict the values associated with spatial or spatiotemporal phenomena. Geostatistics provides a means of exploring spatial data and generating continuous surfaces from selected sampled data points. *See also* spatial statistics.

geosynchronous [ASTRONOMY] Positioned in an orbit moving west to east with an orbital period equal to the earth's rotational period. If a satellite is in a circular geosynchronous orbit in the equatorial plane, its orbit is geostationary. If it is not positioned directly over the equator, the satellite appears to make a figure eight once a day between the latitudes that correspond to its angle of inclination over the equator. *See also* geostationary.

GEOTIFF [DIGITAL IMAGE PROCESSING] A public domain metadata standard that embeds spatial reference information directly into a tagged image file format (TIFF), including map projection, coordinate systems, ellipsoids, and datums. *See also* metadata.

geotypical model [SYMBOLOGY] A symbolic representation for a class of map features, such as government buildings. For example, on a map of the United States, a white building with a dome on top could be used as a geotypical model for all state capitols. *See also* geospecific model.

GIF [GRAPHICS (COMPUTING)] Acronym for *graphics interchange format.* A low-resolution format for image files, commonly used on the internet. It is well-suited for images with sharp edges and reduced numbers of colors. *See also* JPEG.

GIS [GIS TECHNOLOGY] Acronym for *geographic information system.* An integrated collection of computer software and data used to view and manage information about geographic places, analyze spatial relationships, and model spatial processes. A GIS provides a framework for gathering and organizing spatial data and related information so that it can be displayed and analyzed. *See also* analysis, model, spatial analysis, spatial data.

GIScience [SOCIAL CONTEXT OF GIS] An abbreviation of *geographic information science.* A field of science that studies the principles of GIS. GIScience research includes the theoretical and technical basis of GIS, the social implications of its use, and where it intersects or interacts with related disciplines—such as cognitive, social, and information science.

glacial contour [PHOTOGRAMMETRY] A contour that represents the surface of an ice mass or permanent snow field at the date of the imagery used to compile the feature.

global analysis [DATA ANALYSIS] The computation of an output raster where the output value at each cell location may be a function of all the cells in the input raster. *See also* local analysis, spatial function.

global functions *See* global analysis.

global geoid [GEODESY] The slightly undulating, nearly ellipsoidal surface that best fits mean sea level for all the earth's oceans. *See also* geoid, local geoid, mean sea level.

G

global navigation satellite system *See* GNSS.

Global Positioning System *See* GPS.

global spatial data infrastructure [DATA SHARING] Also known by the acronym *GSDI*. A global framework of technologies, policies, standards, and human resources necessary to acquire, process, store, distribute, and improve the use of geospatial data across multiple countries and organizations. *See also* National Spatial Data Infrastructure, SDI.

GlobalID [ESRI SOFTWARE] A type of UUID (Universal Unique Identifier) in which values are automatically assigned by the geodatabase when a row is created. The GlobalID field is necessary for maintaining object uniqueness across replicas. A GlobalID is not editable. *See also* globally unique identifier, identifier.

globally unique identifier [ESRI SOFTWARE] Also known by the abbreviation *GUID*. A string used to uniquely identify an interface, class, type library, component category, or record. *See also* GlobalID, identifier.

globe [CARTOGRAPHY] A sphere on which a map of the earth or a celestial body is represented. Since the earth's natural shape is similar to a sphere, globes distort the earth's features far less than flat maps. *See also* observer, sphere.

GLONASS [GPS] An abbreviation of *Global Navigation Satellite System*. A global geopositioning satellite constellation used for positioning, navigation, and timing (PNT); operated by the Russian government, it is part of the international GNSS constellations spread between several orbital planes. *See also* BeiDou Navigation Satellite System, Galileo, geopositioning, GNSS, GPS, PNT, satellite constellation, satellite navigation.

glyph [SYMBOLOGY]

1. The geometric shape of a character in a font.

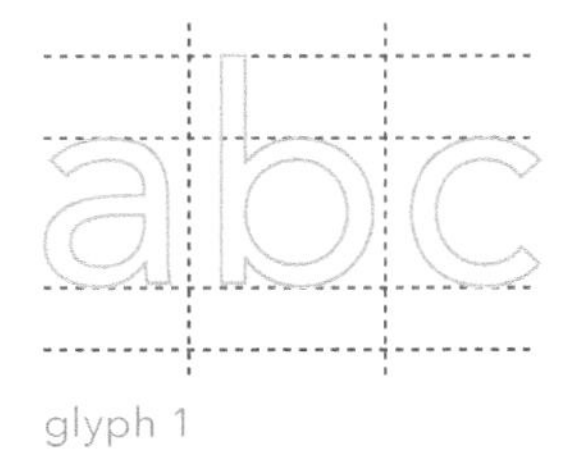

glyph 1

2. A specialized form of map symbol used to convey both local and global structure in multivariate glyph plots.

See also multivariate point symbol.

GM2022 *See* NAPGD2022.

GMT *See* Greenwich mean time.

gnomonic projection [MAP PROJECTIONS] A planar projection, tangent to the earth at one point, projected from the center of the globe. All great circles appear as straight lines on this projection, so that the shortest distance between two points is a straight line. The gnomonic projection is useful in navigation. The gnomonic projection was used by Thales of Miletus, an ancient Greek astronomer and philosopher, to chart the heavens. It is possibly the oldest map projection. *See also* projection.

gnomonic projection

GNSS [GPS] Acronym for *Global Navigation Satellite System.* Global geopositioning satellite constellations, spread between several orbital planes, that are used for positioning, navigation, and timing (PNT) relative to objects on the earth's surface. The GNSS currently includes Chinese BeiDou Navigation Satellite System, European Union Galileo, Russian Global Navigation Satellite System (GLONASS), and U.S. Global Positioning System (GPS). *See also* BeiDou Navigation Satellite System, Galileo, geopositioning, GLONASS, GPS, PNT, satellite constellation.

GNSS time [GPS] Time that is unique to each of the global navigation satellite systems: U.S. GPS (GPST), European Galileo (GST), Russian GLONASS (GLONASST), and Chinese BeiDou Navigation Satellite System (BDT). *See also* GPS time.

gon *See* gradian.

goodness of fit [STATISTICS] In modeling, the degree to which a model predicts observed data; a measure of predictive power. *See also* model.

gore [CARTOGRAPHY] A section of a map area that lies between two lines of longitude. Each gore can be fitted to the surface of a globe to form a complete map with little distortion. *See also* interrupted projection, map.

gore

GPS [GPS] Acronym for *Global Positioning System.* A global geopositioning satellite constellation used for positioning, navigation, and timing (PNT) relative to objects on the earth's surface.

Orbiting satellites transmit signals that allow a GPS receiver anywhere on earth to calculate its own location through trilateration. Developed and operated by the U.S. government for the purpose of logistics, mapping, surveying, and other applications that require precise positioning; it is part of the international GNSS constellations spread between several orbital planes. *See also* BeiDou Navigation Satellite System, Galileo, geopositioning, GLONASS, GNSS, satellite constellation, trilateration.

GPS augmentation [GPS] The improvement of GPS positioning using external information—often integrated into the calculation process—to improve the accuracy, availability, or reliability of the satellite signal. *See also* GPS, QZSS.

GPS time [GPS] A continuous time scale, without leap seconds, defined by the GPS control segment and maintained by atomic clocks onboard monitor stations and satellites. It started (starting epoch) at 0h UTC (coordinated universal time, midnight) 5th–6th January, 1980 (6.d0). GPS time is synchronized with the U.S. Naval Observatory's UTC (USNO) at a 1 microsecond level and kept within 25 ns. *See also* GNSS time.

GPS-aided GEO augmented navigation *See* GAGAN.

grad *See* gradian.

gradian [EUCLIDEAN GEOMETRY] A unit of angular measurement in which the angle of a full circle is 400 gradians and a right angle is 100 gradians. The common abbreviation for gradian is *grad. See also* degree, radian, unit of measure.

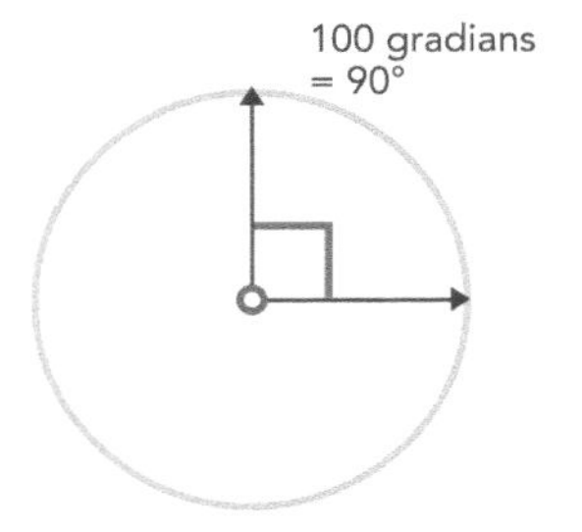

gradian

gradient

1. [GEODESY] A vector used to determine the maximum amount of vertical change on a surface and the direction in which that change occurs. Can also be expressed as the ratio, or percentage, between vertical distance (rise) and horizontal distance (run). A 10 percent gradient rises 10 feet for every 100 feet of horizontal distance.
2. [GEODESY] The inclination of a surface in a given direction.
3. [PHYSICS] The rate at which a quantity such as temperature or pressure changes in value.
4. [GRAPHICS (MAP DISPLAY)] The interpolated transition of thematic color between fixed colors.

See also color ramp, gradient of gravity, slope.

gradient of gravity [GEODESY] The direction of the maximum increase in gravity in a horizontal plane. *See also* gradient.

gradient vector map [MAP DESIGN] A map that uses vector lines to show the maximum slope magnitude and downward direction. *See also* gradient, vector.

graduated color map [MAP DESIGN] A map on which a range of colors indicates a progression of numeric values. For example, increases in population density might be represented by the increased saturation of a single color or temperature differences by a sequence of colors from blue to red. *See also* color map, color ramp.

graduated color map

graduated symbol map [MAP DESIGN] A map with symbols that are scaled proportionately according to the value of the data attribute they represent; each data value is symbolized to show its location in the progression of smaller to larger data values. For example, denser populations might be represented by larger dots or larger rivers by thicker lines. *See also* symbolization.

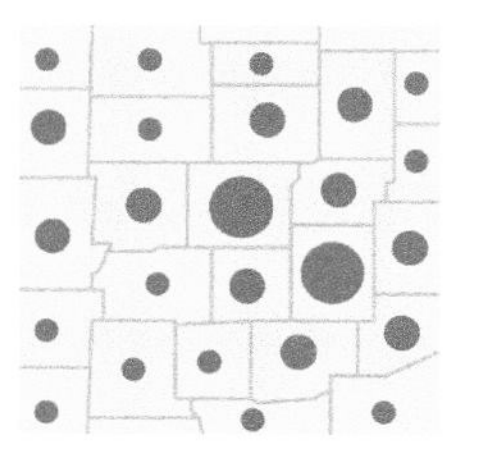

graduated symbol map

granularity

1. [DATA QUALITY] The coarseness or resolution of data. Granularity describes the clarity and detail of data during its capture and visualization.
2. [PHOTOGRAMMETRY] The objective measure of the random groupings of silver halide grains into denser and less dense areas in a photographic image.

See also resolution, visualization.

graph *See* chart.

graphic [GRAPHICS (COMPUTING)] A digital image. *See also* image, symbol.

graticule

1. [CARTOGRAPHY] A network of longitude and latitude lines on a map or chart that relates points on a map to their true locations on the earth.

graticule 1

2. [ASTRONOMY] A glass plate or cell with a grid or cross wires on it that rests in the focal plane of the

eyepiece of a telescope, used to locate and measure celestial objects.
See also latitude, longitude, grid.

graticule alignment of labels [SYMBOLOGY] A label positioning method in which labels are oriented along the graticule of the data frame. This is useful for maps of large areas, for cartographic or stylistic reasons. *See also* label orientation.

GRAV2022 *See* NAPGD2022.

gravimeter [GEODESY] A device used to measure small variations in the earth's gravitational field between two or more points. *See also* densitometer, geodesy.

gravimetric geodesy [GEODESY] The science of deducing the size and shape of the earth by measuring its gravitational field. *See also* densitometer, geodesy.

gravimetric geoid [GEODESY] A geoid model determined from terrestrial, airborne, or satellite gravity measurements, or any combination thereof. *See also* geoid, GEOID2022, hybrid geoid.

gravity model

1. [GEOGRAPHY] A model that assumes that the influence of phenomena or populations on each other varies inversely with the distance between them.
2. [BUSINESS] The idea that the probability of a given consumer visiting and purchasing at a given site is some function of the distance to that site, the site's attractiveness, and the distance and attractiveness of competing sites.

See also attractiveness.

gray level [PHOTOGRAMMETRY] In raster data, a unitless digital value of a pixel, sometimes called a density number (DN), representing a measurement of light, intensity, energy, elevation, temperature, or another physical characteristic of an object. This is a relative value subject to sensor sensitivity, bit depth, and material properties of the object captured.

grayscale

1. [SYMBOLOGY] Shades of gray from white to black.

grayscale 1

2. [GRAPHICS (COMPUTING)] Levels of brightness used to display information on a monochrome display device.

great circle [GEODESY] The largest possible circle that can be drawn on the surface of the spherical earth. Its circumference is that of the sphere, and its center is the center of the earth. All great circles divide the earth into halves. The equator and all lines of longitude are great circles. *See also* equator, small circle.

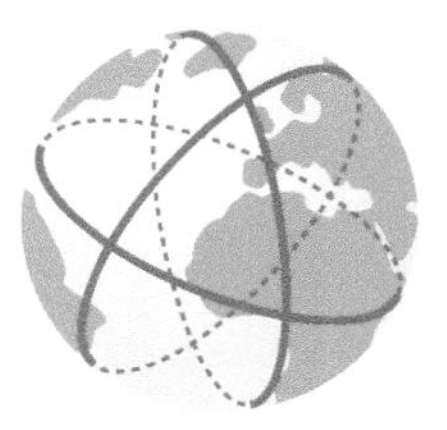

great circle

great circle route [NAVIGATION] The shortest path between two points on the globe. *See also* great circle, navigation.

Greenwich mean time [ASTRONOMY] Also known by the acronym *GMT*. The time at the prime meridian, which runs through the Royal Observatory in Greenwich, England. From 1884 to 1928, Greenwich mean time was the official name (and is still the popular name) for universal time. It sometimes also refers to coordinated universal time (UTC). *See also* coordinated universal time, Greenwich meridian, prime meridian, universal time.

Greenwich meridian [COORDINATE SYSTEMS] The 0-degree meridian from which all other longitudes are calculated; adopted by international agreement in 1884 as the prime meridian. The Greenwich prime meridian runs through the Royal Observatory in Greenwich, a suburb of London, England. *See also* prime meridian, meridian.

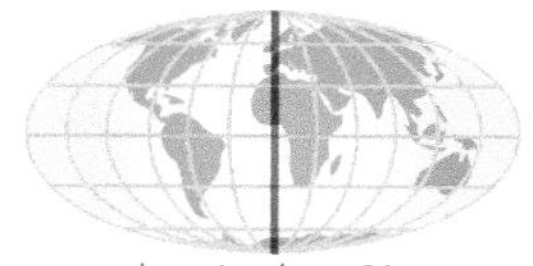

Greenwich meridian

grid [CARTOGRAPHY] In cartography, any network of parallel and perpendicular lines superimposed on a map and used for reference. These grids are usually referred to by the map projection or coordinate system they represent, such as universal transverse Mercator grid. *See also* Esri Grid, raster.

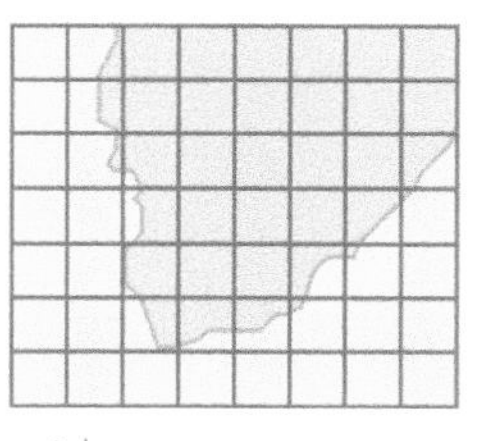

grid

grid azimuth [MAP PROJECTIONS] The horizontal angle from a grid north reference line to a direction line. *See also* direction line, grid north, reference line.

grid cell *See* cell.

grid cell location system [CARTOGRAPHY] A reference system used to identify locations on maps with alphanumeric codes using letters for columns and numbers for rows, or vice versa. *See also* cell.

grid convergence [COORDINATE SYSTEMS] The slight rotation of grid lines from horizontal and vertical at grid coordinate system zone edges. *See also* grid coordinate system.

grid coordinate system [COORDINATE SYSTEMS] A rectangular coordinate system superimposed mathematically on a map projection. *See also* Oregon Lambert system.

grid declination [NAVIGATION] The angular difference between grid north and true north. *See also* grid north, true north.

grid meridian [SURVEYING] Any meridian that is parallel to the central meridian, used when computing points in planar rectangular coordinate systems of limited extent. *See also* meridian.

grid north [CARTOGRAPHY] The northerly direction indicated by the north–south grid lines or ticks used on a map. *See also* true north, magnetic north.

grid reference box [MAP DESIGN] A type of map marginalia that identifies the grid zone and explains the grid used on the map. *See also* map marginalia.

grid stack [DATA STORAGE, ESRI SOFTWARE] A mechanism for storing multivariate raster data, consisting of layers referenced by a geodatabase. A grid stack is treated as a single entity for multivariate analysis. *See also* grid, multivariate analysis.

ground control [GEODESY] A system of points with known positions, elevations, or both, used as fixed references in georeferencing map features, aerial photographs, or remotely sensed images. *See also* georeferencing, photogrammetry.

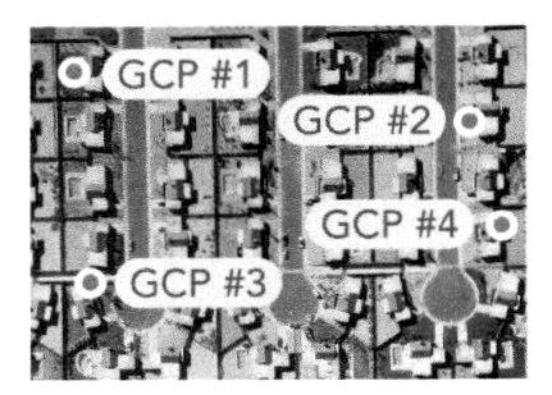

ground control

ground control point *See* control point.

ground distance multiplier [MATHEMATICS] A coefficient used to convert map distance to ground distance when taking slope error into account. *See also* physical distance, slope error.

ground receiving station [REMOTE SENSING] Communications equipment for receiving and transmitting signals to and from satellites such as Landsat. *See also* remote sensing.

ground resolution [GEOGRAPHY] The minimum distance on the ground between two closely located objects distinguishable as separate objects.

ground truth [ACCURACY] The accuracy of remotely sensed or mathematically calculated data based on data actually measured in the field.

group [ESRI SOFTWARE] In ArcGIS organizations, a way for administrators to organize user hierarchies and allocate credits or other resources. Groups typically have shared privileges for collaboration and content exchange. *See also* ArcGIS account, credits, members, organization.

group layer [DATA MANAGEMENT] A collection of several layers that appear and act as a single layer to simplify organization, advanced drawing options, and sharing. *See also* sublayer.

GRS80 *See* Geodetic Reference System of 1980.

GSDI *See* global spatial data infrastructure.

GUID *See* globally unique identifier.

hachure [SYMBOLOGY] A short line on a map that indicates the direction and steepness of a slope. Hachures that represent steep slopes are short and close together; hachures that represent gentle slopes are longer, lighter, and farther apart. *See also* slope.

hachure

hachured contour [SYMBOLOGY] On a topographic map, concentric contour lines drawn with hachures (right-angle ticks) to indicate a closed depression or basinlike features. Added to the downslope (or inside) of the contour. Concentric contour lines drawn without hachure marks indicate a hill. *See also* contour line, depression, topographic map.

hachured contour

half point [NAVIGATION] One of the eight compass points indicating NNE, ENE, ESE, SSE, SSW, WSW, WNW, and NNW.

halftone image [GRAPHICS (MAP DISPLAY)] A continuous tone image photographed through a fine screen that converts it into uniformly spaced dots of varying size while maintaining the gradations of highlight and shadow. The size of the dots varies in proportion to the intensity of the light passing through them. *See also* dot screen, continuous tone image.

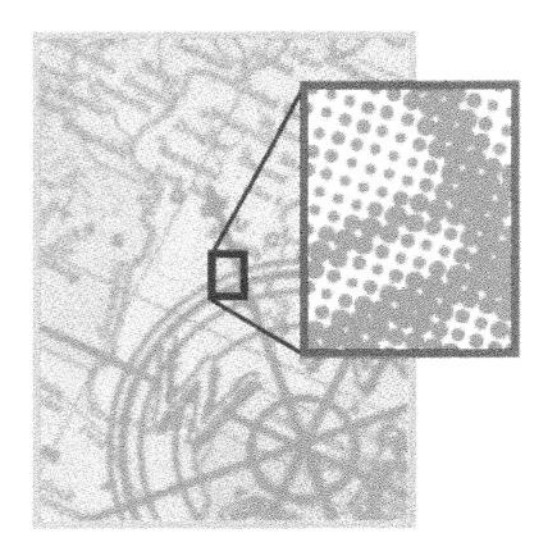

halftone image

Hamiltonian path [NETWORK ANALYSIS] A path through a network that visits each junction in the network only once without returning to its point of origin. Hamiltonian paths are named after the Irish mathematician, physicist, and astronomer William Rowan Hamilton (1805–1865). *See also* traveling salesperson problem.

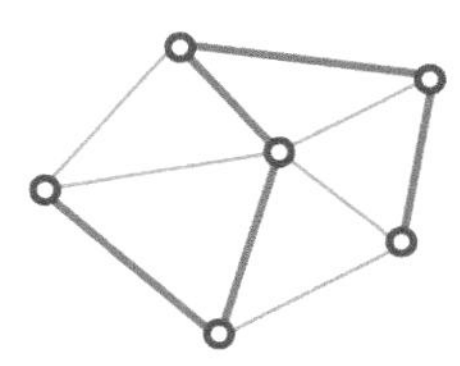

Hamiltonian path

HARN *See* High Accuracy Reference Network.

hatches [LINEAR REFERENCING] A series of vertical line or marker symbols used in linear referencing that store relative geographic locations or positions. Hatches, or hatch layers, are displayed on top of features at an interval specified in route measure units. *See also* route measure.

hatching [LINEAR REFERENCING] A type of labeling used in linear referencing that posts and labels hatches (or line marker symbols) at a regular interval along measured line features. *See also* hatches.

HDOP *See* horizontal dilution of precision.

heading [NAVIGATION] The direction of a moving object, expressed as an angle from a known direction, usually north. *See also* bearing.

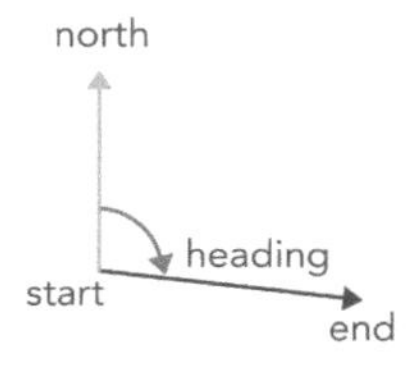

heading

heat map [MAP DESIGN] A symbology style that shows the relative distribution of point density using a continuous color ramp that implies temperature ranging from cool (few points) to hot (many points). *See also* color ramp, density, point.

hectare [CADASTRAL AND LAND RECORDS] A common areal unit used in the measurement of land area that is equal to 10,000 square meters or 2.472 acres. *See also* are.

height [EUCLIDEAN GEOMETRY] The vertical distance between two points, or above a specified datum. *See also* altitude, elevation, slope, z-value.

height above ellipsoid [GEOREFERENCING] Also known by the acronym *HAE*. The measurement of elevation above a theoretical three-dimensional model of the earth's surface based on an ellipsoid rather than a sphere. Most commonly used by WGS84. *See also* ellipsoid height, WGS84.

Helmert transformation [COORDINATE SYSTEMS, SPATIAL ANALYSIS] A type of affine transformation (a geometric transformation) used to transform coordinates between coordinate systems. A Helmert transformation includes translation, rotation, rescaling, and a skew component that allows for the correction of systematic errors between coordinate systems. The Helmert transformation is named for the German mathematician and geodesist Friedrich Robert Helmert

(1843–1917). *See also* affine transformation, rescale, rotate, skew, translate.

hemisphere
1. [ASTRONOMY] Half of a celestial body, such as the earth. For example, the equatorial plane divides the earth into the northern and southern hemispheres.
2. [EUCLIDEAN GEOMETRY] Half of a sphere.

See also sphere.

H

heuristic [MATHEMATICS] In graph theory, a function used to determine the lowest cost or shortest path between two given nodes. *See also* algorithm.

hex code [GRAPHICS (COMPUTING)] A color model that uses a hexadecimal palette. *See also* color model, hexadecimal, RGB.

hexadecimal [MATHEMATICS] A number system using base 16 notation, usually composed of the digits 0–9 and the letters A–F or a–f. *See also* hex code.

hexmap [DATA STRUCTURES] A map made by binning data into hexagonal cells. *See also* data binning.

hierarchy [NETWORK ANALYSIS] An attribute of an element in a network dataset that is used during network analysis to assign priority. For example, in a transportation network, a "road class" hierarchy can be assigned to edges to favor highways instead of local streets. *See also* network attribute, network dataset.

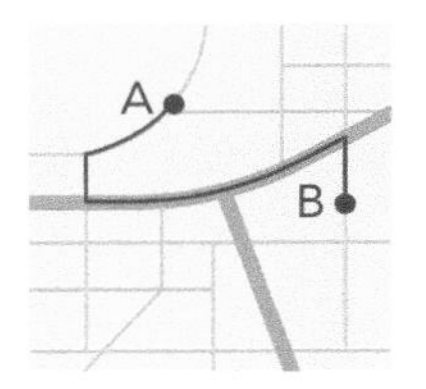

hierarchy

High Accuracy Reference Network [GEODESY] Also known by the acronym *HARN*. A regional or statewide resurvey and readjustment specifically of NAD 1983 control points using GPS techniques. The resurvey date is often included as part of the datum name: NAD 1983 (1991) or NAD91. *See also* control point, GPS, North American Datum of 1983, surveying.

high-pass filter [DIGITAL IMAGE PROCESSING] A spatial filter that blocks low-frequency (longwave) radiation, resulting in a sharpened image. *See also* band-pass filter, low-pass filter, edge enhancement.

hillshading [MAP DESIGN]
1. Shadows drawn on a map to simulate the effect of the sun's rays over varied terrain.
2. A technique used to enhance the three-dimensional appearance of terrain features that create patterns of light and shadow on the landscape. Illumination is from an imaginary light source at a specified azimuth and altitude (typically the upper-left corner of the map). Hillshading provides a sense of visual relief for cartography, and a relative measure of incident light for analysis.

hillshading 2

See also altitude, azimuth, shaded relief image.

hillsign [MAP DESIGN] A rough line drawing of highly stylized hills and mountains, used on both large- and small-scale maps.

hillslope element [GEOGRAPHY] A division of a hillside defined by its position and slope and influenced by geomorphic and hydrologic processes (erosion). *See also* geomorphology, hydrology.

histogram [GRAPHICS (MAP DISPLAY)] A graphic display that shows the set of values in a dataset and the number of features with those values. It can be enhanced to also show the proportion of features that fall into each category. Often used to create range breaks for thematic maps. *See also* natural breaks classification, thematic map.

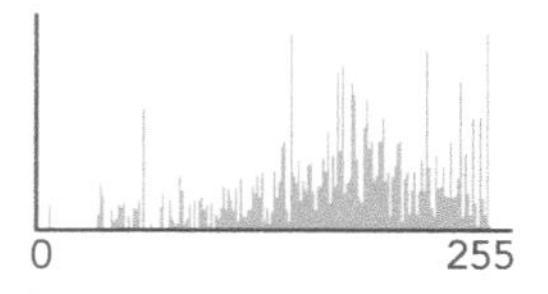

histogram

histogram equalization [DIGITAL IMAGE PROCESSING] The redistribution of pixel values in an image so that each range contains approximately the same number of pixels. A histogram showing this distribution of values would be nearly flat. *See also* digital image processing, pixel.

historic parcel [SURVEYING] In the parcel fabric record, a parcel that has been retired. When parcels become historic they are replaced with new parcels. Historic parcels are also referred to as parent parcels when defining parcel lineage. *See also* parcel fabric.

historical marker [ESRI SOFTWARE] A user-created reference to a time and date stamp.

hole [DATA QUALITY] A small gap in a raster line feature, typically caused by source data errors. *See also* error.

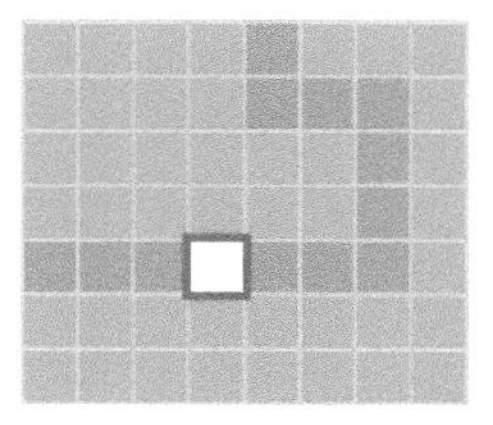

hole

homolatitudes [MAP PROJECTIONS] The two small-circle lines of tangency that are equidistant from the parallel where the map projection is centered. *See also* line of tangency, projection, small circle.

homolosine projection [MAP PROJECTIONS] An equal-area map projection that is a composite of two pseudocylindrical projections—the Mollweide projection for higher latitudes and the sinusoidal projection for lower latitudes. The two projections join at 40°44'11.8" north and south, where the linear scale of the two projections matches. Sometimes called an uninterrupted homolosine projection. *See also* equal-area projection, projection.

horizon

1. [NAVIGATION] The apparent or visible junction of land and sky.
2. [ASTRONOMY] The horizontal plane tangent to the earth's surface and perpendicular to the line through an observer's position and the zenith of that position. The apparent or visible horizon approximates the true horizon only when the point of vision is very close to sea level.
3. [ASTRONOMY] The great circle in which an observer's horizon meets the celestial sphere.
4. [MAP PROJECTIONS] The edge of a map projection.

See also celestial sphere, horizon circle, zenith.

horizon circle [MAP PROJECTIONS] The circle containing all points equidistant from the center of an azimuthal projection. *See also* azimuthal projection.

horizontal accuracy [ACCURACY] The accuracy of a position on the surface of the earth.

horizontal angle [NAVIGATION] The angle formed by the intersection of two lines in a horizontal plane. *See also* navigation, operation codes, zenith angle.

horizontal control [GEODESY] A network of known horizontal geographic positions, referenced to geographic parallels and meridians or to other lines of orientation such as plane coordinate axes. *See also* horizontal geodetic datum.

horizontal control datum *See* horizontal geodetic datum.

horizontal datum *See* horizontal geodetic datum.

horizontal dilution of precision [GEODESY] Also known by the acronym *HDOP*. A measure of the geometric quality of a GPS satellite configuration. HDOP is a factor in determining the relative accuracy of a horizontal position. The smaller the DOP number, the better the geometry. *See also* dilution of precision.

horizontal geodetic datum [GEODESY] A geodetic datum for any extensive measurement system of positions, usually expressed as latitude-longitude coordinates, on the earth's surface. A horizontal geodetic datum may be local or geocentric. If it is local, it specifies the shape and size of an ellipsoid representing the earth, the location of an origin point on the ellipsoid surface, and the orientation of x- and y-axes relative to the ellipsoid. If it is geocentric, it specifies the shape and size of an ellipsoid, the location of an origin point at the intersection of x-, y-, and z-axes at the center of the ellipsoid, and the orientation of the x-, y-, and z-axes relative to the ellipsoid. Examples of local horizontal geodetic datums include the North American Datum of 1927, the European Datum of 1950, and the Indian Datum of 1960; examples of geocentric horizontal geodetic datums include the North American Datum of 1983 and the World Geodetic System of 1984. *See also* geocentric datum, geodetic datum, local datum, North American Datum of 1983, WGS84.

hosted layer [DATA MANAGEMENT] A layer accessible through an ArcGIS account and added to a map or a scene from a service. Examples of the types of layers that can be hosted include, but are not limited to, feature layers, Web Feature Service (WFS) layers, tile layers, imagery layers, elevation layers, tables, and scene layers. *See also* ArcGIS account, content, item, layer, service.

HSV [GRAPHICS (MAP DISPLAY)] A color model that uses *hue, saturation,* and *value.* Hue specifies the perceived color, such as red or green. Saturation specifies the intensity, or how vivid the color appears. Value specifies the brightness, or white intensity, with higher values perceived as lighter. *See also* CMYK, color model, RGB.

hub [NETWORK ANALYSIS] A central node in a network for routing goods to their destinations.

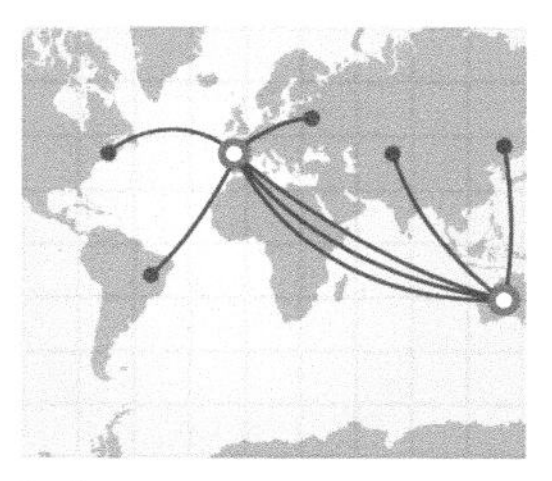

hub

H

hue [PHYSICS] The dominant wavelength of a color, by which it can be distinguished as red, green, yellow, blue, and so on. *See also* saturation, value, intensity, chroma.

human geography [GEOGRAPHY] The field of geography concerning a range of social, cultural, and political aspects of human life as related to their distribution through physical space. *See also* geography.

human landscape [ECOLOGY] The landscape resulting from many generations of human occupancy. *See also* landscape.

hybrid geoid [GEODESY] A gravimetric geoid model constrained to GPS or GNSS observations at leveled benchmarks. In general, the resulting surface consists of three components: a bias, a tilt, and locally varying surface curvature. The most common use of a hybrid geoid model is to transform GPS-derived ellipsoidal heights to a leveling-based regional vertical datum, such as NAVD 88. *See also* benchmark, curvature, ellipsoid height, gravimetric geoid, GNSS, GPS, leveling, NAVD 88, transformation, vertical geodetic datum.

hydrographic chart [NAVIGATION] A nautical chart published for an inland water area.

hydrographic datum [COORDINATE SYSTEMS] A plane of reference for depths, depth contours, and elevations of foreshore and offshore features. *See also* datum, hydrography, hydrographic survey.

hydrographic survey [GEODESY] A survey of a water body, particularly of its currents, depth, submarine relief, and adjacent land. *See also* surveying, hydrography, hydrographic datum.

hydrography [GEODESY] The measurement and description of water features and their related land areas for the purposes of safe marine navigation. *See also* bathymetry.

hydrologic cycle [GEOGRAPHY] The circulation of water from the earth through the atmosphere and back again. Its major stages are evaporation, condensation, precipitation, runoff, transpiration, infiltration, and percolation. *See also* hydrology.

hydrology [GEOGRAPHY] The study of water, its behavior, and its movements across and below the surface of the earth, and through the atmosphere. *See also* drainage, watershed.

hyperspectral imagery [PHOTOGRAMMETRY] An image composed of many narrow wavelength bands across the electromagnetic spectrum, typically greater than 100 bands. Hyperspectral imaging captures narrow-width spectral bands over specific ranges of wavelengths, corresponding with spectral signatures of interest. *See also* band, electromagnetic spectrum.

hypsography [CARTOGRAPHY]
1. The study and representation of elevation and the earth's topography.
2. The representation of relief features on a map.

See also elevation, relief, topography.

hypsometric curve [CARTOGRAPHY] A curve showing the relationship of area to elevation for specified terrain. A hypsometric curve is plotted on a graph on which the x-axis represents surface area and the y-axis represents elevation above or below a datum (normally sea level). The curve shows how much area lies above and below marked elevation intervals. *See also* hypsometry.

hypsometric map [CARTOGRAPHY] A map showing relief, whether by contours, hachures, shading, or tinting. *See also* contour, hachure, hypsometric tinting, relief, shading.

hypsometric tinting [MAP DESIGN] A method of adding color between contour lines to illustrate changes in elevation depth. Each color represents a different range of elevation. Also called layer tinting or hypsometric coloring. *See also* contour line, elevation.

hypsometric tinting

hypsometry [GEODESY]

1. The science that determines the spatial distribution of elevations above an established datum, usually sea level.
2. The determination of terrain relief, by any method.

See also elevation, hypsometric map.

H

ICAO [NAVIGATION, STANDARDS] Acronym for *International Civil Aviation Organization.* A member organization that represents the worldwide body of nations for standardizing flight rules, regulations, and requirements.

ID *See* identifier.

I

ideal camera model [PHOTOGRAMMETRY] In imagery, the mathematical relationship between the coordinates of a point in three-dimensional space and its projection onto the two-dimensional image plane of a pinhole camera. Pinhole describes where the image point, ground point, and perspective center converge colinearly. The perspective geometry is distortion-free. *See also* perspective geometry.

identification key [AERIAL PHOTOGRAPHY] An aid to an image interpreter that shows the typical appearance of a feature on an aerial photograph. *See also* aerial photograph, image interpretation.

identifier [COMPUTING] A unique character string or numeric value associated with a particular object. *See also* GlobalID, globally unique identifier.

identify [ESRI SOFTWARE] A software tool that, when applied to a feature (by clicking it), opens a window showing that feature's attributes.

identity [ANALYSIS/GEOPROCESSING] In geoprocessing, a topological overlay that computes the geometric intersection of two datasets. The output dataset preserves all the features of the first dataset plus those portions of the second (polygon) dataset that overlap the first. For example, a road passing through two counties would be split into two arc features, each with the attributes of the road and the county it passes through. *See also* intersect, union.

identity link [DATA EDITING] An anchor that prevents the movement of features during rubber sheeting. *See also* rubber sheeting.

IDW *See* inverse distance weighted interpolation.

IFR enroute high altitude chart [NAVIGATION] A chart designed for air navigation at or above 18,000 feet above mean sea level for the United States. Provided and maintained by the U.S. FAA. IFR is an acronym for *instrument flight rules. See also* enroute aeronautical chart, FAA, IFR enroute low altitude chart, mean sea level.

IFR enroute low altitude chart [NAVIGATION] A chart that provides aeronautical information for air navigation below 18,000 feet above mean sea level for the United States. Provided and maintained by the U.S. FAA. IFR is an acronym for *instrument flight rules. See also* enroute aeronautical chart, FAA, IFR enroute high altitude chart, mean sea level.

IFSAR [REMOTE SENSING] An abbreviation of *interferometric synthetic aperture radar.* A dual-antenna radar sensor mounted on an airborne or spaceborne platform that collects a remotely sensed radar image, called an interferogram. There is a measured energy shift between the signals received by each antenna, and this interference can be colorized to measure elevation or changes in the topography on the earth's surface. *See also* interferogram, radar interferometry.

IID [PROGRAMMING] An abbreviation of *interface identifier.* A string that provides the unique name of an interface. An IID is a type of globally unique identifier (GUID). *See also* globally unique identifier, identifier.

illumination [CARTOGRAPHY] The light incident on a surface or object, either natural or artificial, as determined by the surface's slope and aspect and by the light source's azimuth and altitude. *See also* azimuth, altitude.

illumination

illumination map [MAP DESIGN] A map that accounts for the location of the sun and any objects that impede the sun's illumination, such as hills, trees, or buildings.

image

1. [DATA CAPTURE] A representation or description of the geographic environment, typically produced by an optical or electronic device or an electromagnetic sensor, such as a camera or a scanning radiometer. Common examples include remotely sensed data (for example, satellite data), scanned data, and photographs.
2. [ESRI SOFTWARE] A raster dataset.

See also aerial photograph, analog image, digital image, raster.

image adjustment [PHOTOGRAMMETRY] The process of adjusting the parameters of image support data to get an accurate transformation between the image and the ground. The image adjustment process is based on a calculated relationship between overlapping images, control points, the camera model, and topography and consists of computing a transformation for a block (or group) of images. With aerial digital data, image adjustment consists of three key components: tie points, ground control points, and aerial triangulation. *See also* control point, tie point, transformation, triangulation.

image basemap [PHOTOGRAMMETRY] A basemap created using geometrically correct (orthorectified) imagery, where distortion caused by terrain relief, elevation, and any other geometric anomaly has been removed, giving the image basemap the geometric integrity of a map. *See also* basemap, orthorectification.

I

image catalog *See* raster catalog.

image chip [PHOTOGRAMMETRY] Small images extracted from a larger source image that contain the feature or object of interest; used to train a deep learning model. *See also* deep learning.

image coordinate

1. [DATA STRUCTURES, ESRI SOFTWARE] A two-digit method to locate a pixel, or cell, within an image file using row and column referencing (x,y); x provides the column number (commonly starting from 0 at the left) and y provides the row number (commonly starting from 0 at the top).
2. [REMOTE SENSING] The coordinates in a physical sensor model originating at the principal point of the sensor focal plane. Units are typically in mm and align with the rows and columns of an output image.

See also raster, pixel.

image coordinate system [COORDINATE SYSTEMS, ESRI SOFTWARE] A spatial reference system that uses the relative geometric coordinate (x,y,z) locations for an image. Once this is established for the focus image, all other feature and image data is transformed to use that coordinate reference.

image data [DATA CAPTURE] Data produced by scanning a surface with an optical or electronic device. Common examples include scanned documents, remotely sensed data (for example, satellite images), and aerial photographs. An image is stored as a raster dataset of binary or integer values that represent the intensity of reflected light, heat, or other range of values on the electromagnetic spectrum. *See also* electromagnetic spectrum.

image division [DIGITAL IMAGE PROCESSING] A digital image processing technique for increasing the contrast between features in an image by dividing the pixel values in the image by the values of corresponding pixels in a second image. Image division is normally used to identify concentrations of vegetation. *See also* digital image processing.

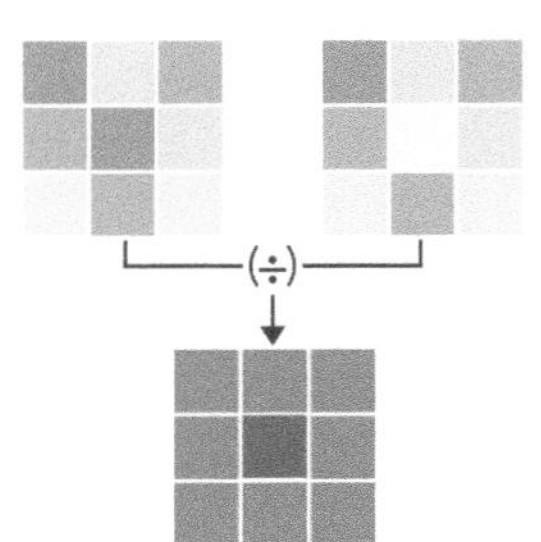

image division

image element [PHOTOGRAMMETRY] The characteristics of an image, including its tone, color, shape, size, pattern, shadow, texture, location, context, height, and date; used for the purpose of detection, identification, or image interpretation. *See also* color, context, date, location, pattern, shadow, shape, size, tone.

image filter [DIGITAL IMAGE PROCESSING] In image processing, a computing technique by which the size, colors, shading, and other characteristics of an image are altered. Typically uses an <n × n> matrix of weights. The resulting calculation to each output pixel is restricted to the size of the matrix, referred to as a kernel. Filters are used primarily in pixel-based analysis where the value of a center pixel is changed to the mean, sum, or other function of all pixel values within the filter. Image filters move systematically across a raster until each pixel has been processed. *See also* kernel, matrix.

image index [REMOTE SENSING] A computed value within a raster dataset that is an arithmetic calculation of different spectral bands used to mitigate illumination effects of the terrain and accentuate phenomena based on spectral signature curves. For example, Normalized Difference Vegetation Index (NDVI) is a common calculation correlated with vegetation vigor [(NIR – Red) / (NIR + Red)]. Many image indices require that the gray levels in bands be transformed to either radiance or reflectance for inputs. *See also* radiance, raster dataset, reflectance.

image interpretation [GEOREFERENCING] The use of photographic and digital images to identify objects and their attributes. *See also* identification key.

image map [MAP DESIGN] A map made by superimposing traditional map symbols over an image base. *See also* image.

I

image pixel [PHOTOGRAMMETRY] In imagery, a pixel is the smallest element (a cell) that composes a raster dataset. A picture element representing a property such as reflectance, elevation, temperature, or energy. *See also* pixel, pixel value.

image scale [DATA CAPTURE] The ratio between a distance in an image and the actual distance on the ground, calculated as focal length divided by the height above mean ground elevation. Image scale can vary in a single image from point to point due to surface relief and the tilt of the camera lens. *See also* image, scale.

image service [ESRI SOFTWARE] A persistent software process that provides access to prerendered images. Some properties of an image service may be defined by the client application. *See also* service, web service.

image space

1. [DATA STRUCTURES] The x,y coordinate space defined by the number of columns and rows in a raster dataset. The origin of image space is commonly the center of the top left pixel of the data and is labeled (0,0). The x-axis corresponds to the number of columns in the raster and the y-axis to the number of rows. For raster data to be used in GIS software, image space must be transformed to a real-world coordinate system through georeferencing.
2. [DIGITAL IMAGE PROCESSING] A method to locate a pixel within a file using row and column referencing (0,0) independent of any sensor, geographic, or other coordinate reference.

See also raster, pixel, coordinate system, georeferencing.

image statistics [DIGITAL IMAGE PROCESSING] In image processing, calculations from the pixel values of each band in an image—often based on the image's histogram—including, but not limited to, the minimum, maximum, mean, and standard deviation of pixel values. Image statistics are required for specific rendering and geoprocessing operations, such as multivariate spatial analysis. *See also* multivariate analysis, pixel value.

image stretch [DIGITAL IMAGE PROCESSING] A technique that expands, compresses, or truncates different areas of an image histogram to enhance or depress certain features or information. This can be done according to statistical parameters or performed interactively piece by piece. *See also* image adjustment, transformation.

image support data [PHOTOGRAMMETRY] Also known by the acronym *ISD*. In imagery, a collection of information identifying image metadata and sensor model information used when processing and displaying imagery products. ISD typically includes interior and exterior orientation data, spectral data, sensor gain and bias, image collection angles, information about the mission, and other pertinent information. *See also* metadata.

image tile layer [DATA MANAGEMENT, MAP DISPLAY] A data layer stored as a series of tiles and typically used in basemaps to provide geographic or topographic context. Image tile layers differ from tile layers in that they are provided by a service rather than a tiling scheme. *See also* basemap, tile service.

image transformation [PHOTOGRAMMETRY, COORDINATE SYSTEMS] In imagery, a mathematical function that changes the image's geometry and resamples its pixels, forming a new image that conforms to a different coordinate reference or map projection system. *See also* coordinate system, projected coordinate system, transformation.

imagery [REMOTE SENSING] The use of images—specifically images captured by sensors, drones, or aircraft—to represent and analyze the spatial data of a surface (such as the surface of the earth). An incomplete list of imagery types includes satellite imagery, aerial photographs, thermal imagery, and multispectral imagery. *See also* aerial photograph, drone imagery, multispectral image, satellite imagery.

imagery layer [REMOTE SENSING] A collection of map cartography based on raster data obtained from a satellite, drone, or other remote sensing device to show an area. These layers can be displayed dynamically or prerendered as cached image tiles from an image service. *See also* raster, tile cache.

impedance [NETWORK ANALYSIS] A measure of the amount of resistance, or cost, required to traverse a path in a network or to move from one element in the network to another. Resistance may be a measure of travel distance, time, speed of travel multiplied by distance, and so on. Higher impedance values indicate more resistance to movement, and a value of zero indicates no resistance. An optimum path in a network is the path of lowest impedance, also called the least-cost path. *See also* least-cost path, stop impedance.

IMU *See* inertial measurement unit.

incident [NETWORK ANALYSIS] A network location used in closest facility analysis. Car accidents, crime scenes, and fires are examples of incidents. *See also* closest facility analysis.

incident energy [PHYSICS] Electromagnetic radiation that strikes a surface. *See also* electromagnetic radiation.

INCITS [STANDARDS] Acronym for *International Committee for Information Technology Standards.* A forum accredited by the American National Standards Institute (ANSI) that creates and maintains information and communications technology standards through the participation and consensus of its industry members. *See also* American National Standards Institute.

independent variable [STATISTICS] One or a set of variables used to model or predict the dependent variable. For example, a prediction of annual purchases for a proposed store (the dependent variable) might include independent variables representing the number of potential customers, distance to competition, store visibility, and local spending patterns. In the regression equation, independent variables appear on the right side of the equal sign and are often referred to as explanatory variables. *See also* dependent variable, regression equation.

indeterminate flow direction [NETWORK ANALYSIS] In network analysis, a flow direction that is unknown or undiscoverable. Indeterminate flow direction occurs when flow direction cannot be determined from the connectivity of the network, the locations of sources and sinks, and the enabled or disabled states of features. *See also* determinate flow direction.

index [COMPUTING] A data structure, usually an array, used to speed the search for records in a database or for spatial features in geographic datasets. In general, unique identifiers stored in a key field point to records or files holding more detailed information.

index contour line [SYMBOLOGY] On a topographic map, a contour line that is drawn with a thicker line symbol than the intermediate contour lines and usually labeled with the elevation that it represents. Depending on the contour interval, every fourth or fifth contour line may be an index contour. *See also* contour line, intermediate contour line.

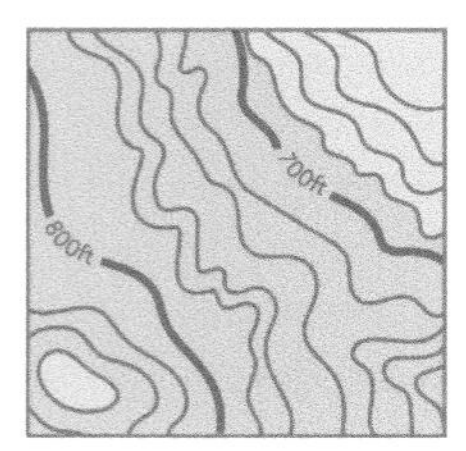

index contour line

index map [CARTOGRAPHY] A schematic map used as a reference for a collection of map sheets, outlining the total area covered along with the coverage extent of, and usually a name or reference for, each map sheet. *See also* map sheet.

index mark [CARTOGRAPHY] The mark at the center of the protractor circle from which angles are measured.

index value [DATA CONVERSION] A complex combination of attributes from one or more sets of spatial data that can be combined in a variety of ways to normalize data. *See also* normalization, spatial data.

Indian Regional Navigation Satellite System *See* NavIC.

indoor positioning system [DATA CAPTURE] Also known by the acronym *IPS*. Dedicated networked devices, sensors, or technology that can create small-area blueprints or find objects within defined indoor spaces.

industry [ORGANIZATIONAL ISSUES] An organization with specific GIS needs. Examples of industries include government, transportation, health care, real estate, and public safety.

inertial measurement unit [REMOTE SENSING] Also known by the acronym *IMU*. A device that measures the linear acceleration (force) and angular (attitude) rates of an object. IMUs are composed of accelerometers and gyroscopes and are used to assist with precision when navigating airborne and spaceborne vehicles and for direct geopositioning of remote sensor data. IMUs are frequently integrated with GPS/GNSS systems and other aiding sensors, such as star trackers in satellites, to determine orientation. *See also* GPS, GNSS.

inferencing [PROGRAMMING] In deep learning, a model that employs a series of mathematical algorithms to analyze and sort data based on training data; sorted data groups are then scored to derive an output. *See also* deep learning, training, training samples.

information space [COGNITION] A geometric representation of relationships between elements in a data domain, in which relative position indicates the degree of similarity between elements. Information spaces are often based on geographic metaphors and are used to provide more intuitive views of a complex, multidimensional data domain. *See also* visualization.

infrared radiation [PHYSICS] The band of the electromagnetic spectrum that has electromagnetic radiation (ER) wavelengths greater than the red end of the visible light spectrum but less than that of microwaves. Its wavelengths span from about 750 nm (reflected light) to 1 mm (thermal radiation). *See also* band, electromagnetic spectrum, longwave infrared, mid-wave infrared, near infrared, shortwave infrared.

infrared satellite image [SATELLITE IMAGING] An image that records thermal infrared (thermal IR) radiation emitted by cloud tops, land, oceans, ice, and snow. *See also* longwave infrared.

infrared scanner *See* infrared sensor.

infrared sensor [DATA CAPTURE] A device that detects infrared radiation and converts it into an electrical signal. *See also* infrared radiation, multispectral sensor, sensor.

infrared weather satellite image map [METEOROLOGY] A map that distinguishes low-, middle-, and high-level clouds by their gray tone or color on a weather satellite image.

infrastructure [DEVELOPMENT] The system of human-made physical structures, for example, roads, bridges, canals, communication towers, hospitals, pipes, and so on, that provide communication, transportation, public services, utilities, or all of the above to a population.

initial point [SURVEYING] A surveyed starting point determined by government land surveyors at the intersection of the baseline and the principal meridian used in establishing land partitioning in the Public Land Survey System. *See also* DLS, PLSS, principal meridian.

input event record [ESRI SOFTWARE] In geocoding, a piece of information such as a customer address and location of an incident. Input event record types vary by application. They include customer addresses, location of the event or incident, location of equipment and facilities, and the monument offset. *See also* event, geocoding, record.

input feature [ANALYSIS/GEOPROCESSING] In geoprocessing, data put into the system for processing, usually specified by a path in a dialog box, script, or at the command line. *See also* feature, geoprocessing.

input table [ANALYSIS/GEOPROCESSING] In geoprocessing, tabular data put into the system for processing, usually specified by a path in a dialog box, script, or at the command line.

inscribed circle [CARTOGRAPHY] The largest circle contained within a given polygon.

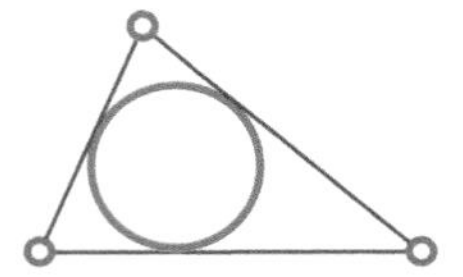

inscribed circle

inset map [MAP DESIGN] A small map set within a larger map. An inset map might show a detailed part of the map at a larger scale or the extent of the existing map drawn at a smaller scale within the context of a larger area. *See also* overview map.

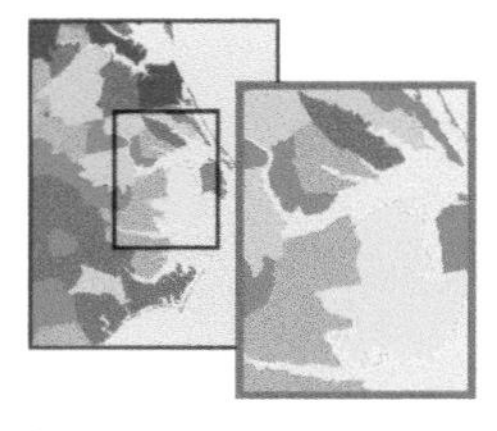

inset map

insolation *See* solar insolation.

inspection method [WAYFINDING] A method that uses linear features or prominent objects to plot map orientation, position, or distance.

integer data *See* discrete data.

integrated feature dataset [DATA MANAGEMENT] In a geodatabase, a feature dataset that stores topologically associated feature classes. *See also* feature, feature class.

integration [DATA TRANSFER] A high degree of interconnection between two or more programs or datasets that share a common schema, ontology, semantic approach, or method, allowing information to be passed between them without being fully processed. *See also* intersect.

intensity [GRAPHICS (COMPUTING)] In the IHS (intensity, hue, saturation) color model, the relative brightness of a color. *See also* saturation, hue, value, chroma.

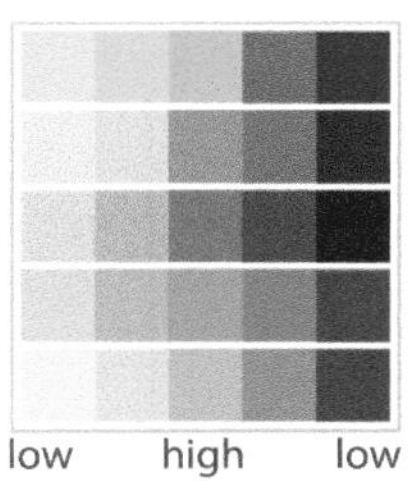

intensity

interactive map [SOFTWARE] A web map that users can manipulate to change the style or format, zoom, search, filter, or view pop-ups. *See also* web map, zoom.

interactive vectorization [DATA CONVERSION] A manual process for converting raster data into vector features that involves tracing raster cells. *See also* vectorization.

interferogram [REMOTE SENSING] A radar image that records interference patterns captured by two antennae a short distance apart. *See also* IFSAR, radar interferometry.

interior orientation [PHOTOGRAMMETRY] In photogrammetry, the internal geometry of a calibrated sensor coordinate system. This consists of the location of the principal point, the calibrated focal length, the x, y coordinate transform from row and column to image x, y in millimeters, and any correction methods required to remove distortion from the ideal sensor model.

intermediate contour line [SYMBOLOGY] On a topographic map, a contour line that is not labeled and is located between index contour lines. *See also* contour, contour line, index contour line.

intermediate data [MODELING] Any data in a process that did not exist before the process existed and that will not be maintained after the process executes. *See also* derived data.

International Chart Series [CARTOGRAPHY] Coordinated by the International Hydrographic Organization, a worldwide system of hydrographic chart standard specifications.

international date line [COORDINATE SYSTEMS] An imaginary line, generally following the meridian of longitude lying 180 degrees east and west of the Greenwich meridian, where the date changes. The time zone east of the international date line is twelve hours ahead of Greenwich mean time; the time zone west of the international date line is twelve hours behind Greenwich mean time. A traveler going west across the date line adds a day; a traveler going east across it subtracts a day. *See also* Greenwich mean time.

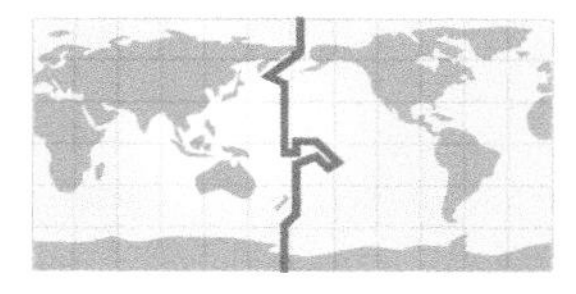

international date line

International Ellipsoid of 1924 [COORDINATE SYSTEMS] The ellipsoid used as the basis for the latitude and longitude coordinates of the European Datum of 1950. *See also* datum, ED50.

international meridian *See* Greenwich meridian.

International Meridian Conference [COORDINATE SYSTEMS] The conference held in October 1884, in Washington, D.C., U.S.A., at which the Greenwich meridian was selected as the international standard for the prime meridian. *See also* Greenwich meridian, prime meridian.

international nautical mile [NAVIGATION]

1. A unit of distance equal to 6,076.1 feet or 1,852 meters (about 1.15 statute miles); used globally for maritime and aviation purposes.
2. A unit of distance equivalent to one minute of latitude on a sphere whose surface area is equal to the surface area of the ellipsoidal earth.

See also ellipsoid.

International Organization for Standardization [STANDARDS] Also known by the acronym *ISO*. A federation of national standards institutes from more than 100 countries that works with international organizations, governments, industries, businesses, and consumer representatives to define and maintain criteria for international standards in specific technical and nontechnical fields.

International Terrestrial Reference Frame *See* ITRF.

International Terrestrial Reference System *See* ITRS.

Internet of Things [INTERNET] Also known by the acronym *IoT*. A conceptual term for devices (such as smart home appliances, security devices, and wearable health monitors) equipped with sensors or other technology that enable communication over a local or public network.

interoperability [INTEROPERABILITY] The capability of components or systems to exchange data with other components or systems or to perform in multiple environments. In GIS, interoperability is required for a GIS user using software from one vendor to study data compiled with GIS software from a different provider.

interpolation

1. [MATHEMATICS] The estimation of surface values at unsampled points based on known surface values of surrounding points. Interpolation can be used to estimate elevation, rainfall, temperature, chemical dispersion, or other spatially based phenomena. Interpolation is commonly a raster operation, but it can also be done in a vector environment using a TIN surface model. There are several well-known interpolation techniques, including spline and kriging.

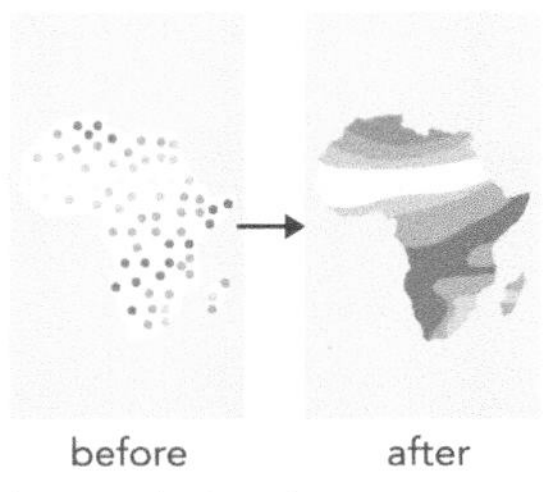

interpolation 1

2. [ESRI SOFTWARE] In the context of linear referencing, the calculation of values for a route between two known measure values.

See also inverse distance weighted interpolation, kriging, natural neighbors, trend surface analysis.

interrupted Goode homolosine projection [MAP PROJECTIONS] A pseudo-cylindrical equal-area map projection created by compositing 12 segments that form 6 interrupted lobes; constructed in 1923 by American geographer John Paul Goode from the uninterrupted homolosine projection. *See also* equal-area projection, homolosine projection, projection.

interrupted projection [MAP PROJECTIONS] A map projection in which the generating globe is segmented to minimize the distortion within any lobe of the projection. *See also* projection, gore.

intersect

1. [ANALYSIS/GEOPROCESSING] A geometric integration of spatial datasets that preserves features or portions of features that fall within areas common to all input datasets.
2. [SPATIAL ANALYSIS] A type of overlay analysis that identifies the overlapping area of two or more layers.

See also identity, overlay, union.

intersection

1. [DATA MANAGEMENT] The point where two lines cross. In geocoding, most often a street crossing.

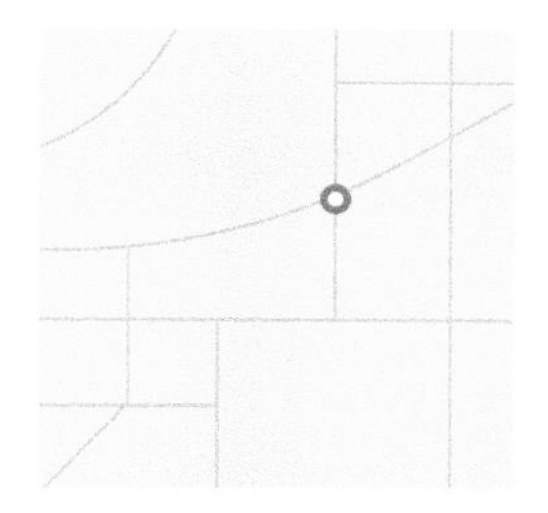

intersection 1

2. [PHOTOGRAMMETRY] The result from calculating the meeting point of two rays from two or more different images, providing the x,y,z ground location of the conjugate image points. Computed using the least squares solution, yielding the location where the rays most closely approach each other.
3. [CARTOGRAPHY] The area of overlap between features.

See also junction, least-squares analysis and adjustment, line-on-line overlay, line-on-point overlay.

intersection connector [GEOCODING] A character used in address data to indicate that an address is located at an intersection. For example, in the address "S. Huntington Dr. & E. Clark Blvd." the ampersand (&) character is the intersection connector. The intersection connector delimits the address into two parts and assigns intersection searches to the address. *See also* address, connector, intersection.

interval data [DATA STRUCTURES] Data consisting of numeric values on a magnitude scale that has an arbitrary zero point (such as temperature in degrees Fahrenheit, calendar years, time of day, and so on).

intrinsic stationarity [SPATIAL STATISTICS (USE FOR GEOSTATISTICS)] In spatial statistics, the assumption that a set of data comes from a random process with a constant mean and a semivariogram that depends only on the distance and direction separating any two locations. *See also* stationarity, semivariogram.

inverse distance [STATISTICS] One divided by distance, often raised to some power ($1/D$ or $1/D^2$, for example), where D is a distance value. By inverting the distance among spatial features, and using that inverted value as a weight, near things have a larger weight or influence than things that are farther away. *See also* inverse distance weighted interpolation.

inverse distance weighted interpolation [MATHEMATICS] An interpolation technique that estimates cell values in a raster from a set of sample points that have been weighted so that the farther a sampled point is from the cell being evaluated, the less weight it has in the calculation of the cell's value. *See also* interpolation, weight.

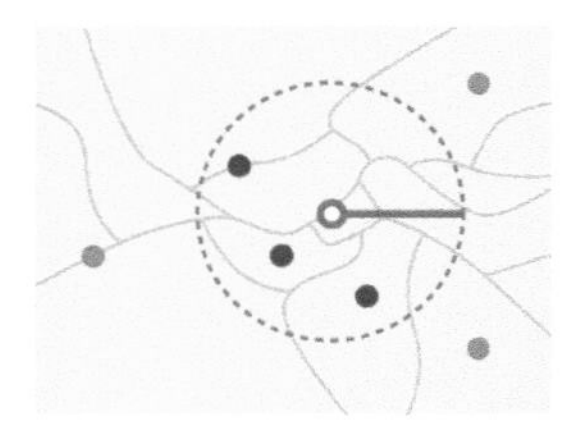

inverse distance weighted interpolation

inverse tangent [MATHEMATICS] The tan–1 of the slope ratio. Sometimes called the arctangent. *See also* slope ratio.

IoT *See* Internet of Things.

IPS *See* indoor positioning system.

IRNSS *See* NavIC.

irregular line profile [CARTOGRAPHY] A profile created for a path on the ground that is not straight (that is, the profile line is irregular). *See also* profile, profile line.

irregular surface area [MAP DESIGN] Computation of the area of a feature that takes the slope of the landscape into account. *See also* slope.

irregular triangular mesh *See* TIN.

irregular triangular surface model *See* TIN.

isanomal [METEOROLOGY] An isoline on a map connecting points of equal difference from a normal value; usually used in meteorology, such as to track average temperature. *See also* isoline.

isarithm [CARTOGRAPHY]

1. An isoline that connects measured values that exist at points, such as temperature or elevation values. Also known as an isometric line.
2. A line connecting points of equal value on a map; an isoline.

See also isoline.

island polygon [SYMBOLOGY] An isolated polygon that does not share boundaries with any other polygons.

ISO *See* International Organization for Standardization.

isobar [CARTOGRAPHY] A line on a weather map connecting places of equal barometric pressure. *See also* isoline.

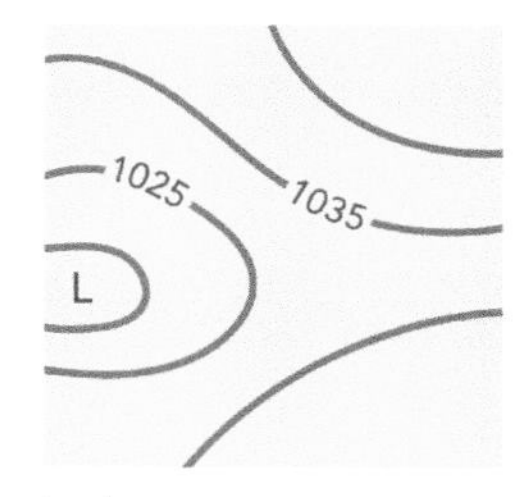

isobar

isochrone [CARTOGRAPHY]
1. A line on a map connecting points of equal elapsed time; especially, travel time to or from a given location.
2. A line on a map connecting points at which an event occurs, or a situation exists, at the same time.

See also isoline.

isogonic line [CARTOGRAPHY] An isoline that connects points of constant angular difference between true and magnetic north.

isogonic line

isohel [CARTOGRAPHY] An isoline that connects points of equal solar radiation.

isohyet [CARTOGRAPHY] A line on a map connecting points of equal precipitation. *See also* isoline.

isoline [CARTOGRAPHY] A line connecting points of equal value on a map. Isolines fall into two classes—isarithms, in which the values actually exist at points, such as temperature or elevation values—and isopleths, in which the values are ratios that exist over areas, such as population per square kilometer or crop yield per acre. *See also* isometric line, isopleth.

isoline interval [CARTOGRAPHY] The distance between isolines.

isometric line [CARTOGRAPHY] An isoline drawn according to known values, either sampled or derived, that can occur at points. Examples of sampled quantities that can occur at points are elevation above sea level, an actual temperature, or an actual depth of precipitation. Examples of derived values that can occur at points are the average of temperature over time for one point or the ratio of smoggy days to clear days for one point. Often used interchangeably with isarithm. *See also* isoline, isopleth.

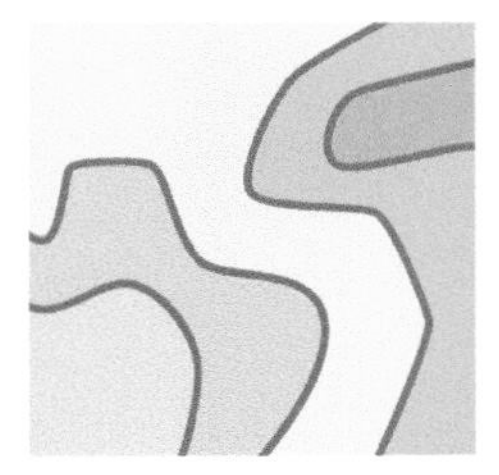
isometric line

isopleth [CARTOGRAPHY] An isoline drawn according to known values that can only be recorded for areas, not points. Examples include population per square mile or the ratio of residential land to total land for an area. *See also* isoline, isometric line.

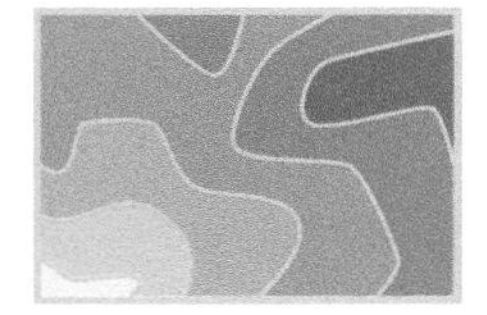
isopleth

isotherm [CARTOGRAPHY] An isoline that connects points of equal temperature. *See also* isoline.

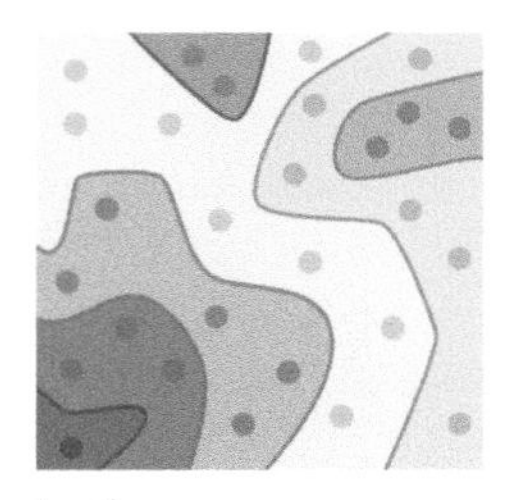
isotherm

isotropic [ANALYSIS/GEOPROCESSING] Having uniform spatial distribution of movement or properties, usually across a surface. *See also* anisotropic, isotropy.

isotropy [SPATIAL STATISTICS (USE FOR GEOSTATISTICS)] A property of a natural process or data where spatial dependence (autocorrelation) changes only with the distance between two locations—direction is unimportant. *See also* anisotropy, autocorrelation, isotropic.

item [ESRI SOFTWARE]

1. A map, layer, or tool that users find, use, or add to a project.
2. In coverages, a field or attribute.

See also content.

ITRF [STANDARDS] Acronym for *International Terrestrial Reference Frame.* A physical realization of a reference system, based on the modeling standards of the ITRS. The ITRF is updated periodically by the International Earth Rotation and Reference Systems Service (IERS) to more closely approximate ITRS standards using the latest technologies. *See also* ITRS, reference system.

ITRS [STANDARDS] Acronym for *International Terrestrial Reference System.* A methodology and set of models for creating an abstract coordinate system, defined and maintained by the International Earth Rotation and Reference Systems Service (IERS). ITRS conventions include a geocentric coordinate system and International System of Units (SI) numerical constants. *See also* geocentric coordinate system, ITRF.

jaggies *See* aliasing.

JavaScript Object Notation [PROGRAMMING] Also known by the acronym *JSON*. An open-standard file and data interchange format used for encoding a variety of geographic data structures. *See also* Esri JSON, GeoJSON.

Jenks' optimization [DATA MANAGEMENT] A method of statistical data classification that partitions numeric data into classes, or class intervals, using an algorithm that minimizes per class average deviation from its mean and maximizes per class deviation from the means of all other classes. In other words, Jenks' optimization seeks to reduce variance within a group and maximize the variance between groups. Named for U.S. cartographer and geographer George Frederick Jenks (1916–1996). *See also* classification, manual classification, natural breaks classification.

JOG *See* joint operations graphic.

joining

1. [ANALYSIS/GEOPROCESSING] Connecting two or more features from different sets of data so that they become a single feature.

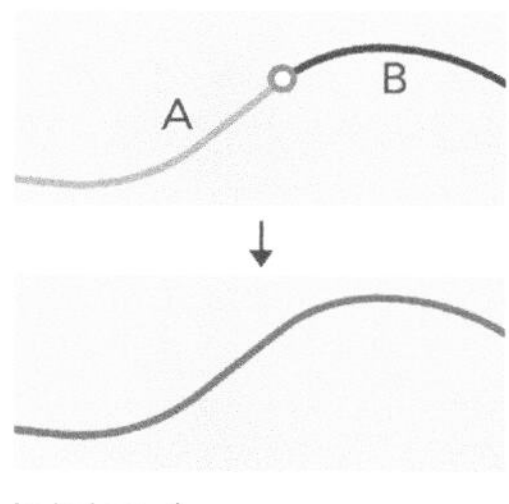

joining 1

2. [DATABASE STRUCTURES, ESRI SOFTWARE] Appending the fields of one table to those of another through an attribute or field common to both tables. A join is typically used to attach more attributes to the attribute table of a geographic layer.

See also link, relate.

joint operations graphic [DEFENSE] Also known by the acronym *JOG*. A 1:250,000-scale topographic map used by militaries worldwide. Joint operations graphics use a common base graphic to facilitate operations involving air, ground, and naval forces. *See also* TLM, VMap.

JPEG [GRAPHICS (COMPUTING)] Acronym for *Joint Photographic Experts Group*. A digital image format that uses lossy compression to reduce file size. Well suited for images that contain graduated colors. An alternate acronym is *JPG*. *See also* GEOTIFF, lossy compression.

JSON *See* JavaScript Object Notation.

junction [NETWORK ANALYSIS]

1. For network data models in a geodatabase, a point at which two or more edges meet.

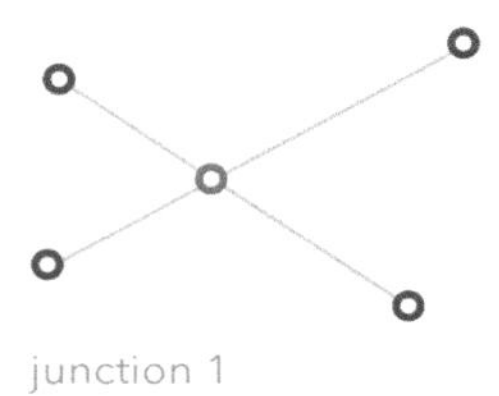

junction 1

2. In a coverage, a node joining two or more arcs.

See also arc, edge, geodatabase, network, node.

junction connectivity policy [NETWORK ANALYSIS] In network datasets, a connectivity policy that defines how a junction may connect to an edge. *See also* connectivity policy, edge connectivity policy, override.

junction element *See* junction.

J

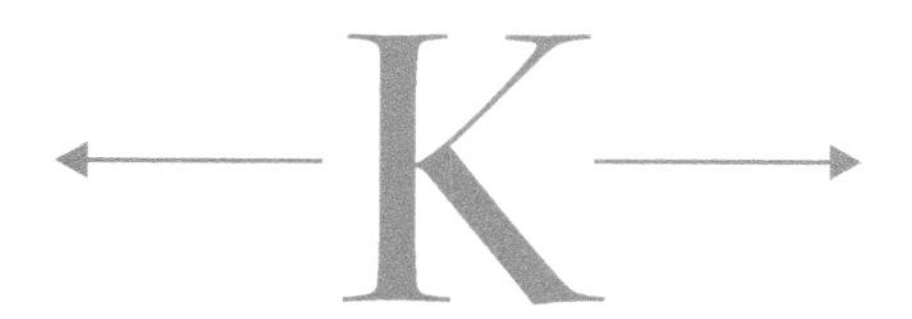

Kappa coefficient [ACCURACY] In map accuracy, a measurement comparing classification results to values assigned by chance. The resulting value ranges between 0 (no agreement between the classified image and the reference data) and 1 (the classified image and the reference data are identical). The higher the Kappa coefficient, the more accurate the classification.

karst topography [GEOGRAPHY] A landscape that is characterized by numerous caves, sinkholes, fissures, and disappearing and underground streams.

kernel [SPATIAL ANALYSIS] A moving matrix of a neighborhood of pixels (commonly three by three) used in the spatial analysis of raster data.

key [DATABASE STRUCTURES] An attribute or set of attributes in a database that uniquely identifies each record. *See also* relational join.

key attribute *See* primary key.

keyframe [VISUALIZATION, ESRI SOFTWARE] In an animation, a snapshot of an object's state that will be transitioned to or from during playback. *See also* animation.

keyword [COMPUTING]

1. A significant word from a document that is used to index or search content.
2. A word searched for in a search command. Keywords are searched for in any order. When defining metadata, users can enter theme and place keywords.

See also index.

kinematic positioning [GPS] Determining the position of an antenna on a moving object such as a ship or an automobile. *See also* static positioning.

king's case [SPATIAL ANALYSIS] In spatial analysis, a case in which contiguity is defined by the four adjacent quadrats sharing a row or column edge with the center quadrat as well as diagonal neighbors, expanding the neighborhood to eight quadrats. Also sometimes called queen's case. *See also* contiguity, quadrat.

knowledge base [ESRI SOFTWARE] A series of tables used for database and cartographic production. Knowledge base tables contain validation rules for feature attribution, data collection, and symbology. *See also* database.

known point [SURVEYING] A surveyed point that has an established x,y coordinate value. Known points are used in survey operations to extend survey computations into a project area. *See also* ground control point, surveying.

Kohonen map [CARTOGRAPHY] A map that uses an artificial neural network algorithm to classify and illustrate associations in complex datasets and reveal multidimensional patterns. A similar set of methods produces maps referred to as self-organizing maps (SOMs). Kohonen maps are named for the Finnish engineer Teuvo Kohonen (1934–2021). *See also* machine learning, neural network.

K

kriging [SPATIAL STATISTICS (USE FOR GEOSTATISTICS)] A spatial interpolation technique used to estimate values at unmeasured locations based on known values at nearby locations. Known values are given weights based on the distance between the measured points, the prediction locations, and the overall spatial arrangement among the measured points. Kriging is unique among interpolation methods in that it provides an easy method for characterizing the variance, or precision, of the predictions. Kriging is based on regionalized variable theory, which assumes that spatial variation in the data being modeled is homogeneous across the surface (that is, the same pattern of variation can be observed at all locations on the surface). Kriging was named for the South African mining engineer Danie G. Krige (1919–2013). *See also* block kriging, interpolation, ordinary kriging, simple kriging, universal kriging.

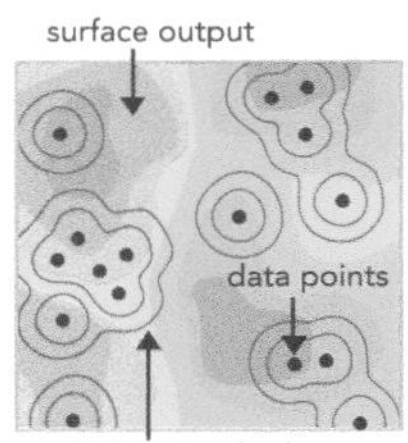

kriging

kurtosis [STATISTICS] A property that indicates how the peaks and tails of a distribution differ from a normal distribution when the peak is sharper or broader or the tails are shorter or longer. This influences the classification technique selected for defining data ranges in a thematic map.

label [CARTOGRAPHY] Text placed on or near a map feature that describes or identifies it. *See also* annotation, tag.

label

label offset [ESRI SOFTWARE] The distance a label should be from the feature it labels on the map. *See also* label placement property.

label orientation [ESRI SOFTWARE] The angle or direction of alignment for feature labels. Labels for features are usually placed horizontally, but they may also be oriented to an angle stored as an attribute, an angle defined by the orientation of the feature geometry, or along the graticule of the data frame. *See also* graticule, graticule alignment of labels.

label placement option *See* label placement property.

label placement property [ESRI SOFTWARE] A software parameter used to define the placement property for a label, which includes offset, placement zone, priority, stacking, and weight.

label point [ESRI SOFTWARE] In a coverage, a feature class used to represent points or identify polygons. When representing points, the x,y location of the point describes the location of the feature. When identifying polygons, the point can be located anywhere within the polygon. *See also* coverage, feature class, point, polygon.

label rule *See* label placement property.

label stacking [ESRI SOFTWARE] The splitting of long labels to place the text on two or more lines. *See also* label placement property.

laddered contour label [SYMBOLOGY] A label placed in-line with the label on adjacent index contour lines so that the elevation values can be easily read from one index contour line to the next. *See also* elevation, index contour line, label.

lag [ESRI SOFTWARE] In the creation of a semivariogram, the sample distance used to group or bin pairs of points. Using an appropriate lag distance can be helpful in revealing scale-dependent spatial correlation. *See also* bin, kriging.

L

Lambert azimuthal equal-area projection [MAP PROJECTIONS] A planar equal-area projection usually restricted to a hemisphere, with the polar and equatorial aspects used most often. Invented by Swiss polymath Johann Heinrich Lambert in 1772. *See also* equal-area projection, projection.

Lambert conformal conic projection [MAP PROJECTIONS] A conic map projection that preserves angles between lines on the map. Two standard parallels are assigned that minimize overall unit scale distortion within the region of interest between these two parallels; distortion increases at further distances. Invented by Swiss polymath Johann Heinrich Lambert in 1772. Lambert conformal conic projections are used for regional maps and aeronautical charts. *See also* projection.

Lambert conformal conic projection

land cover [GEOGRAPHY]
1. The physical and biological cover over the land surface, including water, vegetation, bare soil, and artificial structures.
2. The classification of land according to the vegetation or material that covers most of its surface—for example, pine forest, grassland, ice, water, or sand.

See also classification, land cover map, land use.

land cover map [CARTOGRAPHY] A map that shows the physical and biological land cover. *See also* land cover.

land information system [GOVERNMENT] Also known by the acronym *LIS*. A geographic information system for cadastral and land-use mapping, typically used by local governments. *See also* cadastre, digital elevation model, land use.

land record [CADASTRAL AND LAND RECORDS] A publicly owned and managed system, defined by state or provincial standards, that records real estate ownership, transfers, taxation, and development.

land use [GEOGRAPHY] The classification of land according to what activities take place on it or how humans occupy it—for example, agricultural, industrial, residential, urban, rural, or commercial. *See also* land cover.

land use map [CARTOGRAPHY] A map that shows a limited number of land use or land cover categories. *See also* land cover, land cover map, land use.

landform [GEOGRAPHY] Any natural feature of the land that has a characteristic shape, including major forms such as plains and mountains and minor forms such as hills and valleys. *See also* geomorphology, landmark, terrain.

landmark [GEOGRAPHY]
1. Recognizable natural or cultural features that stand out from the nearby environment and are often visible from long distances, allowing people to establish distance, bearing, or location relative to that feature.
2. A building or location that has historical, architectural, or cultural value.

See also cultural feature, natural feature, topography.

landmark map [CARTOGRAPHY] A map created for navigation using landmarks. *See also* landmark.

Landsat [SATELLITE IMAGING] In remote sensing, a series of polar-orbiting, earth-imaging satellites that collect multispectral imagery used for land cover and land use inventory, geological and mineralogical exploration, crop and forestry assessment, urban development, monitoring short- and long-term land cover change, and more. Developed jointly by NASA and the United States Geological Survey (USGS), Landsat satellites provide a continuous record of medium-resolution coverage of the earth dating back to 1972; the revisit rate is 16 days. *See also* ground receiving station, multispectral, NASA, remote sensing.

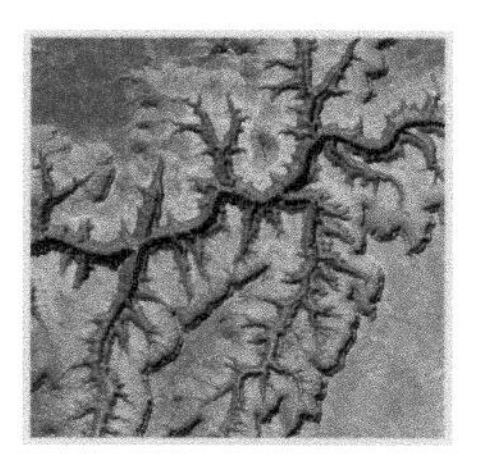

Landsat

landscape
1. [ECOLOGY] An area of land that affects, and is affected by, the ecological processes of interest.
2. [CARTOGRAPHY] A map layout that is wider than it is tall.

See also human landscape.

landscape drawing [MAP DESIGN] An artistically rendered, oblique-perspective map that provides a realistic terrain portrayal. *See also* terrain.

landscape ecology [ENVIRONMENTAL GIS] The study of spatial patterns, processes, and change across biological and cultural structures within areas encompassing multiple ecosystems.

large scale [MAP DESIGN] A map scale that has a large representative fraction, showing features as very small compared to their actual size. Often covers large areas in lower detail. *See also* small scale.

LAS [3D GIS] An industry-standard binary file format that maintains information related to lidar data. *See also* LAS dataset, lidar.

LAS dataset [DATABASE STRUCTURES] A geodataset that references LAS files and surface constraints and assists with examination of LAS files in their native format. *See also* LAS, lidar.

latitude [COORDINATE SYSTEMS] The location of a point on the earth's surface, stated as an angular measurement in degrees, minutes, and seconds north or south of the equator. Lines of latitude are also referred to as parallels. *See also* parallel, longitude.

latitude

L

latitude of center [MAP PROJECTIONS] The latitude value that defines the center, and sometimes the origin, of a projection. *See also* latitude.

latitude of origin [MAP PROJECTIONS] The latitude value that defines the origin of the y-coordinate values for a projection. *See also* longitude of origin.

latitude-longitude [COORDINATE SYSTEMS] A reference system used to locate positions on the earth's surface. Distances east–west are measured with lines of longitude (also called meridians), which run north–south and converge at the north and south poles. Distance measurements begin at the prime meridian and are measured positively 180 degrees to the east and negatively 180 degrees to the west. Distances north–south are measured with lines of latitude (also called parallels), which run east–west. Distance measurements begin at the equator and are measured positively 90 degrees to the north and negatively 90 degrees to the south. *See also* latitude, longitude, meridian, parallel.

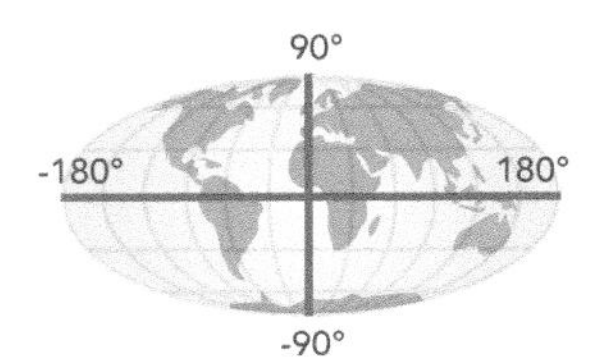

latitude-longitude

lattice [DATA MODELS] A representation of a surface using an array of regularly spaced sample points (mesh points) that are referenced to a common origin and have a constant sampling distance in the x and y directions. Each mesh point contains the z-value at that location, which is referenced to a common base z-value, such as sea level. Z-values for locations between lattice mesh points can be approximated by interpolation based on neighboring mesh points. *See also* raster.

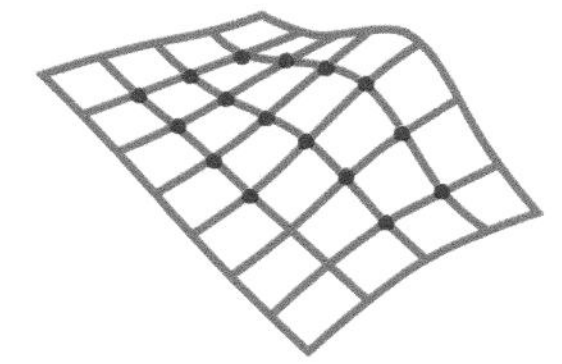

lattice

layer

1. [DATA STRUCTURES] The visual representation of a geographic dataset in any digital map environment. Conceptually, a layer is a slice or stratum of the geographic reality in a particular area and is more or less equivalent to a legend item on a paper map. On a road map, for example, roads, national parks, political boundaries, and rivers might be considered different layers.
2. [ESRI SOFTWARE] A reference to a data source that defines how the data should be shown on a map. Layers can also define additional properties, such as which features from the data source are included.

See also annotation layer, CAD layer, feature layer, network analysis layer, network layer, raster layer, TIN layer.

layout

1. [MAP DESIGN] The arrangement of elements on a map, possibly including a title, legend, north arrow, scale bar, and geographic data.
2. [ESRI SOFTWARE] A presentation document incorporating maps, charts, tables, text, and images.

L-band [GPS] The group of radio frequencies that carry data from GPS satellites to GPS receivers.

LBS *See* location-based services.

least convex hull *See* convex hull.

least-cost path [NETWORK ANALYSIS] The path between two locations that costs the least to traverse, where cost is a function of time, distance, or some other criteria defined by the user. *See also* cost, shortest path.

least-squares analysis and adjustment [SURVEYING] A process based on the theory of probability that estimates new coordinate locations for parcel fabric points. The process uses coordinate geometry (COGO) dimensions on parcel lines, existing coordinates of parcel points, and the network redundancy of connected parcel lines to estimate updated coordinates for parcel points in a best-fit solution. In the parcel fabric, a least-squares analysis can either be used as a consistency check to detect outliers of mistakes in COGO dimensions or as a weighted analysis to estimate updated coordinates for parcel points. When the weighted analysis returns acceptable results, the updated coordinates can be applied to parcel fabric points to adjust them to more spatially accurate positions. *See also* coordinate geometry, parcel fabric.

least-squares corrections [SURVEYING] The final measurement residuals of a least squares analysis and adjustment. *See also* least-squares analysis and adjustment.

legal cadastre [CADASTRAL AND LAND RECORDS] The written records that concern proprietary interests in parcels. Also spelled legal cadaster. *See also* cadastre.

legend [SYMBOLOGY] A description of what each symbol, or graphic, on a map represents. A legend is sometimes called a key. *See also* map element, symbol.

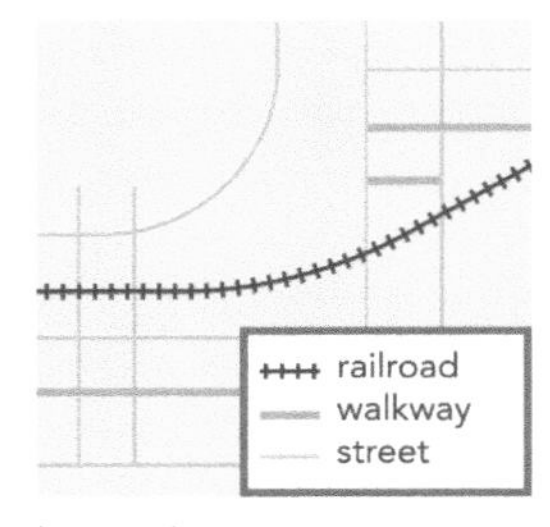

legend

legend disclaimer [ACCURACY] A note in the map legend that indicates something about the map's accuracy.

L

Lehmann system [SYMBOLOGY] A hachuring system named for German geodesist and cartographer Johann Georg Lehmann (1765–1811). *See also* hachure.

lens distortion [PHOTOGRAMMETRY] In imagery, the variance of an image point location within a sensor from its ideal, projected geometric position. It is the result of deviation from rectilinear projection caused by, for example, lens design, camera position, object elevation, and so on. *See also* sensor.

lens falloff [DIGITAL IMAGE PROCESSING] In imagery, the reduction of exposure—or intensity—at the edges of an image compared to its perspective center. Also called vignetting.

lens nodal point [PHOTOGRAMMETRY] In imagery, one of two points on the optical axis of a complex lens that establish parallel alignment. All rays pass between the front and rear nodal points, which are aligned to converge. *See also* axis.

level of confidence *See* confidence level.

level of detail [MAP DESIGN] Also known by the acronym *LOD*. The density and quantity of information provided. *See also* resolution, simplification.

level of measurement [DATA ANALYSIS] A way to characterize the nature of numeric feature data. There are four basic measurement levels: nominal, ordinal, interval, and ratio. *See also* feature data.

level of significance *See* significance level.

leveling [SURVEYING] In surveying, the measurement of the heights of objects and points according to a specified elevation, usually mean sea level. *See also* benchmark, elevation.

LIBID [ESRI SOFTWARE] An abbreviation of *Library Identifier*. A type of GUID consisting of a unique string assigned to a type library. *See also* globally unique identifier.

lidar [REMOTE SENSING] An abbreviation of *light detection and ranging.* An active, optical remote sensing technique that uses rapid pulses of laser light to measure distances and to densely sample the surface of the earth, objects on the earth, and physical infrastructure. Lidar produces highly accurate x,y,z measurements of the positions of objects struck by the light pulses. Some common uses of lidar are to provide topographic and bathymetric maps and assist infrastructure engineering. *See also* radar, sonar.

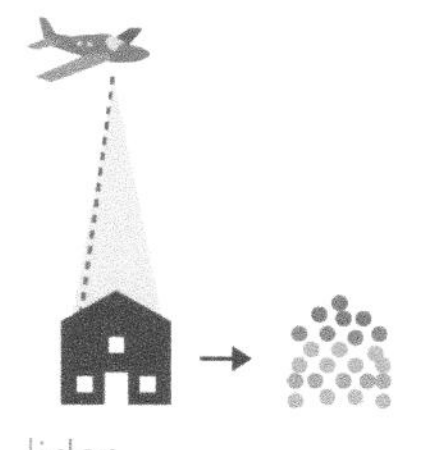

lidar

limited error raster compression [PHOTOGRAMMETRY] In imagery, an open-source raster image compression method that supports rapid encoding and decoding for all common pixel or data types. The main parameter is the maximum encoding error per pixel allowed, to preserve image integrity.

line [EUCLIDEAN GEOMETRY] On a map, a shape defined by a connected series of unique x,y coordinate pairs. A line may be straight or curved. *See also* line feature, polyline.

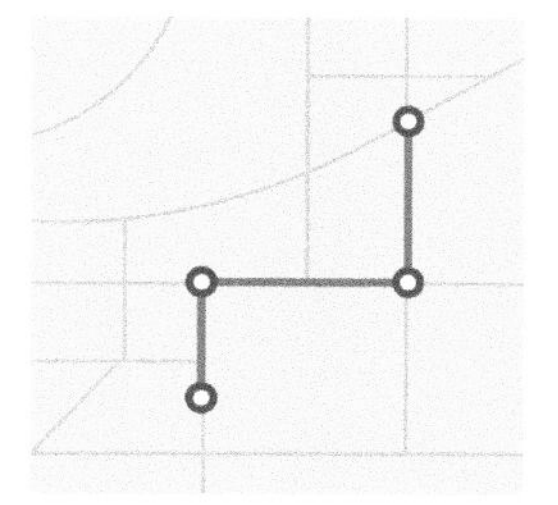

line

line event [LINEAR REFERENCING] In linear referencing, a description of a portion of a route using a from- and to-measure value. Examples of line events include pavement quality, salmon spawning grounds, bus fares, pipe widths, and traffic volumes. *See also* event, linear referencing.

line feature [SYMBOLOGY] A map feature that has length but not area at a given scale, such as a river on a world map or a street on a city map. *See also* feature, polyline feature.

line feature map [CARTOGRAPHY] A map containing symbols showing line features.

line of sight

1. [3D ANALYSIS] A line drawn between two points, an origin and a target, that is compared against a surface to show whether the target is visible from the origin and, if it is not visible, where the view is obstructed.

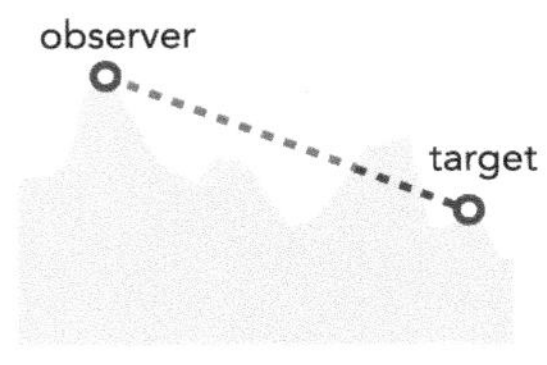

line of sight 1

2. [VISUALIZATION] In a perspective view, the point and direction from which a viewer looks into an image.

See also observer, view.

line of tangency [CARTOGRAPHY] The point at which a line along a secant-case developable surface touches the generating globe. This line on the map is true in scale to the equivalent line on the spherical or ellipsoidal approximation of the earth. *See also* developable surface, homolatitudes.

line pattern [SYMBOLOGY] A set of shapes or lines that create a pattern within a line graphic mark.

line simplification [CARTOGRAPHY] A generalization technique in which vertices are selectively removed from a line feature to eliminate detail while preserving the line's basic shape. *See also* simplification, generalization, line smoothing.

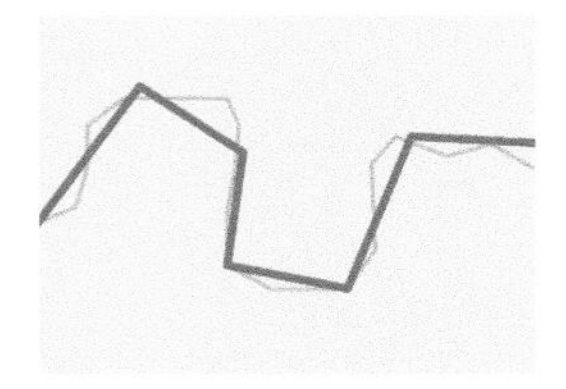

line simplification

line smoothing [DATA EDITING] The process of adding extra points to lines to reduce the sharpness of angles between line segments, resulting in a smoother appearance. *See also* weeding.

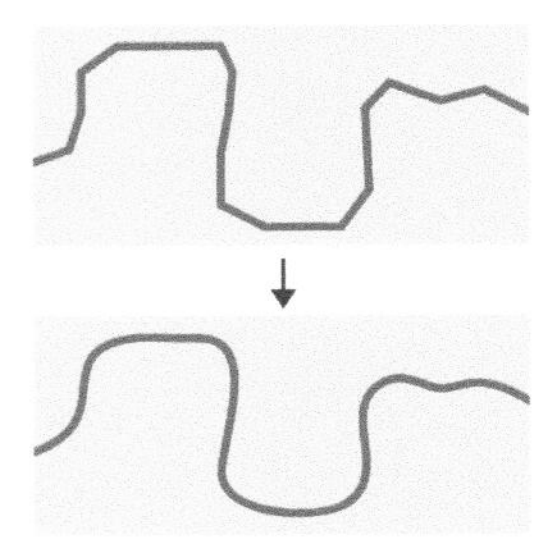

line smoothing

line, sample [REMOTE SENSING] In remote sensing, the physical location of a pixel within the sensor when the image was formed. Sample refers to the index along a linear array; line refers to the sequence of the linear arrays. For a framing camera, sample is equivalent to the column and line to the physical row of the sensor.

linear connections [TRANSPORTATION] Roads, rivers, bus lines, airline routes, and other connections that link one place with another.

linear dimension [SURVEYING] A measurement of the horizontal or vertical dimension of a feature. Linear dimensions may not represent the true distance between beginning and ending dimension points because they do not take angle into account as aligned dimensions do. *See also* aligned dimension.

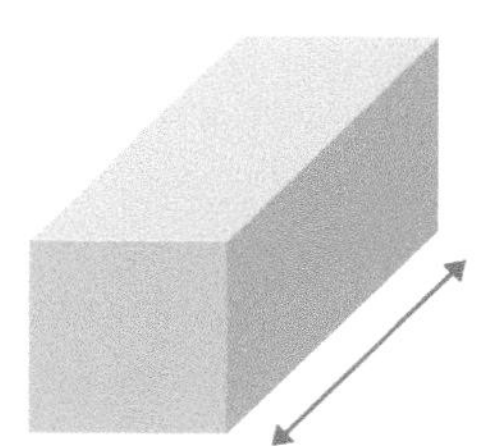
linear dimension

linear feature *See* line feature.

linear interpolation [SPATIAL STATISTICS (USE FOR GEOSTATISTICS)] A method used to predict an unknown value between two known values assuming a constant, or linear, rate of change. *See also* distance, interpolation.

linear referencing [LINEAR REFERENCING] A method for storing geographic data by using a relative position along an already existing line feature; the ability to uniquely identify positions along lines without explicit x,y coordinates. In linear referencing, location is given in terms of a known line feature and a position, or measure, along the feature. Linear referencing is an intuitive way to associate multiple sets of attributes to portions of linear features. An example of a linear reference is saying "seven feet west of mile marker 234 on Route 52." *See also* dynamic segmentation, m-value, point event.

linear surface [CARTOGRAPHY] A surface assumed to increase or decrease in height at a constant or uniform rate between two points.

linear unit [CARTOGRAPHY] The unit of measurement on a plane or a projected coordinate system, often in meters or feet. *See also* angular unit, unit of measure.

line-on-line overlay [LINEAR REFERENCING] In linear referencing, the overlay of two line event tables to produce a single line event table. The new event table can be the logical intersection or union of the input tables. *See also* line-on-point overlay, point-in-polygon overlay.

line-on-point overlay [LINEAR REFERENCING] In linear referencing, the overlay of a line event table and a point event table to produce a single point event table. The new event table can be the logical intersection or union of the input tables. *See also* line-on-line overlay, point-in-polygon overlay.

link

1. [GEOREFERENCING] Connections added between known points in a dataset being georeferenced and corresponding points in the dataset being used as a reference.
2. [COMPUTING] An operation that relates two tables using a common field, without altering either table.
3. [NETWORK ANALYSIS] The linear connection between two nodes in a network.

See also georeferencing, joining, relate, remote sensing, simple relationship.

LIS *See* land information system.

live feed [DATA CAPTURE] A data broadcast, generally in real time, for the purpose of monitoring changing conditions and delivered from sources that collect and process the data. In GIS, live feed data is typically related to weather or traffic. *See also* real time, streaming.

lobe [MAP PROJECTIONS] A section of an interrupted projection. *See also* gore, interrupted Goode homolosine projection.

local analysis [DATA EDITING] The computation of an output raster where the output value at each location is a function of the input value at the same location. *See also* global analysis, spatial function.

L

local datum [GEODESY]

1. A horizontal geodetic datum that serves as a basis for measurements over a limited area of the earth, with its origin at a location on the earth's surface. It uses an ellipsoid whose dimensions conform well to its region of use. It was originally defined for land-based surveys. A local datum in this sense stands in contrast to a geocentric datum. Examples include the North American Datum of 1927 and the Australian Geodetic Datum of 1966.
2. A horizontal or vertical datum used for measurements over a limited area of the earth, such as a nation, a supranational region, or a continent. A horizontal datum that is local in this sense may or may not be geocentric. For example, the North American Datum of 1983 and the Geocentric Datum of Australia 1994 are local in that they are applied to a particular part of the world; they are also geocentric. All vertical datums are local in that there is, at present, no global vertical datum.

See also datum, geocentric datum, horizontal geodetic datum, vertical geodetic datum.

local functions *See* local analysis.

local geoid [CARTOGRAPHY] A surface slightly above or below (usually within two meters) the global geoid elevations. This difference is due to mean sea level at one or more nearby locations being used as the vertical reference datum rather than mean sea level for all the earth's oceans. *See also* geoid, global geoid, mean sea level, vertical datum.

local space rectangular coordinate system [COORDINATE SYSTEMS] A 3D Cartesian coordinate system with an origin at a point either tangent or secant to the earth's ellipsoid, where the axes are defined as Y=North, X=East, and Z is normal to the earth ellipsoid at its origin.

locale [COMPUTING] A parameter that defines a system's language and region.

localization [ORGANIZATIONAL ISSUES] The process of adapting software to the requirements of a different language or culture, including translating user interfaces, documentation, and help systems; customizing features; and accommodating different character sets. *See also* translation.

location

1. [GEOGRAPHY] An identifier assigned to a region or feature.
2. [GEOGRAPHY] A position of an object defined by its coordinate values (x,y,z).
3. [DIGITAL IMAGE PROCESSING] An image element that provides the x,y,z coordinates of an object of interest; used to help identify the object.

See also image element, position.

location query *See* spatial query.

location-allocation [BUSINESS] The process of finding the best locations for one or more facilities that will service a given set of points and then assigning those points to the facilities, considering factors such as the number of facilities available, their cost, and the maximum impedance from a facility to a point. *See also* cost, impedance.

location-based services [LOCATION-BASED SERVICES] Information or a physical service delivered to multiple channels, exclusively based on the determined location of a wireless device. Some location-based applications include emergency services, information services, and tracking services.

locator [GEOCODING, ESRI SOFTWARE] A dataset that stores the address attributes, associated indexes, and rules that define the process for translating nonspatial descriptions of places, such as street addresses, into spatial data that can be displayed as features on a map. A locator contains a snapshot of the reference data used for geocoding, and parameters for standardizing addresses, searching for match locations, and creating output. *See also* address, geocode server.

locator map [CARTOGRAPHY] A map that shows the position of the main mapped area within its state, country, or other area.

locator property [GEOCODING] A parameter in a locator that defines the process of geocoding. *See also* locator.

locator style [GEOCODING] A template on which a locator is built. Each template accommodates a specific address and reference data format and geocoding parameters. *See also* locator.

locomotion [WAYFINDING] The movements of a person following a route. Locomotion is the physical component of navigation. *See also* navigation, wayfinding.

LOD *See* level of detail.

logarithm [MATHEMATICS] The power to which a fixed number (the base) must be raised to equal a given number. The three most frequently used bases for logarithms are base 10, base e, and base 2. *See also* exponent.

logical expression [MATHEMATICS] A string of numbers, constants, variables, operators, and functions that returns a value of true or false. *See also* binary, Boolean expression, expression.

logical network [ESRI SOFTWARE] An abstract representation of a network, implemented as a collection of hidden tables. A logical network contains edge, junction, and turn elements, the connectivity between them, and the weights necessary for traversing the network. It does not contain information about the geometry or location of its elements; this information is one of the components of a network system.

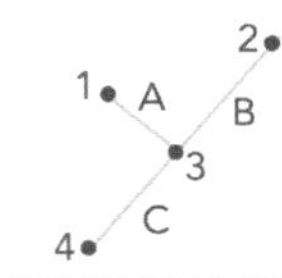

edge ID	connected junction
A	1,3
B	2,3
C	3,4

logical network

logical operator [MATHEMATICS] An operator used to compare logical expressions that returns a result of true or false. Examples of logical operators include less than (<), greater than (>), equal to (=), and not equal to (<>). *See also* operator, logical expression.

logical query [ANALYSIS/GEOPROCESSING] The process of using mathematical expressions to select features from a geographic layer based on their attributes—for example, select all polygons with an area greater than 16,000 units or select all street segments named Green Apple Run. *See also* expression.

logical selection *See* logical query.

long lot [CARTOGRAPHY] A narrow ribbon-shaped parcel used during the French settlement of North America. A long lot's boundaries ran back from the waterfront as roughly parallel lines, allowing a settler to have a dock on the river, a home on the natural levee formed by the river, and a narrow strip of farmland that often ended at the edge of a marsh or swamp. Also called arpent sections or French arpent land grants.

long transaction [DATABASE STRUCTURES] An edit session in an enterprise geodatabase that may last from a few minutes to several months. Long transactions are managed by the enterprise geodatabase traditional versioning mechanism.

longitude [COORDINATE SYSTEMS]

1. The location of a point on the earth's surface, stated as an angular measurement in degrees, minutes, and seconds east or west of the Greenwich prime meridian. All lines of longitude are great circles that intersect the equator and pass through the north and south poles.

longitude 1

2. The numbering system used for meridians.

See also great circle, latitude.

longitude of center [MAP PROJECTIONS] The longitude value that defines the center, and sometimes the origin, of a projection. *See also* latitude of center, longitude.

longitude of origin [MAP PROJECTIONS] The longitude value that defines the origin of the x-coordinate values for a projection. *See also* latitude of origin.

longitudinal distance [COORDINATE SYSTEMS] The distance between meridians. *See also* longitude, meridian.

long-range variation [SPATIAL STATISTICS (USE FOR GEOSTATISTICS)] In a spatial model, coarse-scale variation that is usually modeled as the trend. *See also* trend.

longwave infrared [PHYSICS] Also known by the abbreviation *LWIR* or *TIR*. The band of the electromagnetic spectrum with wavelengths in the 8–14 mm range. This is the passive thermal infrared band (TIR), consisting completely of emitted thermal energy. For imaging, sensors can obtain a completely passive LWIR image of warm objects occurring naturally in a landscape without requiring illumination from the sun. *See also* band, electromagnetic spectrum, infrared radiation, mid-wave infrared, near infrared, shortwave infrared, wavelength.

loop traverse *See* closed loop traverse.

loose coupling [MODELING] A relatively unstructured relationship between two software components or programs that work together to process data, which requires little overlap between methods, ontologies, class definitions, and so on. *See also* tight coupling.

LORAN [NAVIGATION] An abbreviation of *long-range navigation.* A radio navigation system that allows a receiver to determine its position using low-frequency signals transmitted by fixed beacons.

lossless compression [DATA TRANSFER] A method of reducing the size of stored data without changing any of the values, but only at a low compression ratio (typically 2:1 or 3:1). In GIS, lossless compression is often used to compress continuous image data when the pixel values of the raster will be used for analysis or deriving other data products. *See also* JPEG, lossy compression.

lossy compression [DATA TRANSFER] A method of reducing the size of a stored dataset that provides high compression ratios (typically in the range of 10:1 to 100:1) but does not preserve all the information in the original data. In GIS, lossy compression is used to compress raster datasets that can be used as background images or categorical rasters but are not suitable for analysis or deriving other data products. *See also* lossless compression, PNG.

lot [CADASTRAL AND LAND RECORDS] A special type of parcel within a legal subdivision that is recorded on a map. *See also* parcel, plat, subdivision.

low-altitude aerial photograph [AERIAL PHOTOGRAPHY] An aerial photograph taken from a height ranging from just above the ground to around 15,000 feet above the surface.

low-pass filter [REMOTE SENSING] A spatial filter that blocks high-frequency (shortwave) radiation, resulting in a smoother image. *See also* band-pass filter, high-pass filter.

loxodrome *See* rhumb line.

loxodromic curve [GEODESY] The rhumb-line path on the earth that crosses each meridian at the same acute angle, thus forming a spiral that converges on the north or south pole. *See also* rhumb line.

loxodromic curve

machine learning [PROGRAMMING] The use and development of computer programs to adapt without following explicit code through algorithms and statistical models that help the model continually analyze and grow from patterns observed in obtained data. Machine learning is a branch of artificial intelligence (AI). *See also* AI, deep learning, neural network.

magnetic azimuth [GEOGRAPHY] The horizontal angle measured from a magnetic north reference line to a direction line. *See also* direction line, horizontal angle, magnetic north, reference line.

magnetic bearing [NAVIGATION] A bearing measured relative to magnetic north. *See also* bearing.

magnetic declination [GEOGRAPHY] The angle between magnetic north and true north observed from a point on the earth. Magnetic declination varies from place to place, and changes over time, in response to changes in the earth's magnetic field. Sometimes referred to as compass variation on navigational charts. *See also* declination.

magnetic inclination [GEOGRAPHY] The angle between the compass needle and the horizontal plane. Also called the magnetic dip.

magnetic north [GEOGRAPHY] The direction of a compass needle when it is aligned with the earth's magnetic field from a point on the earth's surface. Magnetic north follows a great circle toward the magnetic north pole. *See also* grid north, true north.

magnetic poles [GEOGRAPHY] The positions on the earth's surface at which the geomagnetic field is vertical—that is, perpendicular—to the ellipsoid. *See also* magnetic north.

magnetometer [PHYSICS] An instrument used to measure variations in the strength and direction of the earth's magnetic field.

major axis [EUCLIDEAN GEOMETRY] The longer axis of an ellipse or spheroid. *See also* minor axis.

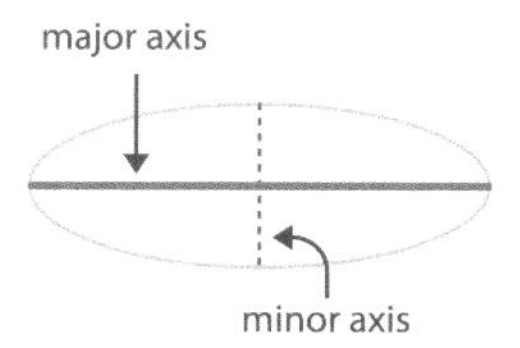

major axis

majority resampling [SPATIAL STATISTICS (USE FOR GEOSTATISTICS), ESRI SOFTWARE] A technique for resampling raster data in which the value of each cell in an output is calculated, most commonly using a 2 × 2 neighborhood of the input raster. Majority resampling does not create any new cell values, so it is useful for resampling categorical or integer data, such as land use, soil, or forest type. Majority resampling acts as a type of low-pass filter for discrete data, generalizing the data and filtering out anomalous data values. *See also* resampling, low-pass filter.

managed raster catalog [DATA STRUCTURES] A raster catalog in which the raster datasets are copied to a location assigned by a geodatabase. When a row is deleted from a managed raster catalog, the data is deleted as well. *See also* unmanaged raster catalog.

manual classification [CARTOGRAPHY] A method of data classification used with web maps where the mapmaker defines arbitrary range breaks. Considerations typically include characteristics of the data distribution, ease of legend interpretation, and the resulting map's communication efficacy. *See also* classification method, class intervals, distribution, equal-interval classification, Jenks' optimization, natural breaks classification.

many-to-many relationship [DATABASE STRUCTURES] An association between two linked or joined tables in which one record in the first table may correspond to many records in the second table, and vice versa. *See also* cardinality, joining, link, many-to-one relationship, one-to-many relationship, one-to-one relationship.

many-to-one relationship [DATABASE STRUCTURES] An association between two linked or joined tables in which many records in the first table may correspond to a single record in the second table. *See also* joining, many-to-many relationship, one-to-many relationship, one-to-one relationship.

map

1. [GEOGRAPHY] A spatial representation of a location.

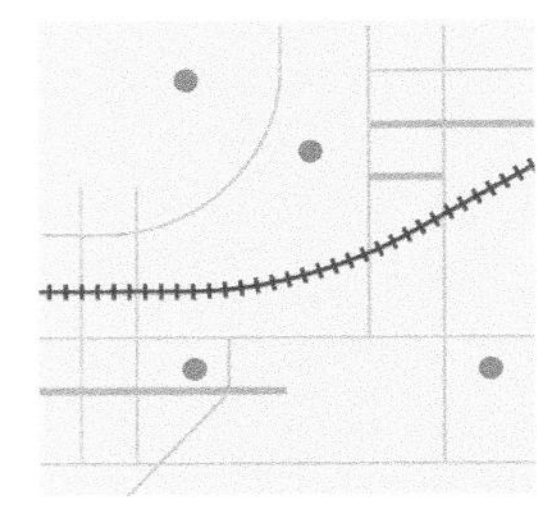

map 1

2. [CARTOGRAPHY] A collection of graphic symbols used to represent a place.
3. [ESRI SOFTWARE] A document used to display and work with geographic data. A map contains one or more layers and various supporting map elements, such as a scale bar.

See also atlas, legend.

map accuracy

1. [ACCURACY] The degree to which a map value represents geographic phenomena at a given scale.
2. [STANDARDS] When discussing the National Map Accuracy Standards, a maximum allowable percentage of points exceeding an error tolerance.

See also geographic phenomena, National Map Accuracy Standards.

map algebra [DATA ANALYSIS] Algebraic language defining a syntax to create map themes by applying mathematical operations and analytical functions to existing map themes. In a map algebra expression, the operators are a combination of mathematical, logical, or Boolean operators (+, >, AND, tan, and so on) and spatial analysis functions (slope, shortest path, spline, and so on), and the operands are spatial data and numbers. *See also* expression, logical operator.

map analysis [SPATIAL ANALYSIS] The use of maps to make geographic counts and measurements and to look for spatial structure, such as patterns, associations, and correspondence. *See also* analysis.

map annotation [ESRI SOFTWARE] Text or graphics stored within the map data frame in an annotation group. Map annotation may be manually entered or generated from labels and can be individually selected, positioned, and modified. *See also* annotation, label.

map collar *See* map surround.

map comparison [SPATIAL ANALYSIS] The process of visually or quantitatively examining the spatial associations between features on different maps. *See also* spatial association.

map critique [MAP DESIGN] An external review of the map design. *See also* map design.

map design [CARTOGRAPHY] The systematic process of arranging and assigning meaning to elements on a map for the purpose of communicating geographic knowledge.

map display [GRAPHICS (MAP DISPLAY)] A graphic representation of a map on a computer screen. *See also* electronic atlas.

map dividers [CARTOGRAPHY] A technical instrument that is used to mark out incremental distances on a map or chart. The act of using map dividers is often called walking. Map dividers are sometimes known as a divider caliper or a compass.

map element [MAP DESIGN] In digital cartography, a distinctly identifiable graphic or object in the map or page layout. For example, a map element can be a title, scale bar, legend, or other map-surround element. The map area itself can be considered a map element; or an object within the map can be referred to as a map element, such as a roads layer or a school symbol. *See also* compass rose, legend, map marginalia, map surround.

map extent [CARTOGRAPHY] The limit of the geographic area shown on a map, usually defined by a rectangle. In a dynamic map display, the map extent can be changed by zooming and panning. *See also* extent.

map feature *See* feature.

map generalization [CARTOGRAPHY] Decreasing the level of detail on a map so that it remains uncluttered when its scale is reduced. *See also* generalization.

map image layer [GRAPHICS (COMPUTING)] A layer composed of map images that is generated dynamically by a server or from tiles during a map visualization or a feature query. Map image layers reference workspaces or datasets that are registered with a federated server. A map image layer can be distinguished from a tile layer, which is pregenerated from raster image or vector tiles. *See also* map view, tile, tile layer.

M

map interpretation [SPATIAL ANALYSIS] The use and analysis of maps to understand the spatial relationships of geographic features. Map interpretation draws on a combination of personal knowledge, fieldwork, expert documents and interviews, and other maps and images. *See also* map analysis.

map marginalia [MAP DESIGN] Information displayed within the mapped area or in the margin of the map that helps explain or support the map. Examples include titles, legends, scale bars, or other indicators of scales, north arrows or other indicators of direction, and sometimes inset or locator maps. Sometimes called marginalia for short. *See also* compass rose, legend, scale bar.

map measurer [CARTOGRAPHY] A technical instrument consisting of a wheel and one or more circular distance dials used to measure the length of curved lines on maps. Also known as a curvimeter, meilograph, or opisometer.

map precision [ACCURACY] A measure that compares repeated observations, or sets of observations, of the same geographic area. Precision is a measure of the random error. *See also* precision.

map projection *See* projection.

map projection families [MAP PROJECTIONS] A method of organizing the range and variety of map projection types into a limited number of groups based on shared properties or attributes. *See also* conic projection family, cylindrical projection family, planar projection family.

map query *See* spatial query.

map reading [CARTOGRAPHY] The activity of viewing a map in a way that allows the viewer to make sense of or gain information from it. Map reading involves interpreting the meanings of codes and cartographic representations used on the map. *See also* map.

map scale *See* scale.

map series [CARTOGRAPHY] A collection of maps usually addressing a particular theme. *See also* map, map sheet.

map service [ESRI SOFTWARE] A persistent software process that provides access to map images for display in a client application. The images can be rendered dynamically for a specific extent or prerendered and cached in a tile grid as static images. A map service can also provide access to the underlying feature layer data used to create the map images. *See also* map, service, web map, web service.

map sheet

1. [CARTOGRAPHY] A single map or chart in a map series, such as any one of the approximately 57,000 USGS 7.5-minute topographic maps of the United States and its territories.
2. [PRINTING] A printed map.

See also digital elevation model, index map, map, map series.

map style *See* style.

map surround [MAP DESIGN] Any of the supporting objects or elements that help a reader interpret a map. Typical map surround elements include the title, legend, north arrow, scale bar, border, source information and other text, and inset maps. *See also* map element.

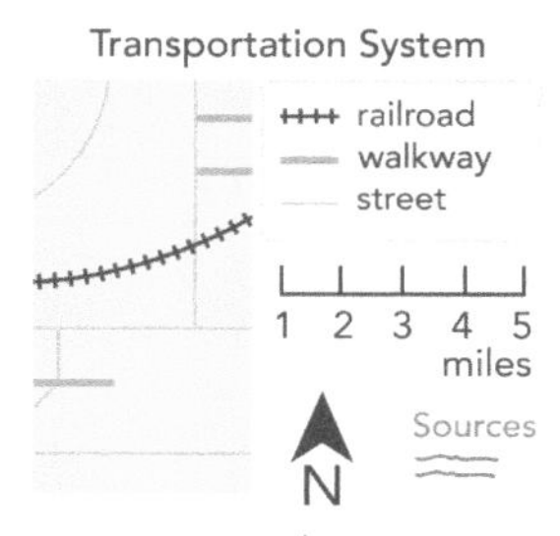

map surround

map symbol *See* symbol.

map topology [GRAPHICS (MAP DISPLAY)] A temporary set of topological relationships between coincident parts of simple features on a map, used to edit shared parts of multiple features. *See also* map display, map element.

map unit [STANDARDS] The ground unit of measurement, for example, feet, miles, meters, or kilometers, in which coordinates of spatial data are stored. *See also* minimum map unit, unit of measure.

map view [MAP DISPLAY] A user interface in a web map that displays map layers, controls the extent (area of the map that is visible), and typically supports user interactions such as pan and zoom. A map view is also used to access the underlying map layer data. *See also* extent.

mapping key *See* classification schema.

mapping methods [CARTOGRAPHY] Commonly used techniques for making maps of quantitative data, such as choropleth mapping, dot density mapping, dasymetric mapping, and stepped-surface mapping. *See also* choropleth map, dasymetric mapping, dot density map, stepped-surface map.

mapping period [CARTOGRAPHY] The time between initial data collection and final map creation.

marginalia *See* map surround.

M

marine navigation [NAVIGATION] The process of planning, recording, and controlling the movement of a watercraft. *See also* air navigation, navigation.

marker symbol [SYMBOLOGY] A symbol used to represent a point location on a map. *See also* fill symbol.

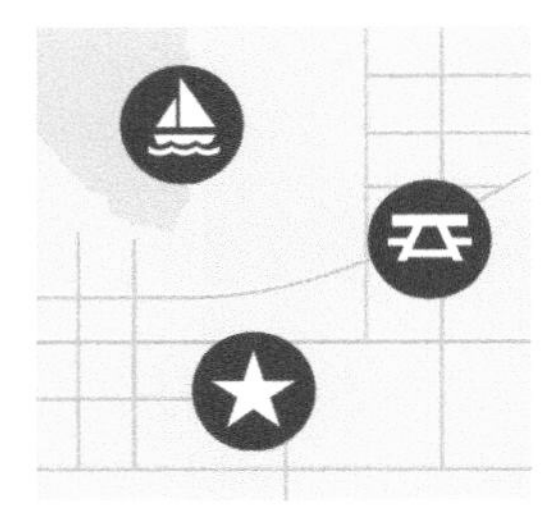

marker symbol

M

market area [BUSINESS] A geographic zone containing the people who are likely to purchase a firm's goods or services. *See also* market penetration analysis, site prospecting, store market analysis.

market penetration analysis [SPATIAL ANALYSIS, BUSINESS] A process that determines the percentage of a market area being reached based on the number of customers within an area divided by the total population in that area. *See also* market area, site prospecting, store market analysis.

mask

1. [CARTOGRAPHY] In digital cartography, a means of covering or hiding features on a map to enhance cartographic representation. For example, masking is often used to cover features behind text to make the text more readable.
2. [ESRI SOFTWARE] A means of identifying areas to be included in analysis. Such a mask is often referred to as an analysis mask and may be either a raster or feature layer.

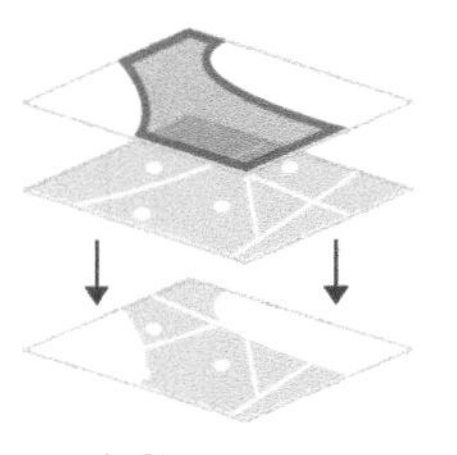

mask 2

See also variable depth masking.

mass point [SPATIAL STATISTICS (USE FOR GEOSTATISTICS)] An irregularly distributed sample point, with an x-, y-, and z-value, used to build a triangulated irregular network (TIN). Ideally, mass points are chosen to capture the more important variations in the shape of the surface being modeled. Also known as spot heights. *See also* TIN.

match score [GEOCODING] In geocoding, a value assigned to all potential candidates for an address match. The match score is based on how well the location found in the reference data matches with the address data being searched. *See also* address matching, geocoding.

matching [GEOCODING] In geocoding, the process of linking a record, such as an address, to a set of reference data. The matched record in the reference data is used to determine the location of the input address. *See also* address matching, geocoding, reference data.

mathematical expression [MATHEMATICS] A combination of numbers, symbols, and operators that represents a mathematical relationship or operation. Mathematical expressions can be simple or complex (involving multiple terms and operations). A common mathematical expression used in GIS is the formula for calculating the Euclidean distance between two points in 2D space: $d = \sqrt{((x2 - x1)^2 + (y2 - y1)^2)}$

Where:

- (x1, y1) and (x2, y2) are the coordinates of two points, and
- d represents the distance between them.

See also expression, operand, operator.

mathematical function [SPATIAL ANALYSIS] A function that applies a mathematical operation to the values of a single input raster. There are four groups of mathematical functions available: logarithmic, arithmetic, trigonometric, and powers. *See also* function, spatial analysis.

mathematical model [ESRI SOFTWARE, SURVEYING] In surveying, a set of relations between measurements and unknown coordinates. *See also* coordinates, model, surveying.

mathematical operator *See* operator.

MATRF2022 [MODELING] Acronym for *Mariana Terrestrial Reference Frame of 2022.* A geometric reference frame of the National Spatial Reference System (NSRS) used to define the geodetic latitude, geodetic longitude, and ellipsoidal height of points on the Mariana tectonic plate. The coordinates defined in MATRF2022 are time-dependent. *See also* NAPGD2022, National Spatial Reference System, reference frame.

matrix [MODELING] A row and column grid of cells or pixels used to store raster data. *See also* array, raster.

MAUP *See* modifiable areal unit problem.

max extent [MAP DISPLAY] The maximum bounding rectangle (in x,y coordinates) of an on-screen map. Users cannot zoom out beyond the max extent. *See also* extent.

maximum elevation figure [NAVIGATION] A number on an aeronautical chart that indicates the altitude above mean sea level an aircraft must maintain to clear all terrain features and obstructions such as towers and antennas. The number indicates the elevation of the tallest feature or obstruction. *See also* aeronautical chart, mean sea level.

maximum slope [MATHEMATICS] A path that does not exceed a specified slope angle. *See also* slope angle.

m-coordinate *See* m-value.

mean [MATHEMATICS] The average for a set of values, computed as the sum of all values divided by the number of values in the set. *See also* median, normal distribution.

mean center [SPATIAL STATISTICS (USE FOR GEOSTATISTICS)] A single x,y coordinate calculated as the average of the x-coordinates and y-coordinates for all features in a study area. *See also* mean, median center.

mean center of population [SPATIAL STATISTICS (USE FOR GEOSTATISTICS)] The average spatial location of distributed (population) features. Conceptually, the point on which a map would balance if all the features in the dataset weighed the same. *See also* mean center.

mean high water [NAVIGATION] Also known by the acronym *MHW*. The datum used to define high water to support harbor and river navigation. *See also* datum, marine navigation.

mean low water [CARTOGRAPHY] Also known by the acronym *MLW*. The average of all the low tide levels recorded over the 19-year Metonic cycle; one of two datums used in North America to define low water and used on Canadian nautical charts as the datum for sounding values. *See also* datum, Metonic cycle, sounding.

mean lower low water [CARTOGRAPHY] Also known by the acronym *MLLW*. The arithmetic average of the lower of the two daily low tides recorded over the 19-year Metonic cycle; one of two datums used in North America to define low water as the datum for sounding values; the official U.S. National Ocean Service nautical chart datum. *See also* datum, Metonic cycle, sounding.

mean sea level

1. [GEODESY] Also known by the acronym *MSL*. The average of all low and high tides at a particular starting location over the 19-year Metonic cycle.
2. [CARTOGRAPHY] The datum used to give the numeric elevation at individual survey points on aeronautical charts, topographic maps, engineering plans, and other large-scale maps.

See also datum, elevation, Metonic cycle.

mean stationarity [SPATIAL STATISTICS (USE FOR GEOSTATISTICS)] In geostatistics, a property of a spatial process in which a spatial random variable has the same mean value at all locations. *See also* stationarity.

measure *See* route measure.

measure location fields [LINEAR REFERENCING] In linear referencing, either one or two fields in a table that describe the position of an event along a route. *See also* event, route.

measure value *See* m-value.

measured grid *See* grid.

M

measurement [DATA ANALYSIS] The quantity associated with features, reported as an amount, intensity, or magnitude. *See also* distance, frequency, intensity.

measurement error [SURVEYING] In surveying, the noise that is expected in every measurement. It occurs because the observer makes estimates and uses measuring equipment that is unpredictable in an environment that is also unpredictable. *See also* noise.

measurement residual [SURVEYING] The difference between a measured quantity and its theoretical true value as determined during each iteration of a least-squares analysis and adjustment. *See also* least-squares analysis and adjustment.

measures of central tendency [DATA ANALYSIS] Statistics that describe the central or typical value for a distribution, including the mean, median, and mode. *See also* descriptive statistics, measures of spread.

measures of spread [DATA ANALYSIS] Statistics that describe how similar or varied the set of values are for a distribution; they include the range, percentiles, absolute deviation, variance, and standard deviation. *See also* descriptive statistics, measures of central tendency.

median [MATHEMATICS] The middle value of a set of values when they are ordered by rank. Half the values in a set are higher than the median, and half are lower. When there are two middle values (if the set has an even number of elements) the median is the mean of these two values. *See also* distribution, mean.

median center [SPATIAL STATISTICS (USE FOR GEOSTATISTICS)]

1. A location representing the shortest total distance to all other features in a study area.
2. The location of a single x,y coordinate value that represents the median x-coordinate value and the median y-coordinate value for all features in a study area.

See also median, mean, mean center.

medium scale [MAP DESIGN] In a three-way grouping of map scales, a mid-range scaled map. Typically, when the numeric value of the representative fraction 1/x is between 1:250,000 and 1:1,000,000, the map is considered medium scale. *See also* large scale, small scale.

M

members

1. [PROGRAMMING] Refers collectively to the properties and methods, or functions, of an interface or class.
2. [ESRI SOFTWARE] ArcGIS users who are part of an organization. Used for group management.

See also ArcGIS account, class, group, organization.

mensuration [PHOTOGRAMMETRY] In imagery, a measurement technique using relative geometric relationships to determine object properties—such as line length (distance), building height, surface area, and object volume—from a series of collected points, lines, and angles from the image or from image metadata. For example, the height of a building can be measured based on the length of its shadow using sun angle information in the image metadata. *See also* measurement, metadata.

mental map [MENTAL MAPS] A person's perception of a place. A mental map may include the physical characteristics of a place, such as boundaries of a neighborhood, or the attributes of a place, such as a neighborhood's perceived unsafe areas. A mental map is primarily a psychological construct, although it may also be rendered as an actual map. *See also* spatial cognition.

Mercator projection [MAP PROJECTIONS] A tangent-case cylindrical conformal projection; the only projection on which all rhumb lines are straight lines on the map. Named for Flemish geographer and cartographer Geert de Kremer (Gerardus Mercator) (1512–1594). *See also* conformal projection, cylindrical projection, tangent projection, transverse Mercator projection, Web Mercator projection.

mereing [SURVEYING] Establishing a boundary relative to ground features present at the time of a survey. *See also* boundary, surveying.

merging [ANALYSIS/GEOPROCESSING] Combining features from multiple data sources of the same data type into a single, new dataset. *See also* appending.

meridian [GEODESY] An imaginary north-south great circle on the earth that passes through the poles, often used synonymously with longitude. Can be used to define locations east or west of the prime meridian. *See also* great circle, longitude, prime meridian.

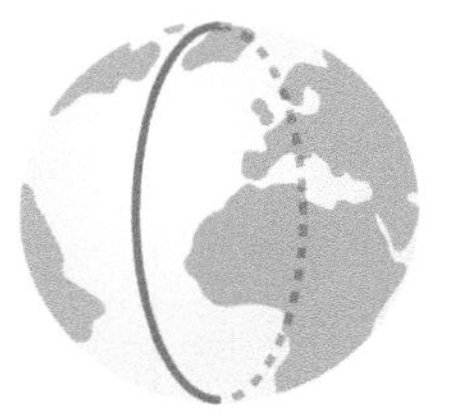

meridian

meridional arc [GEODESY] An arc of a great circle between two parallels measured along a meridian. *See also* arc, great circle.

metadata [DATA MANAGEMENT] Contextual information about a dataset, map, or app. Metadata may include date-time, origin, availability, distribution information, and other relevant properties. Metadata is used for content reliability. *See also* ancillary data, index.

metes and bounds [SURVEYING] A surveying method in which the limits of a parcel are identified as relative distances and bearings from landmarks. Metes and bounds surveying often results in irregularly shaped areas. *See also* bearing, distance, landmark, surveying.

Metonic cycle [GEOREFERENCING] A period of almost exactly 19 years or 235 lunar months, at the end of which the phases of the moon begin to occur in the same order and on the same days as in the previous cycle. Named for Meton of Athens (circa 440 BC). *See also* mean sea level.

metric system [MATHEMATICS] A system of measurement initially based on a fraction of the earth's polar circumference. *See also* English unit, unit of measure.

metropolitan statistical area [FEDERAL GOVERNMENT] A geographic entity defined by the U.S. Office of Management and Budget for use by federal statistical agencies, including the U.S. Census Bureau. A metropolitan statistical area is based on the concept of a core area with a large population nucleus, plus adjacent communities having a high degree of economic and social integration with that core area. According to the 1990 standards, to qualify as a metropolitan statistical area, the area must include at least one city or urbanized area with 50,000 or more inhabitants and a total metropolitan population of at least 100,000 (75,000 in New England). *See also* core-based statistical area, micropolitan statistical area.

MGRS *See* Military Grid Reference System.

Michibiki *See* QZSS.

microdensitometer [GRAPHICS (MAP DISPLAY)] A densitometer that can read densities in minute areas, used particularly for studying spectroscopic and astronomical images. *See also* densitometer.

micrometer [PHYSICS] An instrument for measuring minute lengths or angles. *See also* distance, geometry, stereometer.

micron [PHYSICS] One millionth of a meter, represented by the symbol µm. Microns are used to measure wavelengths in the electromagnetic spectrum. *See also* electromagnetic spectrum, unit of measure, wavelength.

micropolitan statistical area [GOVERNMENT] A geographic region containing at least one urban area with a population between 10,000 and 50,000, defined by the U.S. Office of Management and Budget for use by federal statistical agencies, including the U.S. Census Bureau. Micropolitan statistical areas include adjacent communities having a high degree of economic and social integration with the core area. *See also* core-based statistical area, metropolitan statistical area.

mid-wave infrared [PHYSICS] A band of the electromagnetic spectrum that has wavelengths in the 3–5 micron range. This band is in the higher frequency thermal range of the electromagnetic spectrum and measures emitted thermal energy. *See also* electromagnetic spectrum, infrared radiation, remote sensing.

migratory route map [WILDLIFE] A map that shows the geographic route along which birds or other animals customarily migrate. Also called a migration map.

mileage map [CARTOGRAPHY] A distance cartogram on which routes between major locations are shown by straight lines labeled with the distance along each route. *See also* cartogram.

Military Grid Reference System [GEODESY, DEFENSE] Also known by the acronym *MGRS*. A grid cell location system used by North American Treaty Organization (NATO) nations. Shortens the lengthier numeric coordinates and numeric grid zone specifications typical of other grid systems by providing single alphanumeric substitutions for strings of alphanumeric characters. *See also* grid cell location system.

military symbol [DEFENSE] A graphic representation of a unit, equipment, installation, activity, control measure, or tactical task relevant to military operations that is used for planning or to represent the common operational picture on a map, display, or overlay. *See also* modifier.

mimetic symbol [SYMBOLOGY] A symbol that imitates or closely resembles the thing it represents, such as an icon of a picnic table that represents a picnic area. *See also* arbitrary symbol, symbol.

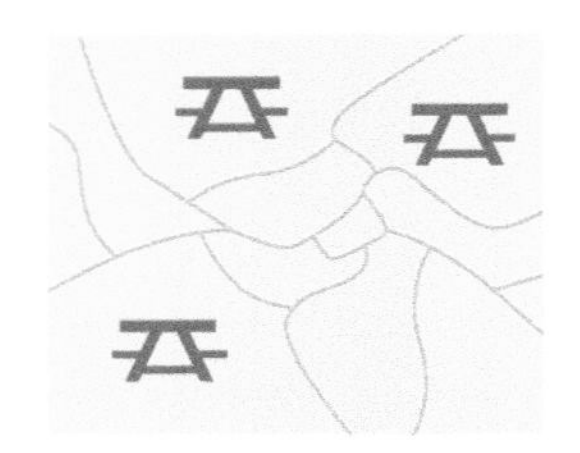

mimetic symbol

min/max scale [ESRI SOFTWARE] The smallest and largest scales at which a layer is visible on a map. Scale ranges are used to prevent detailed layers from displaying when zoomed out and to prevent general layers from displaying when zoomed in.

minimum bounding rectangle [ESRI SOFTWARE] A rectangle, oriented to the x- and y-axes, that bounds a geographic feature or a geographic dataset. It is specified by two coordinate pairs: xmin, ymin and xmax, ymax. *See also* extent.

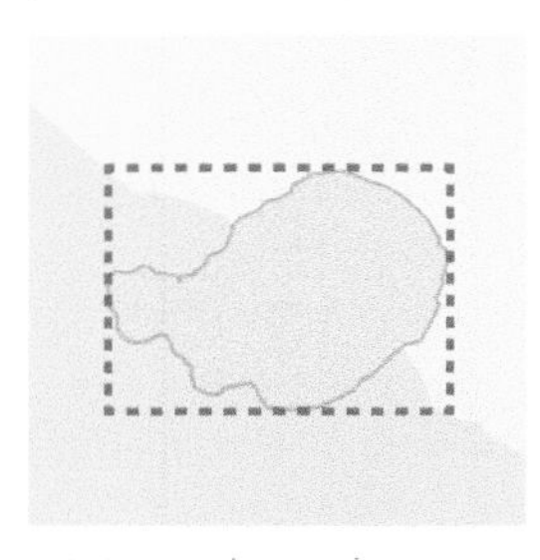

minimum bounding rectangle

minimum candidate score [GEOCODING] In geocoding, the minimum score a potential match record requires to be considered a candidate. *See also* address matching, geocoding.

minimum map unit

1. [MAP DESIGN] For a given scale, the size in map units below which a narrow feature can be reasonably represented by a line and an area by a point.
2. [PHOTOGRAMMETRY] In imagery, the smallest area identified as an object or class in the output dataset, expressed in ground units.
3. [GRAPHICS (MAP DISPLAY)] The smallest allowable size for a group of cells to be seen as a single entity in a raster image.

See also map unit.

minimum match score [ESRI SOFTWARE] In geocoding, the minimum score a match candidate needs to be considered a match in batch geocoding. *See also* address matching, geocoding.

minor axis [EUCLIDEAN GEOMETRY] The shorter axis of an ellipse or spheroid. *See also* major axis.

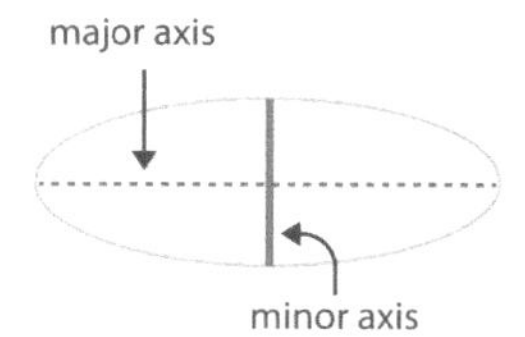

minor axis

minute

1. [GEODESY] An angle equal to 1/60 of a degree of latitude or longitude and containing 60 seconds.
2. [MATHEMATICS] An angle equal to 1/60 of a degree of arc.

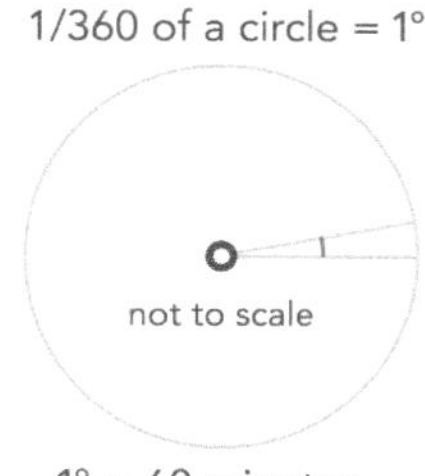

minute 2

See also second.

minute of angle [MATHEMATICS] Also known by the acronym *MOA*. A unit of angular measurement used for a subtended angle. One minute of angle (1/60th of a degree) subtends approximately 1 inch at 300 feet, 2 inches at 600 feet, and so on. *See also* minute.

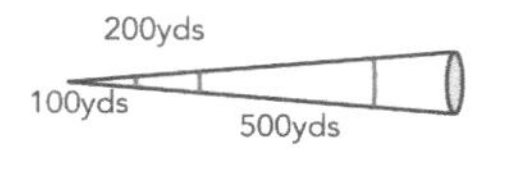

minute of angle

misclosure *See* closure error.

mixed pixel [REMOTE SENSING] In remote sensing, a pixel whose digital number represents the average of several spectral classes within the area that it covers on the ground, each emitted or reflected by a different type of material. Mixed pixels are common along the edges of features. *See also* edge, pixel, remote sensing.

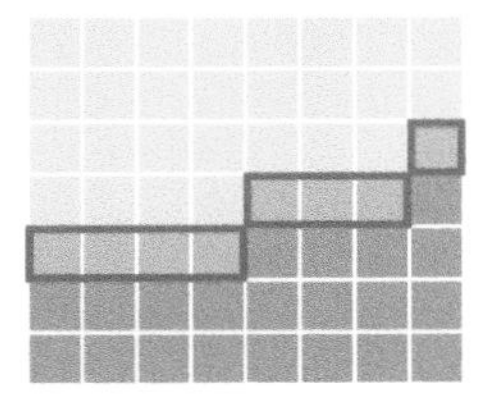
mixed pixel

model

1. [DATA MODELS] An abstraction of reality used to represent objects, processes, or events.
2. [MODELING] A set of rules and procedures for representing a phenomenon or predicting an outcome.
3. [DATA MODELS] A data representation of reality, such as the vector data model.

See also digital elevation model, vector data model.

M

modifiable areal unit problem [SPATIAL ANALYSIS] Also known by the acronym *MAUP*. A statistical bias that can occur during spatial analysis of aggregated data that causes differing results although the same analysis is applied to the same data. MAUP takes two forms: the scale effect and the zone effect.

The scale effect exhibits different results when the scale of the aggregation units is changed. For example, analysis using data aggregated by county will differ from analysis using data aggregated by census tract. Often the resulting difference is valid based on the distinctions of scale.

The zone effect is observed when the scale of analysis is fixed, but the shape of the aggregation units is changed. For example, analysis using data aggregated into one-mile grid cells will differ from analysis using one-mile hexagon cells. The result is a problem because it is an analysis, at least in part, of the aggregation scheme rather than of the data itself.

See also aggregation, analysis, boundary effect, scale, spatial analysis.

modifier [DEFENSE, ESRI SOFTWARE] Text or graphics that display around a military symbol, or a value that changes the appearance of a symbol. In some military specifications, attributes are referred to as modifiers. *See also* military symbol.

MODIS [SATELLITE IMAGING] An abbreviation of *moderate resolution imaging spectroradiometer*. A bundle of remote sensing equipment housed on two NASA satellites, Terra and Aqua. These satellites record multiple images of the earth's surface in various wavelengths and resolutions during each two-day orbit. *See also* Aqua, NASA, remote sensing imagery, sensor.

Mollweide projection [MAP PROJECTIONS] An elliptical equal-area projection that most commonly uses the equator as the standard parallel and the prime meridian as the central meridian. A Mollweide projection typically preserves relative area proportions, making it suitable for global distribution or thematic maps. The Mollweide projection was invented in

1805 by German mathematician Karl B. Mollweide. *See also* central meridian, equal-area projection, equator, prime meridian.

monochromatic

1. [PHYSICS] Related to a single wavelength or a very narrow band of wavelengths.
2. [GRAPHICS (COMPUTING)] A color scheme made up of lighter and darker shades of the same color.

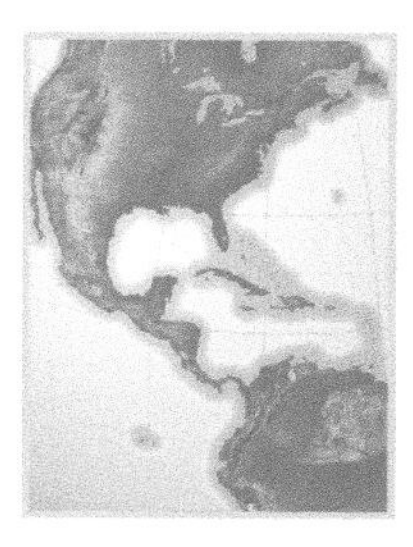

monochromatic 2

See also density slicing, grayscale, wavelength.

monochromatic imagery [PHOTOGRAMMETRY] A single-band, grayscale image. *See also* grayscale.

Monte Carlo method [MODELING] An algorithm for computing solutions to problems that contain many variables by performing iterations with different sets of random numbers until the best solution is found. The Monte Carlo method is usually applied to problems too complex for analysis by anything but a computer. *See also* stochastic model.

monument *See* survey monument.

morphology

1. [GEOGRAPHY] The structure of a surface.
2. [SPATIAL ANALYSIS] The study of structure or form.

See also shape.

mosaic

1. [DIGITAL IMAGE PROCESSING] A raster dataset composed of two or more merged raster datasets—for example, one image created by merging several individual images or photographs of adjacent areas.

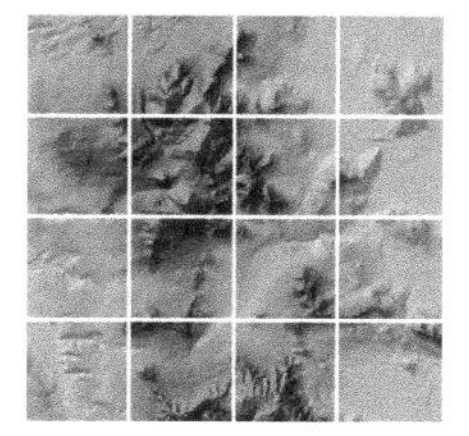

mosaic 1

2. [VISUALIZATION] Maps of adjacent areas with the same spatial reference and scale whose boundaries have been matched and dissolved.
3. [GRAPHICS (MAP DISPLAY)] When discussing photos or images, an image composed of multiple individual photographs or images of adjacent areas merged into a single photograph or image without gaps. For photographs, it is called a photomosaic, and for images, it is called an image mosaic. When discussing landscapes, the entire set of landscape patches that cover a region.

See also edgematching, merging.

mosaic dataset [PHOTOGRAMMETRY] A collection of images stored in a geodatabase and viewed, managed, and analyzed either as a single, mosaicked image or as individual images. *See also* raster dataset.

motion imagery [DIGITAL IMAGE PROCESSING] In imagery, a series of spatially and temporally related images—such as video—that can be analyzed collectively, visualized as animations, and analyzed for change over time. Typically sensor-based, these can also include object data, 3D scenes, and support temporal data (or trend) analysis.

MSA *See* metropolitan statistical area.

MSAS [GPS] Acronym for *Multi-functional Satellite Augmentation System.* A satellite-based augmentation system (SBAS) that uses differential correction to supplement the reliability and accuracy of signals from the GPS system in the Japanese region. MSAS was developed and is operated by Japan's Ministry of Land, Infrastructure, Transport and Tourism and Civil Aviation Bureau (JCAB). *See also* differential correction, GPS, GPS augmentation.

MSS *See* multispectral sensor.

MTSAT *See* MSAS.

multichannel receiver [REMOTE SENSING] A receiver that tracks several satellites at a time, using one channel for each satellite. *See also* multiplexing channel receiver.

multidimensional data [DATA MODELS] Data that is composed of multiple dimensions, such as space and time. For example, a temperature dataset could have dimensions of latitude, longitude, altitude, and time. *See also* dimension.

multidirectional hillshading [MAP DESIGN] A form of hillshading that involves the use of multiple imaginary light sources primarily to the map's west and north sides so that terrain with north–south and east–west orientation also receives sufficient shading. *See also* hillshading.

Multi-functional Satellite Augmentation System *See* MSAS.

multimodal network [NETWORK ANALYSIS] A network in which two or more types of transportation modes (such as walking, riding a train, or driving a car) are modeled. In a network dataset, multiple connectivity groups are required to create a multimodal network. *See also* connectivity, network.

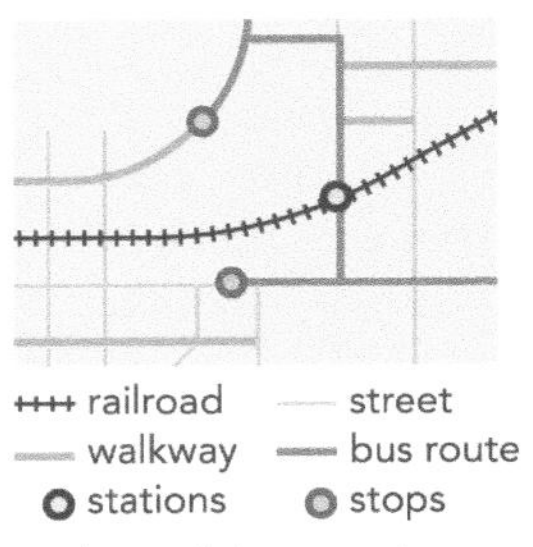

multimodal network

multipart feature [ESRI SOFTWARE] A digital representation of a place or thing that has more than one part but is defined as one feature because it references one set of attributes. In a layer of states, for example, the state of Hawaii could be considered a multipart feature because its separate geometric parts are classified as a single state. A multipart feature can be a point, line, or polygon. *See also* feature.

multipart feature

multipatch [ESRI SOFTWARE] A type of geometry composed of planar 3D rings and triangles, used in combination to model objects that occupy discrete area or volume in 3D space. Multipatches may represent geometric objects, such as spheres and cubes, or real-world objects, such as buildings and trees. *See also* multipatch feature, ring.

multipatch feature [ESRI SOFTWARE] A real-world geographic feature modeled using multipatch geometry. *See also* multipatch.

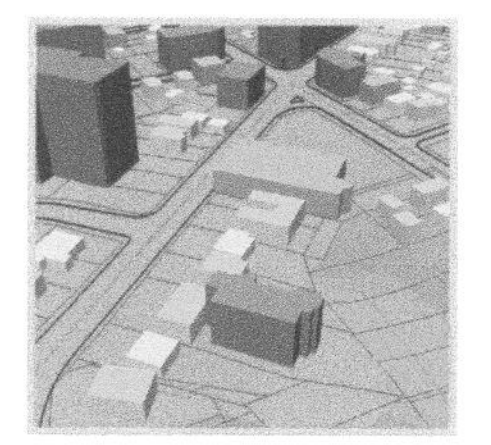

multipatch feature

multipath error [SATELLITE IMAGING] Errors caused when a satellite signal reaches the receiver from two or more paths, one directly from the satellite and the others reflected from nearby buildings or other surfaces. Signals from satellites low on the horizon will produce more error. *See also* error, remote sensing.

multiple display map [GRAPHICS (MAP DISPLAY)] A mapping method for multivariate data in which several individual map displays are used. These displays can either be of constant format (small multiples) or complementary formats. *See also* complementary-formats display.

multiple regression [STATISTICS] Regression in which the dependent variable is measured against two or more independent variables. *See also* regression.

multiplexing channel receiver [REMOTE SENSING] A receiver that tracks several satellite signals using a single channel. *See also* multichannel receiver.

multipoint [ESRI SOFTWARE] A geometric element defined by an unordered set of x,y coordinate pairs. *See also* multipoint feature.

multipoint feature [ESRI SOFTWARE] A digital map feature that represents a place or thing that has neither area nor length at a given scale, and that is treated as a single object with multiple locations. For example, the entrances and exits to a prairie dog den might be represented as a multipoint feature. A multipoint feature is associated with a single record in an attribute table. *See also* feature.

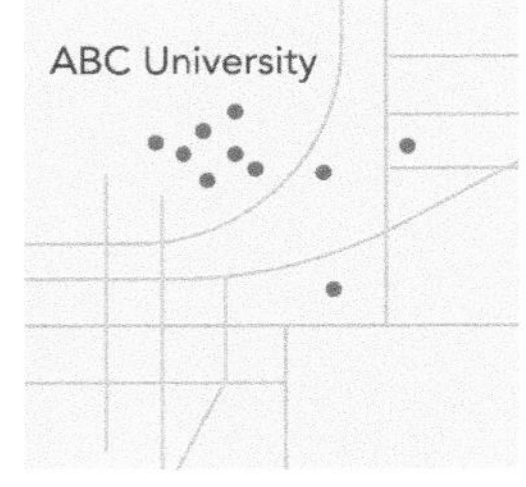

multipoint feature

multipoint feature class [3D GIS] A feature class that can store many points per shape or row, saving storage space and improving read-write performance. *See also* feature class, multipoint feature.

multipurpose cadastre [CADASTRAL AND LAND RECORDS] The body of land records containing information about parcels and their attributes, including slope and aspect, soil characteristics, drainage, vegetation cover, number of residents, building construction, access road width and surface material, utility service, and zoning restrictions. Also spelled *multipurpose cadaster. See also* cadastre.

multiscale maps [MAP DESIGN] A series of maps at varying scales, each depicting an amount and type of information appropriate to the particular scale. *See also* large scale, medium scale, scale, small scale.

multispectral [PHYSICS] Related to two or more frequencies or wavelengths in the electromagnetic spectrum. *See also* electromagnetic spectrum, multispectral image.

multispectral image [DIGITAL IMAGE PROCESSING] An image or imagery that shows information gathered in spectral bands across the electromagnetic spectrum; typically composed of 4–15 bands from a set of targeted wavelengths at chosen locations, from ultraviolet to thermal (400nm–12,500nm). *See also* multispectral, color composite.

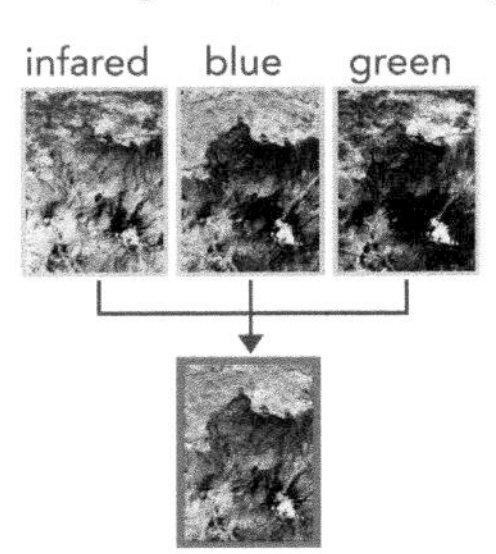

multispectral image

multispectral imaging [PHOTOGRAMMETRY] A process, such as sensor technology, that captures image data within discrete and somewhat narrow wavelength ranges across the electromagnetic spectrum. Typically measures several spectral bands using a subset of targeted wavelengths at chosen locations, for example, 400–1100 nm in increments of 20 nm. *See also* CIR film, electromagnetic spectrum.

multispectral scanner
See multispectral sensor.

multispectral sensor [SATELLITE IMAGING] A device carried on satellites and aircraft that records energy from multiple portions of the electromagnetic spectrum. *See also* electromagnetic spectrum, sensor.

multitemporal [PHOTOGRAMMETRY] In imagery, a characteristic of a collection of raster datasets (images) or other features that have multiple time or date stamps. *See also* temporal data.

multivariate analysis [STATISTICS] Any statistical analysis method for evaluating the relationship between two or more variables. *See also* analysis, variance.

multivariate map [SYMBOLOGY] A map that shows more than one theme. It may show a composite of two or more related themes, or several attributes of a feature may be incorporated into the same symbol. *See also* theme, symbol.

multivariate point symbol [SYMBOLOGY] A symbol that shows multiple attributes of a point feature, such as angled arrows indicating the location and direction of recorded wind speeds. *See also* glyph.

multivariate symbol mapping [SYMBOLOGY] A mapping method in which multivariate data is shown on a single map through the use of multivariate symbols, most often point symbols. Mapmakers can display two or more variables in a single symbol, or they can segment a symbol to show the relative magnitudes of subcategories of feature attributes. *See also* mapping methods, multivariate analysis, multivariate point symbol.

multiversioned view [ESRI SOFTWARE] A view that uses stored procedures and triggers to access a specified version of data in a single business table in the geodatabase. A multiversioned view includes all the records in the business table that have been selected and merged with records from the delta tables. The schema of a multiversioned view is identical to that of the business table on which it is based. *See also* view.

m-value
1. [LINEAR REFERENCING] In linear referencing, a measure value that is added to a line feature. M-values are used to measure the distance along a line feature from a vertex (a known location) to an event.
2. [ESRI SOFTWARE] Vertex attributes that are stored with x,y point coordinates. Every type of geometry (point, polyline, polygon) can have attributes for every vertex.

See also distance, line feature, vertex.

M

NAD 1927 *See* North American Datum of 1927.

NAD 1983 *See* North American Datum of 1983.

NAD27 *See* North American Datum of 1927.

NAD83 *See* North American Datum of 1983.

nadir

1. [AERIAL PHOTOGRAPHY] In aerial photography and satellite imagery, the point on the ground vertically beneath the perspective center of the camera lens or scanner's detectors, or the point of convergence (vanishing point) for all vertical rays within a scene.

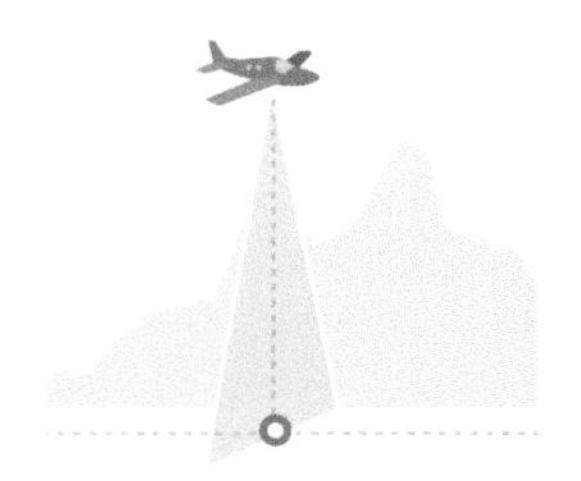

nadir 1

2. [ASTRONOMY] In astronomy, the point on the celestial sphere directly beneath an observer. Both the nadir and zenith lie on the observer's meridian; the nadir lies 180 degrees from the zenith and is therefore unobservable.

See also zenith, meridian, celestial sphere.

NAICS *See* North American Industry Classification System.

NaN [MATHEMATICS] Acronym for *not a number.*

NAPGD2022 [COORDINATE SYSTEMS] Acronym for *North American–Pacific Geopotential Datum of 2022.* The geopotential basis for the vertical and gravity components of the National Spatial Reference System (NSRS) in the United States, including vertical datum, the geoid, gravity, deflections of the vertical, and other quantities related to the earth's gravity field. NAPGD2022 replaces NAVD 88 and all other vertical datums throughout the U.S. and its territories. NAPGD2022 is composed of four interrelated, time-dependent products:

- A global model of Earth's geopotential field (GM2022)
- Regional gridded geoid undulation models (GEOID2022)
- Regional gridded deflection of the vertical models (DEFLEC2022)
- Regional gridded surface gravity models (GRAV2022)

N

See also geoid, National Spatial Reference System, NAVD 88, North American Datum of 1983.

NASA [FEDERAL GOVERNMENT] Acronym for *National Aeronautics and Space Administration.* The United States government agency responsible for civil space exploration, aeronautics, and space scientific research.

National Elevation Dataset [GEOGRAPHY, GOVERNMENT] Also known by the acronym *NED.* A digital elevation model of one arcsecond resolution distributed by the U.S. Geological Survey covering the continental United States, Hawaii, and the U.S. island territories; the resolution for Alaska's grid is two arcseconds. *See also* arcsecond, digital elevation model.

National Geodetic Survey *See* U.S. National Geodetic Survey.

National Geodetic Vertical Datum of 1929 [COORDINATE SYSTEMS, GOVERNMENT] Also known by the acronym *NGVD29.* A datum defined by the observed heights of mean sea level at 26 North American tide gauges, as well as the elevations of benchmarks resulting from more than 60,000 miles (96,560 kilometers) of leveling across the continent, totaling more than 500,000 vertical control points. Established in 1929 by the U.S. Coast and Geodetic Survey as the surface against which elevation data in the United States is referenced. *See also* datum, NAVD 88.

National Geospatial-Intelligence Agency [FEDERAL GOVERNMENT] Also known by the acronym *NGA.* The United States government agency whose mission is to collect and provide timely, relevant, and accurate geospatial intelligence in support of national security; formerly called the National Imagery and Mapping Agency (NIMA).

National Hydrography Dataset [GEOLOGY, GOVERNMENT] A dataset compiled by the U.S. Geological Survey using streamgage data to assign streamflow values to streamline features in a GIS. This dataset represents the nation's drainage network and includes features such as rivers, streams, canals, lakes, ponds, coastlines, and dams. *See also* stream flow, streamgage.

National Land Cover Database [REMOTE SENSING, GOVERNMENT] A Landsat-based, 30-meter resolution, land cover database of the United States maintained by the U.S. Geological Survey. *See also* Landsat, U.S. Geological Survey.

National Map Accuracy Standards [ACCURACY, GOVERNMENT] Standards for accuracy in published maps that are created and maintained by the U.S. Geological Survey. *See also* U.S. Geological Survey.

National Ocean Service [FEDERAL GOVERNMENT] Also known by the acronym *NOS.* The United States government agency, under the administration of NOAA, whose responsibility is to preserve and enhance the nation's coastal resources and ecosystems. *See also* NOAA.

National Oceanic and Atmospheric Administration *See* NOAA.

National Park Service [FEDERAL GOVERNMENT] Also known by the acronym *NPS*. The United States government agency within the U.S. Department of the Interior that administers national parks, monuments, historic sites, and recreational areas in the country.

National Spatial Data Infrastructure [DATA SHARING, GOVERNMENT] Also known by the acronym *NSDI*. A federally mandated framework of spatial data that refers to U.S. locations, as well as the means of distributing and using that data effectively. Initiated and coordinated by the FGDC, the NSDI encompasses policies, standards, procedures, technology, and human resources for organizations to cooperatively produce and share geographic data. The NSDI is developed by the federal government; state, local, and tribal governments; the academic community; and the private sector. *See also* Federal Geographic Data Committee.

National Spatial Reference System [COORDINATE SYSTEMS] Also known by the acronym *NSRS*. A consistent coordinate system that defines latitude, longitude, height, scale, gravity, and orientation throughout the United States. The NSRS is currently based on NAD83 and NAVD 88. It is defined and managed by the U.S. National Geodetic Survey. In addition to the NOAA CORS Network (NCN), the NSRS includes a network of permanently marked stations; an up-to-date, consistent national shoreline; and a set of accurate models describing the geophysical processes that affect spatial measurements. *See also* CORS, NAVD 88, NOAA, North American Datum of 1983, U.S. National Geodetic Survey.

National Topographic System [MAP DESIGN, GOVERNMENT] Also known by the acronym *NTS*. The map standards used by Natural Resources Canada to define topographic map coverage of Canada. *See also* Natural Resources Canada, topographic map.

NATRF2022 [MODELING] Acronym for *North American Terrestrial Reference Frame of 2022*. A geometric reference frame of the National Spatial Reference System (NSRS) used to define the geodetic latitude, geodetic longitude, and ellipsoidal height of points on the North American tectonic plate. The coordinates defined in NATRF2022 are time-dependent. *See also* NAPGD2022, National Spatial Reference System, reference frame.

natural breaks classification [DATA MANAGEMENT] A method of manual data classification that seeks to partition data into class intervals based on natural groups in the data distribution. Natural breaks are gaps in a distribution of data. For example,

N

on a histogram, natural breaks occur at the low points of valleys. Breaks are assigned in the order of the size of the valleys, with the largest valley being assigned the first natural break. *See also* classification, Jenks' optimization.

natural feature [GEOGRAPHY] A naturally occurring feature represented on a map, such as a mountain, lake, forest, or river. *See also* cultural feature, feature.

natural neighbors [MATHEMATICS] An interpolation method for multivariate data in a Delaunay triangulation. The value for an interpolation point is estimated using weighted values of the closest surrounding points in the triangulation. These points, the natural neighbors, are the ones the interpolation point would connect to if inserted into the triangulation. *See also* Delaunay triangulation, interpolation, triangulation.

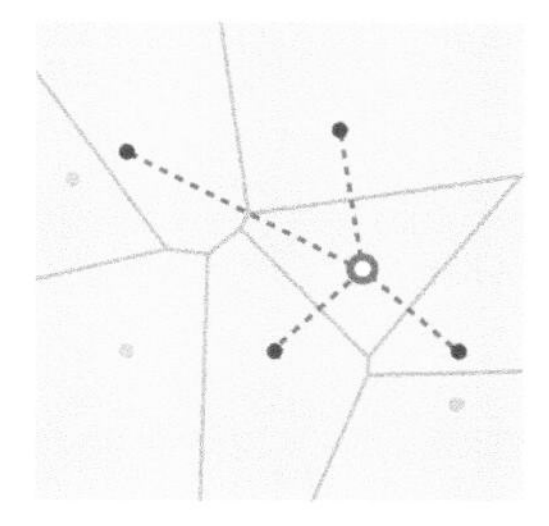

natural neighbors

Natural Resources Canada [FEDERAL GOVERNMENT] Also known by the abbreviation *NRCan.* The Canadian government national mapping agency responsible for the Canadian topographic map series. *See also* National Topographic System, topographic map.

Natural Resources Conservation Service [FEDERAL GOVERNMENT] Also known by the acronym *NRCS.* The United States government agency—a branch of the U.S. Department of Agriculture—that works with private landowners to help them conserve, maintain, and improve their natural resources. Formerly the Soil Conservation Service (SCS).

natural variation error [ACCURACY] A type of error due to natural diversity in the phenomena or natural fluctuations in the system. *See also* error, precision, uncertainty.

nautical chart [NAVIGATION] A map created specifically for water navigation. *See also* hydrographic chart.

NAVD 88 [COORDINATE SYSTEMS] Acronym for *North American Vertical Datum of 1988.* The vertical control datum for heights used for surveying in the United States. NAVD is a leveling network on the North American continent with its origin at Pointe-au-Père (Father Point), Rimouski, Quebec, Canada. It is referenced to the International Great Lakes Datum of 1985 mean sea level. NAVD 88 uses Helmert orthometric heights, which define the location of the geoid from leveling and gravity data. *See also* datum, geoid, mean sea level, orthometric height.

NavIC [GPS] A combined abbreviation and acronym for *Navigation with Indian Constellation,* formerly known as the *Indian Regional Navigation Satellite System (IRNSS).* A regional navigation satellite system (RNSS) used for positioning, navigation, and timing (PNT) in the Indian region; operated by the Indian government. *See also* GAGAN, GPS, GPS augmentation, PNT, satellite constellation.

navigate [NAVIGATION] To interactively change the observer's or target's position. *See also* animation, wayfinding.

navigation [NAVIGATION]
1. A two-part activity that involves planning a route to get from one place to another and then following that route in the field. Navigation is composed of wayfinding and locomotion.
2. The activity of guiding a ship, plane, or other vehicle to a destination, along a planned or improvised route, according to reliable methods.

See also locomotion, route, sextant, wayfinding.

navigation chart [NAVIGATION] A map or chart designed for use with air or marine navigation. *See also* air navigation, marine navigation.

Navigation with Indian Constellation *See* NavIC.

Navstar [GPS] A combined abbreviation and acronym for *navigation satellite timing and ranging.* The original name of the U.S. GPS system. *See also* GPS.

NCN *See* CORS.

near infrared [PHYSICS] Also known by the acronym *NIR.* A band of the electromagnetic spectrum that has wavelengths from 0.75 to 1.4 μm, varying slightly based on the technology and application. Near infrared is reflected light and is used in remote sensing analysis of vegetation (high reflectivity) and in detecting water bodies (high absorption). *See also* electromagnetic spectrum, infrared radiation.

near ultraviolet [PHYSICS] A band of the electromagnetic spectrum that has wavelengths from 300 nm to 400 nm (slightly shorter than visible light). *See also* band, electromagnetic spectrum, wavelength.

near visible [PHYSICS] The near ultraviolet and near infrared portions of the electromagnetic spectrum. Neither near ultraviolet nor near infrared light is visible to the human eye but are just outside the range of visible light. Near ultraviolet light has longer wavelengths (between visible light and x-rays); near infrared light has shorter wavelengths (between visible light and microwaves). *See also* electromagnetic spectrum, near infrared, near ultraviolet.

nearest neighbor [SPATIAL ANALYSIS] The feature that is in closest proximity to something.

N

nearest neighbor resampling [MATHEMATICS] A technique for resampling raster data in which the value of each cell in an output raster is calculated using the value of the nearest cell in an input raster. Nearest neighbor assignment does not change any of the values of cells from the input layer; for this reason, it is often used to resample categorical or integer data (for example, land use, soil, or forest type), or radiometric values, such as those from remotely sensed images. *See also* bilinear interpolation, cubic convolution, neighborhood functions, resampling.

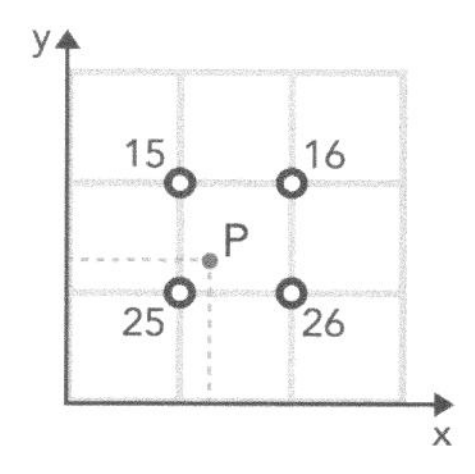

nearest neighbor resampling

near-polar orbit [GPS] A satellite orbit inclined about 10 degrees from the polar axis.

neatline [CARTOGRAPHY] The border delineating and defining the extent of geographic data on a map. It demarcates map units so that, depending on the map projection, the neatline does not always have 90-degree corners. In a properly made map, it is the most accurate element of the data; other map features may be moved slightly or exaggerated for generalization or readability, but the neatline is never adjusted. *See also* cartography, map, map extent.

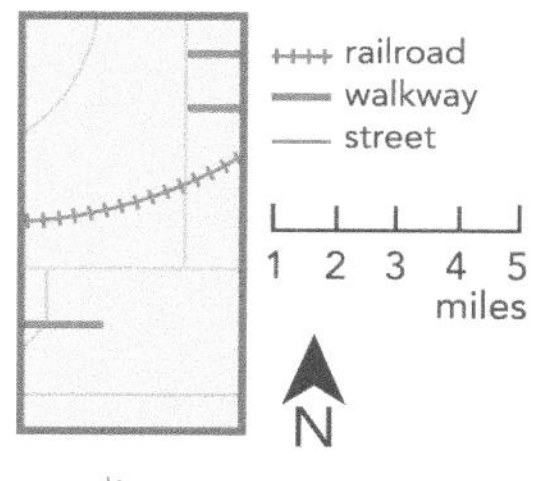

neatline

NED *See* National Elevation Dataset.

neighborhood *See* filter.

neighborhood functions [MATHEMATICS] Methods of defining new values for locations using the values of other locations within a given distance or direction. *See also* natural neighbors, nearest neighbor resampling, proximity analysis.

neighborhood statistics *See* focal analysis.

network [NETWORK ANALYSIS] An interconnected set of points and lines that represent possible routes from one location to another. For example, an interconnected set of lines representing a city streets layer is a network. *See also* link, node, traveling salesperson problem.

network analysis [NETWORK ANALYSIS] Any method of solving network problems—such as traversability, rate of flow, or capacity—using network connectivity. *See also* network.

network analysis layer [ESRI SOFTWARE] A composite layer that contains the properties and network analysis classes used in the analysis of a network problem, and the results of the analysis. *See also* layer, network layer.

network attribute [ESRI SOFTWARE] A type of attribute associated with a network element in a network dataset. Network attributes are used to help control flow through a network. All network elements in a network dataset have the same set of attributes. *See also* attribute, cost, descriptor, hierarchy, restriction.

network dataset [ESRI SOFTWARE] A collection of topologically connected network elements (edges, junctions, and turns) that are derived from network sources, typically used to represent a linear network, such as a road or subway system. Each network element is associated with a collection of network attributes. Network datasets are typically used to model undirected flow systems. *See also* network attribute, network element.

network dataset

N

network element [ESRI SOFTWARE] A component in a network dataset: an edge, junction, or turn. All elements in a network dataset share the same set of network attributes. Network elements are used to model topological relationships in undirected flow networks such as traffic flow systems. Network elements are generated from point, line, and turn features. When the network dataset is built, point features become junctions, line features become edges, and turn features become turn elements. *See also* network.

network flow map [NETWORK ANALYSIS] A flow map that shows interconnectivity between places. The flow lines typically describe transportation or communication linkages. *See also* flow line, flow map, network.

network hierarchy [NETWORK ANALYSIS] A network in which nodes can contain more than one level of connectivity. A single place may, for example, be connected by roads with different surface or traffic flow characteristics. *See also* network.

network layer [ESRI SOFTWARE] A layer that references a network dataset. In a geodatabase, a network dataset is a collection of network elements (edges, junctions, and turns) that are derived from network sources. *See also* layer, network analysis layer, network dataset.

network location [NETWORK ANALYSIS] A geographic position in a network system.

network solver *See* solver.

network source [ESRI SOFTWARE] Feature classes in a geodatabase that are used to generate and define a network dataset. *See also* network dataset, network element.

network transformation [NETWORK ANALYSIS, COORDINATE SYSTEMS] The process of optimizing network data (such as road or utility networks, their associated attributes, and connectivity information) to analyze or visualize the data within a network model. Common network transformations include routing, analysis, optimization, and simulation. *See also* closest facility analysis, network analysis, optimization, route analysis, spatial analysis.

neural network [PROGRAMMING] A shortened form of *artificial neural network*. A branch of machine learning designed to solve the problems typically solved by human beings, such as recognizing patterns and making predictions from past performance. Neural networks calculate a number of weighted inputs to generate an output and "learn" to generate better outputs by adjusting the weights and thresholds applied to those inputs. *See also* AI, deep learning, machine learning.

new store analysis [ESRI SOFTWARE, BUSINESS] A process that finds a potential location for a new store by calculating the centroid of a group of customers. *See also* centroid, store market analysis, store prospecting.

NGA *See* National Geospatial-Intelligence Agency.

NGS *See* U.S. National Geodetic Survey.

NGVD 1929 *See* National Geodetic Vertical Datum of 1929.

NGVD29 *See* National Geodetic Vertical Datum of 1929.

NIIRS [PHOTOGRAMMETRY] Acronym for *National Imagery Interpretability Rating Scale*. In imagery, a subjective scale of image quality that defines levels of interpretability in terms of features that can be effectively resolved and identified. Consists of 10 levels, from 0 (worst quality) to 9 (best quality). NIIRS depends on ground sample distance (GSD), clarity, and the multispectral characteristics of the sensor system. *See also* NITF, sensor.

NITF [PHOTOGRAMMETRY, GOVERNMENT] Acronym for *National Imagery Transmission Format*. A digital file container format that holds an image and its associated information and metadata, including security classification; this is the preferred format for U.S. imagery disseminated within the Department of Defense or Intelligence Community. *See also* NIIRS.

NMEA [STANDARDS] Acronym for *National Marine and Electronics Association.* A nonprofit association composed of manufacturers, distributors, dealers, educational institutions, and others interested in peripheral marine electronics occupations. The NMEA created electrical interface and data protocol for communications between marine instrumentation that has been adopted by the GPS industry as a standard. *See also* GPS.

NOAA [FEDERAL GOVERNMENT] Acronym for *National Oceanic and Atmospheric Administration.* The United States government agency with a mission to understand and predict changes in the earth's environment and manage coastal and marine resources to meet the nation's economic, social, and environmental needs.

NOAA CORS Network *See* CORS.

NoData [DATA CAPTURE] In raster data, pixels that do not contain a recorded value. NoData is not equivalent to zero value; rather, it indicates that no measurements have been collected for that pixel. *See also* NaN, null value.

node

1. [DATA STRUCTURES] In a TIN, one of the three corner points of a triangle, topologically linked to all triangles that meet there. Each sample point in a TIN becomes a node in the triangulation that may store elevation z-values and tag values.

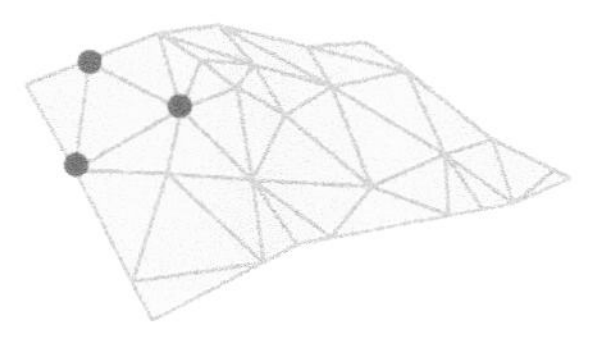

node 1

2. [ESRI SOFTWARE] In a geodatabase, the point representing the beginning or ending point of an edge, topologically linked to all the edges that meet there.
3. [NETWORK ANALYSIS] A point at which a line ends or where two or more lines come together or intersect in a network.
4. [MATHEMATICS] In graph theory, any vertex in a graph.

See also edge, TIN, triangle.

noise

1. [REMOTE SENSING] In remote sensing, any disturbance in a frequency band.
2. [DATA QUALITY] Any irregular, sporadic, or random oscillation in a transmission signal.
3. [TELECOMMUNICATIONS] Random or repetitive events that interfere with communication.
4. [DATA QUALITY] In a raster, irrelevant or meaningless cells that exist due to poor scanning or imperfections in the original source document.

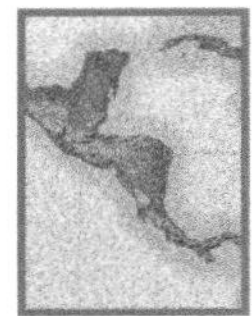

noise 4

See also frequency, PRN, signal.

nominal data [DATA STRUCTURES] A method of organizing data values into categories that have an equivalent, or nonhierarchical, value. In GIS, nominal data is used to distinguish different types of features within a map theme. For example, a group of polygons colored to represent different soil types. Also known as categorical data or nominal-level data. *See also* ordinal data.

nominal-level data [VISUALIZATION] Data that consists of categories used to distinguish different types of features within a map theme. *See also* theme.

noncontiguous cartogram [MAP DESIGN] A cartogram on which the shape of geographic areas is maintained at the expense of proximity and contiguity with neighboring geographic units. *See also* cartogram, contiguous.

nonparametric statistics [ANALYSIS/GEOPROCESSING] A statistical analysis approach that makes no assumptions about probability distributions or other a priori aspects about data. The number and nature of the parameters are flexible and not fixed in advance.

nonsimple polygon [DATA EDITING] A polygon that violates topological integrity by crossing its own boundary (usually by making a small loop). *See also* boundary, polygon.

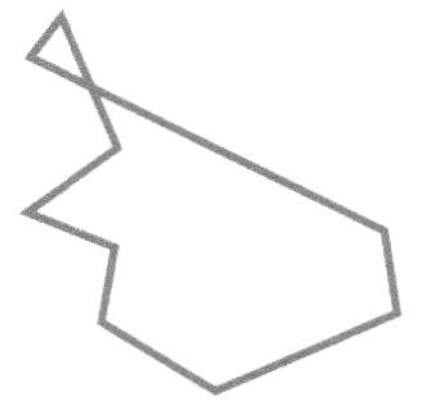
nonsimple polygon

nonspatial data [DATA MANAGEMENT] Data without inherently spatial qualities, such as attributes. *See also* attribute data.

nontopographic profile [MAP DESIGN] A profile for a continuous statistical surface, such as temperature or precipitation, rather than a terrain surface. *See also* continuous surface, profile, terrain surface.

normal aspect [MAP PROJECTIONS] The most commonly used aspect for each of the three map projection families based on a developable surface. For planar projections, the normal aspect is polar; for conic projections, it is oblique (somewhere between the equator and a pole); and for cylindrical projections, it is equatorial. *See also* conic projection, cylindrical projection, map projection families, planar projection.

normal distribution [STATISTICS] A theoretical frequency distribution of a dataset in which the distribution of values can be graphically represented as a symmetrical bell curve. Normal distributions are typically characterized by a clustering of values near the mean, with few values departing radically from the mean. There are as many values on the left side of the curve as on the right, so the mean and median values for the distribution are the same. Some 68 percent of the values are plus or minus one standard deviation from the mean; 95 percent of the values are plus or minus two standard deviations; and 99 percent of the values are plus or minus three standard deviations. *See also* distribution, mean, median, standard deviation.

normal form [DATABASE STRUCTURES] A set of guidelines for designing table and data structures in a relational database. When followed, normal form guidelines prevent data redundancy, increase database efficiency, and reduce consistency errors. A database is said to be in first normal form (1NF), second normal form (2NF), third normal form (3NF), and so on, depending on the level of normal form guidelines followed in its design. In practice, 3NF is commonly used, but higher levels are rarely used. *See also* first normal form, second normal form, third normal form.

normal probability distribution
See normal distribution.

normalization

1. [DATA MANAGEMENT] The process of organizing, analyzing, and cleaning data to increase efficiency for data use and sharing. Normalization usually includes data structuring and refinement, redundancy and error elimination, and standardization.
2. [STATISTICS] A process in which data attributes are adjusted to facilitate comparison of values in different units and scales. For spatial data, normalization is also used to adjust for differences in the sizes of data collection units. A special case of normalization is standardization.

See also data collection unit, standardization.

North American Datum of 1927
[COORDINATE SYSTEMS] Also known by the acronym *NAD 1927*. The primary local horizontal geodetic datum and geographic coordinate system used to map the United States during the middle part of the twentieth century. NAD 1927 is referenced to the Clarke spheroid of 1866 and an origin point at Meades Ranch, Kansas. Features on USGS topographic maps, including the corners of 7.5-minute quadrangle maps, are referenced to NAD 1927. It has gradually been replaced by the North American Datum of 1983. *See also* North American Datum of 1983, Clarke ellipsoid of 1866.

North American Datum of 1983
[COORDINATE SYSTEMS] Also known by the acronym *NAD83*. The horizontal datum and spatial reference system for geometric positioning used in Canada, Greenland, and the United States. NAD83 uses the Geodetic Reference System 1980 (GRS80) ellipsoid as a reference surface and the Bureau International de l'Heure Terrestrial System of 1984 (BTS-84) for orientation. NAD83

is defined to remain constant over time for points on the North American tectonic plate. *See also* geocentric datum, Geodetic Reference System of 1980, North American Datum of 1927.

North American Industry Classification System [STATISTICS, BUSINESS] Also known by the acronym *NAICS.* A system for classifying individual businesses by their types of economic activity. Developed collaboratively by Statistics Canada, Mexico's Instituto Nacional de Estadística, Geografía y Informatica (INEGI), and the U.S. Economic Classification Policy Committee (ECPC) to standardize the business statistical data produced by the three countries. NAICS is used by federal and local government agencies, trade associations, private businesses, and other organizations in North America to identify industry-related statistical data. NAICS replaced the Standard Industrial Classification (SIC) system in 1997. *See also* Standard Industrial Classification codes.

North American Vertical Datum of 1988 *See* NAVD 88.

North American-Pacific Geopotential Datum of 2022 *See* NAPGD2022.

north arrow [SYMBOLOGY] A map symbol that shows the direction of north on the map, showing how the map is oriented. *See also* map element, orientation, symbol.

north celestial pole [ASTRONOMY] The location in the sky directly above the true northern polar axis. *See also* north pole.

north pole [NAVIGATION] The point in the northern hemisphere at which the earth's axis of rotation meets the surface of the earth. It defines latitude 90 degrees north, as well as the direction of true north. *See also* north celestial pole, south pole.

northing [COORDINATE SYSTEMS]

1. The distance north of the origin that a point in a Cartesian coordinate system lies, measured in that system's units.

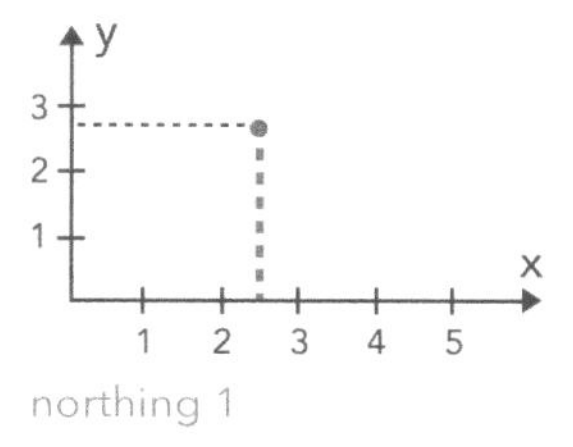

northing 1

2. The y-coordinate distance measured north from the origin in a grid coordinate system.

See also easting.

NOTAM *See* Notice to Airmen.

Notice to Airmen [NAVIGATION] Also known by the abbreviation *NOTAM.* An advisory bulletin containing information about the National Airspace System. It typically contains time-sensitive information between publishing cycles or corrections to published documents and charts. *See also* navigation, Notice to Mariners.

Notice to Mariners [NAVIGATION] Also known by the acronym *NTM*. A periodic update to existing nautical charts, issued by maritime authorities. *See also* navigation, Notice to Airmen.

NSDI *See* National Spatial Data Infrastructure.

NSDI Clearinghouse Network [DATA SHARING] A community of digital spatial data providers that maintains NSDI Clearinghouse Nodes as part of the U.S. National Spatial Data Infrastructure. *See also* Federal Geographic Data Committee, National Spatial Data Infrastructure.

NSDI Clearinghouse Node [DATA SHARING] An internet server that hosts a collection of metadata and data maintained and stored on a server by a data provider. An NSDI Clearinghouse Node provides information about geographic data within the data provider's areas of responsibility. Nodes must host FGDC-compliant metadata and data and use a common access protocol per the U.S. National Spatial Data Infrastructure. *See also* clearinghouse, National Spatial Data Infrastructure, NSDI Clearinghouse Network.

N

NSRS *See* National Spatial Reference System.

NTM *See* Notice to Mariners.

nugget [SPATIAL STATISTICS (USE FOR GEOSTATISTICS)] A parameter of a covariance or semivariogram model that represents independent error, measurement error, or microscale variation at spatial scales that are too fine to detect. The nugget effect is seen as a discontinuity at the origin of either the covariance or semivariogram model. *See also* covariance, error, random noise, semivariogram.

null hypothesis [STATISTICS] A statement that essentially outlines an expected outcome when there is no pattern, no relationship, or no systematic cause or process at work; any observed differences are the result of random chance alone. The null hypothesis for a spatial pattern is typically that the features are randomly distributed across the study area. Significance tests help determine whether the null hypothesis should be accepted or rejected. *See also* confidence level, p-value, significance level.

null value [SPATIAL STATISTICS (USE FOR GEOSTATISTICS)] The absence of a recorded value for a field. A null value differs from a value of zero in that zero may represent the measure of an attribute, whereas a null value indicates that no measurement has been taken. *See also* NoData.

numerical error [DATA QUALITY] A type of error introduced when processing numeric data. This can occur when numbers are rounded or when faulty computer processing compromises the data. *See also* error.

numerically lossless compression *See* lossless compression.

object [DATA MODELS] In GIS, a digital representation of a spatial or nonspatial entity. Objects usually belong to a class of objects with common attribute values and behaviors. *See also* attribute, object class.

object class

1. [ESRI SOFTWARE] In a geodatabase, a collection of nonspatial data of the same type or class. While spatial objects (features) are stored in feature classes in a geodatabase, nonspatial objects are stored in object classes.
2. [DATA MODELS, ESRI SOFTWARE] A table in a geodatabase used to store a collection of objects with similar attributes and behavior. Objects with no locational information are stored as rows or records in object classes. Spatial objects, or features, are stored as rows in feature classes, which are a specialized type of object class in which objects have an extra attribute to define their geographic location.

See also feature class, geodatabase.

object model diagram [PROGRAMMING] A graphical representation of the types in a library and their relationships. *See also* object.

object view [COGNITION] A philosophical view of geographic space in which space is seen as empty except when occupied by objects. In this view, every spatial location is either something (an object) or nothing. *See also* field view.

ObjectID [ESRI SOFTWARE] A system-managed value that uniquely identifies a record or feature. More commonly known as an item ID. *See also* item, object.

oblate ellipsoid

1. [COORDINATE SYSTEMS] A figure used to approximate the true shape and size of the earth for mapping purposes. The shape of the earth approximates an oblate ellipsoid with a flattening ratio of 1 to 298.257. Often shortened to ellipsoid.

oblate ellipsoid 1

2. [MATHEMATICS] A slightly flattened sphere of precisely known dimensions obtained by rotating an ellipse 180 degrees about its polar axis.

See also ellipsoid, prolate ellipsoid, spheroid.

oblate spheroid *See* oblate ellipsoid.

O

oblique image [REMOTE SENSING] In imagery, an image collected at an off-nadir, or nonvertical, angle that is typically greater than 3 degrees and less than 70 degrees.

oblique photograph [AERIAL PHOTOGRAPHY] An aerial photograph taken with the axis of the camera held at an angle between the horizontal plane of the ground and the vertical plane perpendicular to the ground. A low oblique image shows only the surface of the earth; a high oblique image includes the horizon. *See also* aerial photograph.

oblique photograph

oblique projection [MAP PROJECTIONS]

1. A planar or cylindrical projection whose point of tangency is neither on the equator nor at a pole.

oblique projection 1

2. A conic projection whose axis does not line up with the polar axis of the globe.
3. A cylindrical projection whose lines of tangency or secancy follow neither the equator nor a meridian.

See also conic projection, cylindrical projection, line of tangency, planar projection, point of tangency.

oblique-perspective map [MAP DESIGN] A map that gives an oblique, aerial perspective of the landscape. These maps have a more three-dimensional look than planimetric maps. *See also* oblique-perspective projection, planimetric map.

oblique-perspective projection [MAP PROJECTIONS] A map projection with the center point of projection at any point other than the point or lines of tangency of the projection surface. *See also* line of tangency, point of tangency.

observer [ESRI SOFTWARE] The position of the camera in a scene or globe. *See also* camera.

observer offset [ESRI SOFTWARE] The height of the observer point above a surface used in analysis when calculating lines of sight and viewsheds. *See also* camera, line of sight, observer, viewshed.

occlusion [REMOTE SENSING] In imagery, the occurrence when an object blocks another object from view within an image, such as a tree overhanging a roof, or building lean obscuring the street below.

OD cost matrix *See* origin-destination cost matrix.

off-nadir

1. [REMOTE SENSING] In satellite imagery, the number of degrees that the sensor varies from a vertical view when pointing directly downward from the platform. Also called angle of obliquity or elevation angle.

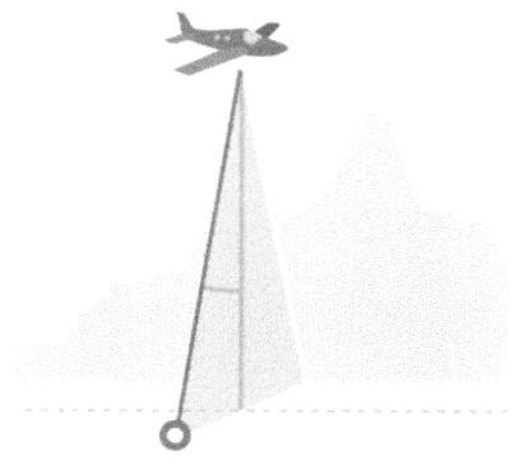

off-nadir 1

2. [PHOTOGRAMMETRY] In imagery, the angle between the vertical from the sensor's perspective center to a point directly below the optics—or nadir—and another point on the ground.

See also nadir, zenith.

off-nadir view [REMOTE SENSING] In imagery, the view when a sensor is not pointed down vertically but rather from an oblique angle. *See also* off-nadir.

offset

1. [CARTOGRAPHY] In cartography, the displacement or movement of features so that they do not overlap when displayed at a given scale. For example, a road can be offset from a river if the symbols are wide enough that they overlap.
2. [SYMBOLOGY] In symbology, the shift of the origin or insertion point of a symbol in an x and/or y direction.
3. [ESRI SOFTWARE] A change in or the act of changing the z-value for a surface or features in a scene by a constant amount or by using an expression. Offsets can be applied to make features draw just above a surface.

See also symbology, z-value.

offset distance *See* label offset.

OGC *See* Open Geospatial Consortium.

OGIS *See* Open Geodata Interoperability Specification.

OID *See* ObjectID.

on the fly [COMPUTING] Assembled, created, presented, or calculated dynamically during a transaction—such as a web page search or data display query. *See also* live feed.

one-to-many relationship [DATABASE STRUCTURES] An association between two linked or joined tables in which one record in the first table corresponds to many records in the second table. *See also* many-to-one relationship, one-to-one relationship, many-to-many relationship, joining.

one-to-one relationship [DATABASE STRUCTURES] An association between two linked or joined tables in which one record in the first table corresponds to only one record in the second table. *See also* joining, many-to-many relationship, many-to-one relationship, one-to-many relationship.

ontology [SOCIAL CONTEXT OF GIS] In computer science, a data model that represents a domain by detailing the entities that comprise it and the semantic relationships between them. Ontologies generally include individuals, classes, attributes, and relations. *See also* data model, loose coupling, tight coupling.

opacity *See* transparency.

Open Geodata Interoperability Specification [INTEROPERABILITY] Also known by the acronym *OGIS*. A specification, developed by the Open Geospatial Consortium, to support interoperability of GIS in a heterogeneous computing environment. *See also* interoperability, Open Geospatial Consortium.

Open Geospatial Consortium [STANDARDS] Also known by the acronym *OGC*. An international consortium of companies, government agencies, and universities participating in a consensus process to develop publicly available geospatial and location-based services. Interfaces and protocols defined by OpenGIS specifications support interoperability and seek to integrate geospatial technologies with wireless and location-based services. *See also* Digital Geographic Information Exchange Standard, Defence Geospatial Information Working Group, Open Geodata Interoperability Specification.

OpenGIS Consortium *See* Open Geospatial Consortium.

OpenLS [DATA SHARING] An abbreviation of *OpenGIS Location Services*. A protocol, designed to work across many different wireless networks and devices, that allows seamless access to multiple content repositories and service frameworks.

operand [MATHEMATICS] A data value or the symbolic representation of a data value in an expression. Operands may be numbers, character strings, functions, variables, parenthetical expressions in the body of a larger expression, and so on. Symbolic representations of operands, such as variables and functions, are evaluated before they are operated upon by the operators in the expression. For example, in the expression "1 + 2", the operands are 1 and 2, and are operated upon by the + (plus) operator, which adds the operands together and returns the value 3. *See also* operator, operator precedence.

operation codes [SURVEYING] In surveying, an alphanumeric or numeric value included in an instrument vendor's data collector file format. Operation codes are used to describe such elements as new instrument setups and numeric values for horizontal angles, zenith angles, slope distance measurements, height of instrument, and height of target. *See also* surveying.

operator [MATHEMATICS] The symbolic representation of a process or operation performed against one or more operands in an expression, such as "+" (plus, or addition) and ">" (greater than). When evaluated, operators return a value as their result. If multiple operators appear in an expression, they are evaluated in order of their operator precedence. *See also* operator precedence, operand.

operator precedence [MATHEMATICS] The order in which operators are evaluated in an expression; operators with a higher precedence are evaluated before those with a lower precedence. If all operators in an expression have the same precedence, they are evaluated in the order in which they appear, from left to right. Parentheses may be used to override operator precedence; portions of an expression within parenthesis are evaluated first, and parenthetical expressions may be nested. *See also* operand, operator.

optical center *See* visual center.

optimization [GIS TECHNOLOGY] The process of fine-tuning data, software, or processes to increase efficiency, improve performance, and produce the best possible results. *See also* genetic algorithm, gradient.

optimized route *See* traveling salesperson problem.

orbital inclination [ASTRONOMY] The angle between a reference plane, such as the equatorial plane, and the axis of direction of the orbiting object. *See also* axis, equatorial plane.

orbital period [ASTRONOMY] The time it takes an orbiting object—such as a satellite, a moon, or the earth—to make one complete orbit around another object—such as the earth or the sun.

ordering [NETWORK ANALYSIS] The process of assigning a numeric connectivity value to the links in a network. *See also* connectivity, link, network.

orders of hierarchy [NETWORK ANALYSIS] The levels of links that are used to describe the hierarchical nature of a network. *See also* hierarchy, link, network.

ordinal data [DATA STRUCTURES] A method of organizing data values into a ranked hierarchy with no fixed interval between the hierarchical levels. A common example of ordinal data is a Likert scale, which assigns a numeric value on an ordinal scale to questionnaire responses such as "Agree," "Somewhat agree," "Neutral," "Somewhat disagree," and "Disagree." A group of polygons colored lighter to darker to represent less to more densely populated areas is an example of ordinal data in GIS. *See also* nominal data.

ordinary kriging [SPATIAL STATISTICS (USE FOR GEOSTATISTICS)] A kriging method in which the weights of the values sum to unity. It uses an average of a subset of neighboring points to produce a particular interpolation point. *See also* kriging.

ordinate [EUCLIDEAN GEOMETRY] In a rectangular coordinate system, the distance of the y-coordinate along a vertical axis from the horizontal or x-axis. For example, a point with the coordinates (7,3) has an ordinate of 3. *See also* abscissa.

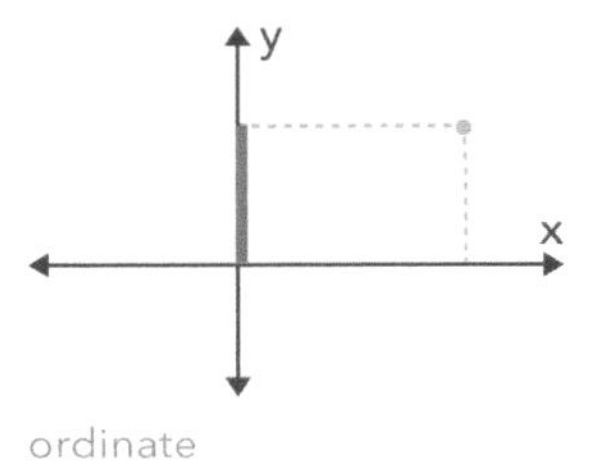

ordinate

ordination method [DATA ANALYSIS] A type of exploratory data analysis complementary to data clustering in multivariate analysis. Ordination orders objects that are characterized by values on multiple variables (multivariate objects) so that similar objects are near each other, and dissimilar objects are farther from each other. Relationships between the objects, on each of several axes (one for each variable), are then characterized numerically or graphically. Examples of ordination methods include principal components analysis (PCA), correspondence analysis (CA), and so on. *See also* principal component analysis.

Ordnance Survey [CARTOGRAPHY] The national mapping and cartographic agency of the United Kingdom. Now a civilian organization, the Ordnance Survey is one of the world's largest producers of maps and was the first national mapping organization in the world to complete a large-scale program of digital mapping. *See also* National Geospatial-Intelligence Agency.

Ordnance Survey National Grid [COORDINATE SYSTEMS] The grid coordinate system used in England, Scotland, and Wales. *See also* Ordnance Survey.

Oregon Lambert system [COORDINATE SYSTEMS] A single grid coordinate system that replaces Oregon, USA's two state plane coordinate zones with the same central meridian but a different parallel of origin. *See also* central meridian, grid coordinate system, state grid, state plane coordinate system, UTM, WTM.

organization [ESRI SOFTWARE] A collection of ArcGIS users who belong to the same school, business, or other association. Organizations and groups are overseen by administrators, who manage organizational functions such as sharing and storage. *See also* ArcGIS account, group, members.

orientation

1. [GEOGRAPHY] An object's position or relationship in direction with reference to points of the compass.
2. [NAVIGATION] A technique for orienting a map. Also known as a range estimation.
3. [ESRI SOFTWARE] The method by which horizontal angle readings for TPS measurements are converted into azimuths.

See also azimuth, bearing, position, TPS measurement.

orienteering [NAVIGATION] A sport that requires the use of a map and compass to navigate from point to point. *See also* orienteering compass.

orienteering compass [NAVIGATION] A compass consisting of a compass card, north arrow, and rotating magnetic needle on which the housing is mounted on a transparent rectangular base that facilitates use with a map. *See also* orienteering.

origin

1. [COORDINATE SYSTEMS] A fixed reference point in a coordinate system from which all other points are calculated, usually represented by the coordinates (0,0) in a planar coordinate system and (0,0,0) in a three-dimensional system. The center of a projection is not always its origin.

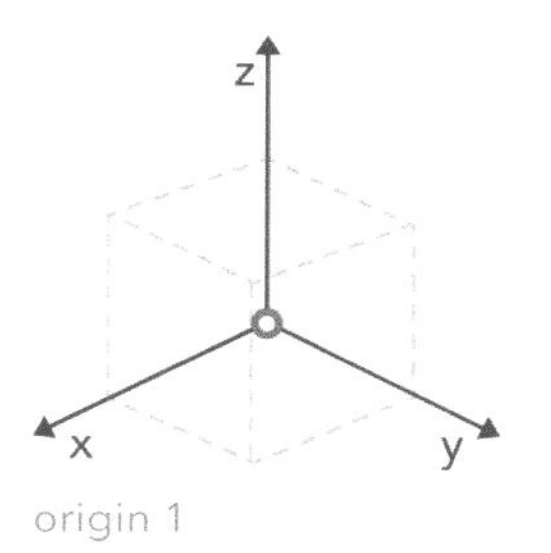

origin 1

2. [ESRI SOFTWARE] The primary object in a relationship, such as a feature class containing measurement points.
3. [NETWORK ANALYSIS] A network location used in origin-destination cost matrix analysis that specifies a starting location.

See also destination.

origin-destination cost matrix [ESRI SOFTWARE] A type of network analysis that computes a table containing the total impedance from each origin to each destination. Additionally, it ranks the destinations that each origin connects to in ascending order of the time it takes to travel from that origin to each destination. *See also* destination, origin.

		destinations			
		A	B	C	D
origins	A	0	5	7	8
	B	5	0	16	11
	C	7	16	0	20

origin-destination cost matrix

ortho mosaic [DIGITAL IMAGE PROCESSING] A photogrammetrically orthorectified image product mosaicked from an image collection where the geometric distortion has been corrected and the imagery has been color balanced to produce a seamless mosaic dataset. *See also* orthorectified image.

orthocorrection *See* orthorectification.

orthogonal

1. [EUCLIDEAN GEOMETRY] Intersecting at right angles.

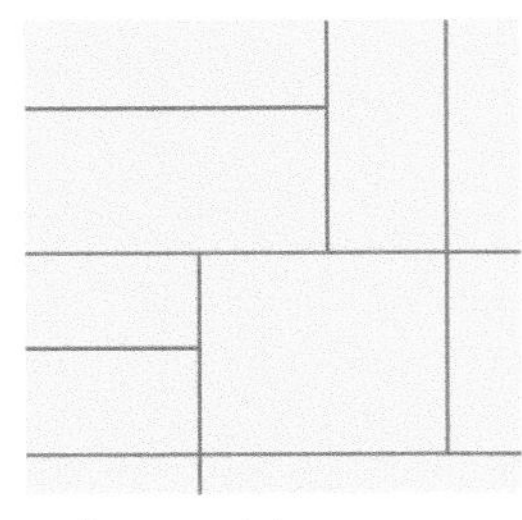

orthogonal 1

2. [COORDINATE SYSTEMS] A coordinate system with axes at 90-degree angles, such as a Cartesian coordinate system.
3. [COORDINATE SYSTEMS] Data that has uncorrelated parameters or a statistically independent coordinate reference system.

See also intersection, orthogonal offset, rectilinear.

orthogonal offset [EUCLIDEAN GEOMETRY] A line that is perpendicular to another line at its point of tangency, often used to measure distance from a line to a separate point that does not lie along the original line. *See also* distance, orthogonal, unit of measure.

orthogonal offset

orthographic projection [MAP PROJECTIONS] A planar projection based on a light source at an infinite distance from the generating globe so that all rays are parallel; how the earth would appear if viewed from a distant planet. *See also* generating globe, orthographic view, planar projection, projection.

orthographic projection

orthographic view [3D ANALYSIS] In 3D analysis, a perspective that allows viewing of data in a scene as a two-dimensional plane seen from above. There is no perspective foreshortening in an orthographic view, so scale is constant across the entire display. *See also* orthographic projection.

orthoimage *See* orthorectified image.

orthometric height [REMOTE SENSING] The vertical distance from a point of interest to a reference surface (or geoid), which is a geopotential surface corresponding to mean sea level. *See also* geoid.

orthomorphic projection *See* conformal projection.

orthophanic projection [MAP PROJECTIONS] A map projection that appears to be a natural and correct representation of the earth.

orthophoto [AERIAL PHOTOGRAPHY] A digitized aerial photograph that has been geometrically corrected to remove distortions—such as from camera orientation or elevation differences. An orthophoto has the same scale throughout and can be used as a map. *See also* aerial photograph, geometric correction, georectification.

orthophoto

orthophotomap [MAP DESIGN] A conventional map, such as a topographic map, that uses an orthophoto as the base. *See also* map, orthophoto, topographic map.

orthophotoquad [AERIAL PHOTOGRAPHY] An orthophotograph that has been formatted as a USGS 1:24,000 topographic quadrangle with little or no cartographic enhancement. *See also* digital orthophoto quadrangle, digital orthophoto quarter quadrangle, orthophoto.

orthophotoscope [AERIAL PHOTOGRAPHY] A photomechanical or optical-electronic device that creates an orthophotograph by removing geometric and relief distortion from an aerial photograph. *See also* distortion, orthophoto.

orthorectification [SATELLITE IMAGING] The process of geometrically correcting an image to remove relief distortions, sensor artifacts, earth curvature, and other perspective distortions and align the image with coordinates on the ground, restoring geometric integrity. Ground control points, tie points, and elevation data are used to correct perspective and terrain distortion in aerial, drone, and satellite imagery. *See also* geometric correction, rectification.

O

orthorectified image [REMOTE SENSING] An image that has been aligned to a coordinate reference system and has been corrected for relief displacement, sensor artifacts, earth curvature, and other perspective distortions. The resulting image has the geometric integrity of a map. Also known as an orthoimage. *See also* orthophoto.

OSGB36 [COORDINATE SYSTEMS] Acronym for *Ordnance Survey Great Britain 1936*. A datum developed as the national coordinate system for topographic mapping in Great Britain. Latitude and longitude coordinates in this datum are based on the Airy 1830 ellipsoid. *See also* Airy 1830 ellipsoid, Ordnance Survey.

outlier

1. [STATISTICS] An unusual or extreme data value in a dataset. In data analysis, outliers can potentially have a strong effect on results and must be analyzed carefully to determine whether they represent valid data.
2. [GEOLOGY] In geology, a feature that lies apart from the main body or mass to which it belongs: for example, a rock or stratum that has been separated from a formation by erosion.

See also spike.

outline [DATA MODELS] The path that follows the boundary of an object. Outlines are also called strokes. *See also* boundary.

O

outline vectorization [DATA CONVERSION] A vectorization method that generates lines along the borders of connected cells. It is typically used for vectorizing scanned land-use and vegetation maps. *See also* centerline vectorization.

outline vectorization

overlap [AERIAL PHOTOGRAPHY] The portion of the ground duplicated on successive aerial photographs along a flight line. *See also* aerial photograph, flight line.

overlapping rings [SPATIAL ANALYSIS] A method of defining the rings in an analysis so that the values inside the rings are cumulative. For example, if you had an analysis with three concentric rings and 10 households in each, the total number of households for ring 1 would be 10, the total for ring 2 would be 20 (ring 1 + ring 2), and the total for ring 3 would be 30 (ring 1 + ring 2 + ring 3). *See also* ring.

overlay [ANALYSIS/GEOPROCESSING]

1. A spatial operation in which two or more maps or layers registered to a common coordinate system are superimposed for the purpose of showing and analyzing the relationships between features that occupy the same geographic space.
2. In geoprocessing, the geometric intersection of multiple datasets to combine, erase, modify, update, or analyze features in a new output dataset.

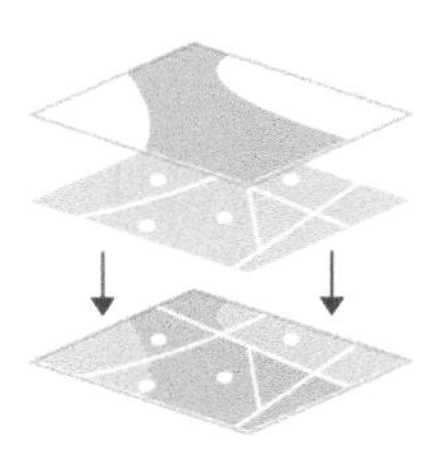

overlay 2

See also spatial analysis.

overlay events *See* event overlay.

overprinting [CARTOGRAPHY] In cartography, portraying cartographic updates on a map by printing new or modified information over the original cartography, usually in a distinctive color.

override [ESRI SOFTWARE]

1. In network datasets, a type of junction connectivity policy in which the way junctions connect to other junctions is not based on the existing edge connectivity policy; junctions "override" the edge connectivity policy.
2. An exception made to a property of a feature representation's representation rule so that the feature is drawn differently than others sharing the same rule.

See also junction connectivity policy.

overshoot [DATA STRUCTURES] The portion of an arc digitized past its intersection with another arc. *See also* dangling arc.

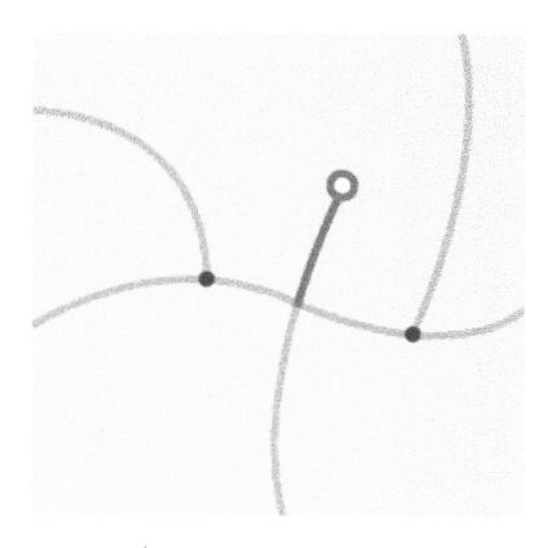

overshoot

overview [DATA MANAGEMENT] A lower-resolution image created to increase display speed and reduce CPU usage when viewing a mosaicked image from a mosaic dataset. *See also* mosaic dataset.

overview map [CARTOGRAPHY] A generalized, smaller-scale map that shows the limits of another map's extent along with its surrounding area. *See also* inset map.

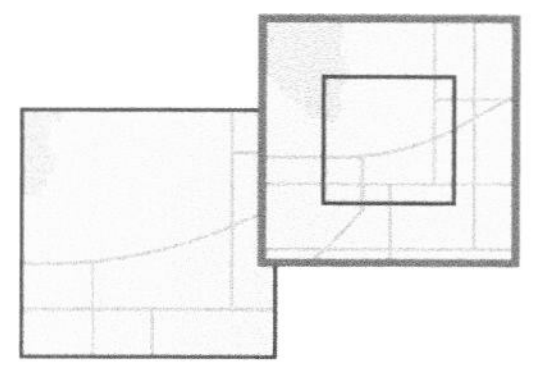

overview map

overview terrain [3D GIS] The coarsest representation of the terrain dataset, intended for fast drawing at small scales. *See also* scale, terrain, terrain dataset.

owner [ESRI SOFTWARE] In ArcGIS organizations, the person or organization responsible for creating and maintaining data or a map. The owner determines settings such as content permissions. *See also* ArcGIS account, group, organization.

pan [MAP DISPLAY] To shift a map image relative to the display window without changing the viewing scale. *See also* zoom.

pan sharpening [DIGITAL IMAGE PROCESSING] An abbreviation of *panchromatic sharpening*. A method of radiometric or image color transformation in which a higher-resolution, panchromatic image is fused with a lower-resolution, multispectral image dataset; used to increase spatial resolution and improve the ability to resolve finer details in a multispectral image. *See also* multispectral image, resolution.

panchromatic [REMOTE SENSING] Sensitive to light of all wavelengths in the visible spectrum. *See also* color, multispectral, wavelength.

panchromatic image [REMOTE SENSING] An image that measures a single, broad band of wavelength that spans across a wide range of the electromagnetic spectrum; for instance, it may span the entire visible spectrum, or a portion of the visible and the near infrared. Because it is a single, wide band, what is measured is target brightness; objects shown appear in shades of gray. *See also* electromagnetic spectrum, panchromatic.

panchromatic sharpening *See* pan sharpening.

paneled map [MAP DESIGN] A map spliced together from smaller maps of neighboring areas.

paneled map

parallax [PHOTOGRAMMETRY] The apparent displacement of an object's position relative to the point of observation; the basis for the perception of depth in stereo imagery. *See also* remote sensing, stereometer.

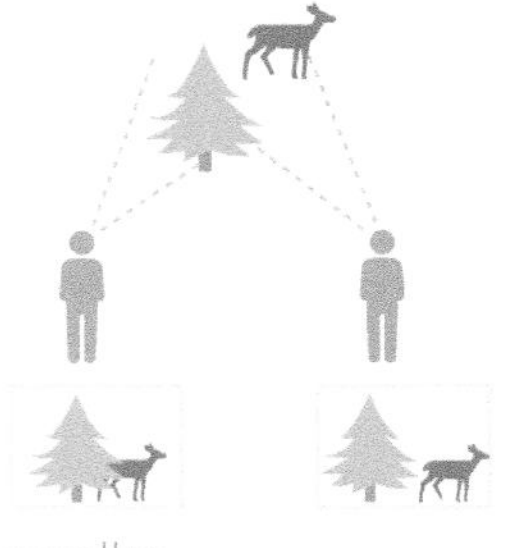

parallax

parallax bar *See* stereometer.

parallel [GEODESY] An imaginary east–west line encircling the earth, parallel to the equator and connecting all points of equal latitude that can be used to define locations north or south of the equator. Also, the representation of this line on a globe or map. *See also* latitude.

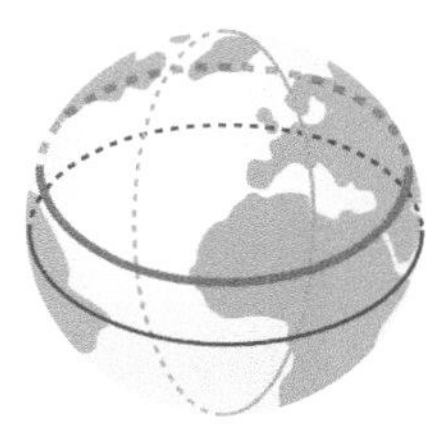

parallel

parallelepiped classifier [PHOTOGRAMMETRY] An algorithm that is used to label unknown pixels in an image based on statistical minimum and maximum values as a surrogate for variance. *See also* variance.

parameter

1. [MAP PROJECTIONS] One of the variables that define a specific instance of a map projection or a coordinate system. Parameters differ for each projection and can include central meridian, standard parallel, scale factor, or latitude of origin.
2. [MATHEMATICS] A variable that determines the outcome of a function or operation.

See also coordinate system, mathematical function, projection.

parametric curve [MATHEMATICS] A curve that is defined mathematically rather than by a series of connected vertices. A parametric curve has only two vertices, one at each end. *See also* Bézier curve.

parametric measure [STATISTICS] A statistical calculation in which certain assumptions are made about the parameters of the population from which a sample is taken. Most commonly, the assumption that data is normally distributed.

parcel

1. [CADASTRAL AND LAND RECORDS] A piece or unit of land, defined by a series of measured straight or curved lines that connect to form a polygon. There are some implications of landownership. Commonly also called a tract.

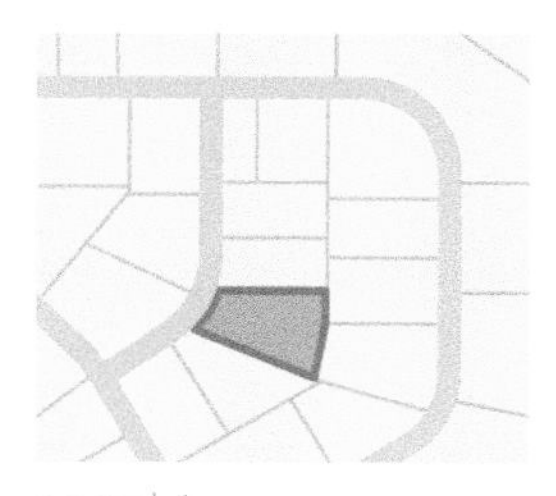

parcel 1

2. [SURVEYING, ESRI SOFTWARE] A group of features representing an area or volume of land or water. In the parcel fabric, a two-dimensional parcel is composed of a polygon feature and line and point features. The parcel polygon represents the land area, the parcel lines are COGO-enabled and store the boundary measurements, and the parcel points represent the

P

parcel corners. A parcel can be used to represent rights, restrictions, or responsibilities. Parcels can also be used to model administrative boundaries or cadastral frameworks. *See also* lot, plan, PLSS, surveying.

parcel fabric [CADASTRAL AND LAND RECORDS] A digital collection of parcel polygons and attributes spanning a municipal domain that form the framework for managing and editing administrative boundaries. *See also* boundary, connection line, parcel.

parcel fabric record [CADASTRAL AND LAND RECORDS] A polygon feature representing the footprint of a land parcel transaction. Parcel fabric record polygons match the cumulative shape of all the parcels created or retired by the transaction. Parcel features (points, lines, polygons) are associated to the parcel fabric record that created them. Historic parcel features are also associated to the record that retired them. Examples of parcel fabric records are plans, plats, deeds, records of survey, or similar. *See also* feature, historic parcel, polygon feature.

parcel identification number *See* parcel PIN.

parcel map [CADASTRAL AND LAND RECORDS] A geographic record of landownership at the local level. Used as a reference for surveying, tax assessment, and other municipal concerns. *See also* boundary survey.

parcel PIN [CADASTRAL AND LAND RECORDS] PIN is an acronym for *parcel identification number*. In parcel fabric records, a unique identifier for a parcel. The format of an identifier is defined by the government's organization and may contain a combination of alphanumeric values. *See also* identifier, parcel fabric record.

parcel polygon [CADASTRAL AND LAND RECORDS, ESRI SOFTWARE] A closed shape that stores COGO dimensions obtained from the recorded, legal document. *See also* connection line, parcel.

parcel type [CADASTRAL AND LAND RECORDS] The way parcels are categorized in the parcel fabric. For example, lot, tax, ownership, subdivision, and so on, are all parcel types. Parcel types are composed of a polygon and a line feature class; there can be one or many parcel types in a parcel fabric. *See also* classification, parcel, parcel fabric, surveying.

parity [MATHEMATICS] The even or odd property of an integer. In address matching, parity is used to locate a geocoded address on the correct side of the street (such as odd numbers on the left-hand side, even numbers on the right). *See also* address matching.

parse [EDUCATION] A method of syntax analysis. For example, to analyze symbols, data structures, or parts of speech. For example, to break a sentence into a sentence diagram, identifying the grammatical components. *See also* schema.

partial address support [GEOCODING] The ability to return a list of geocoding candidates based on incomplete address information. For example, if a city name but no country is entered in a partial address support search, the result list contains cities whose names match the name entered. *See also* address, candidate, geocoding.

partial sill [SPATIAL STATISTICS (USE FOR GEOSTATISTICS)] A parameter of a covariance or semivariogram model that represents the variance of a spatially autocorrelated process without any nugget effect. In the semivariogram model, the partial sill is the difference between the nugget and the sill. *See also* sill, nugget.

partial-hachuring system [MAP DESIGN] A hachuring method on north-oriented maps in which hachures are eliminated on the northwest sides of hills to improve the three-dimensional impression of relief and make terrain features easier to identify. Developed in the mid-19th century by Swiss cartographer Guillaume Henri Dufour (1787–1875), this method largely replaced the Lehmann system. *See also* hachured contour, Lehmann system.

passive remote sensing [REMOTE SENSING] A remote-sensing system, such as an aerial photography imaging system, that detects only energy naturally reflected or emitted by an object. *See also* active remote sensing, remote sensing.

passive sensors [REMOTE SENSING] Imaging sensors that can only receive radiation, not transmit it. *See also* passive remote sensing, sensor.

patch [3D GIS] A single triangular face inside a multipatch geometry. In most cases, many patches (faces) are used together to create a complex 3D model. Examples include geometric shapes, such as spheres, cubes, and tubes; geographic features, such as buildings, cars, and light poles; and other boundary representations, such as isosurfaces, used to represent geologic structures or environmental plumes. Patches in a multipatch geometry may or may not include an image (texture) displayed on them. *See also* 3D model, face, multipatch.

path

1. [NETWORK ANALYSIS] The connecting lines, arcs, or edges that join an origin to a destination.
2. [DATA MODELS, ESRI SOFTWARE] A geometric element from which polylines and polygons are constructed. A path is a sequence of connected, nonintersecting segments, with no two segments having the same start point or the same endpoint.

See also segment.

path distance analysis [SPATIAL ANALYSIS, ESRI SOFTWARE] A description of each cell's least accumulative cost relationship to a source or a set of sources, accounting for surface distance, horizontal cost factors, and vertical cost factors. *See also* cost, path, spatial analysis.

pathfinding [NETWORK ANALYSIS] The process of calculating the optimal path between an origin and a destination point or points in a network. *See also* destination, network, origin.

PATRF2022 [MODELING] Acronym for *Pacific Terrestrial Reference Frame of 2022.* A geometric reference frame of the National Spatial Reference System (NSRS) used to define the geodetic latitude, geodetic longitude, and ellipsoidal height of points on the Pacific tectonic plate. The coordinates defined in PATRF2022 are time-dependent. *See also* NAPGD2022, National Spatial Reference System, reference frame.

P

pattern [DIGITAL IMAGE PROCESSING] An image element that provides information about the spatial and spectral arrangement or configuration of objects; used to identify features of interest, land use, or land cover. *See also* image element.

pattern arrangement [GRAPHICS (MAP DISPLAY)] The order, usually random or uniform, of a set of shapes or lines that create a pattern within a graphic mark.

pattern orientation [GRAPHICS (MAP DISPLAY)] The orientation of a pattern arrangement. *See also* pattern arrangement.

pattern recognition [DIGITAL IMAGE PROCESSING] In image processing, the computer-based identification, analysis, and classification of objects, features, or other meaningful regularities within an image. *See also* digital image processing.

P-code [GPS] The PRN code used by United States and allied military GPS receivers. *See also* PRN code, civilian code.

PDF [DATA STRUCTURES] Acronym for *Portable Document Format.* A proprietary file format that creates text-based, formatted files for distribution to a variety of operating systems. *See also* CSV.

PDOP *See* dilution of precision.

peak

1. [GEOGRAPHY] The highest point of a mountain or hill.

peak 1

2. [MODELING] In modeling, a point on a surface around which all slopes are negative.

See also pit.

percent slope [EUCLIDEAN GEOMETRY] The slope ratio expressed as a percentage (multiplied by 100). A 45-degree slope and a 100-percent slope are the same. Also known as percent rise and slope percentage. *See also* slope.

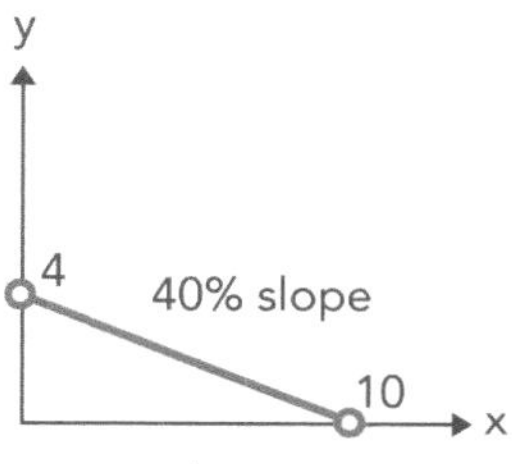

percent slope

perigee [ASTRONOMY] In an orbit path, the point at which the object in orbit is closest to the center of the body being orbited. *See also* apogee.

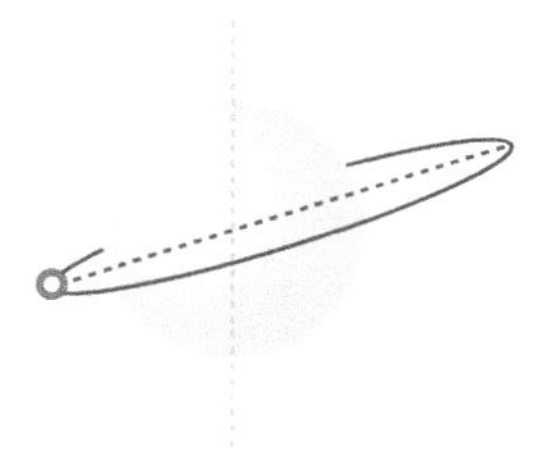
perigee

perimeter [CARTOGRAPHY] The continuous line that forms the external boundary of a feature. *See also* boundary.

perspective geometry [PHOTOGRAMMETRY] A method of re-creating or rendering forms to produce realistic two-dimensional images of three-dimensional objects placed at different distances from the viewer using mathematical concepts, such as lines of focus, a linear horizon, and vanishing point algorithms. *See also* horizon, vanishing point.

perspective view [3D GIS] A projection mode in 3D applications that allows the viewer to control map navigation of the scene or globe from a specified location. *See also* view, projection.

perspective-view map [3D GIS] A map created by draping imagery or hillshading over an underlying fishnet map in oblique view. *See also* fishnet map, hillshading, perspective view.

persuasion map [MAP DESIGN] A map used to promote a specific sociopolitical perspective or ideology. *See also* propaganda map.

phenology [BIOLOGY] The study of periodic events in biological life cycles, specifically how season, climate, and habitat affect plants and animals.

phenomena *See* geographic phenomena.

photogeology [AERIAL PHOTOGRAPHY] The science of interpreting and mapping geologic features from aerial photographs or remote-sensing data. *See also* aerial photograph, photogrammetry, remote sensing.

photogrammetrist [PHOTOGRAMMETRY] A person trained to compile reliable data from aerial photographs. *See also* photogrammetry.

photogrammetry [PHOTOGRAMMETRY] The science of making reliable measurements of physical objects and the environment by measuring and plotting electromagnetic radiation data from aerial photographs and remote sensing systems against land features identified in ground control surveys. Generally done to produce planimetric, topographic, and contour maps. *See also* aerial photograph, electromagnetic radiation, planimetric map.

photomap [AERIAL PHOTOGRAPHY] An aerial photograph or photographs, referenced to a ground control system and overprinted with map symbology. *See also* aerial photograph, symbology.

photometer [PHYSICS] An instrument that records the intensity of light by converting incident radiation into an electrical signal and measuring it. *See also* spectrophotometer.

P

photon [PHYSICS] A quantum (smallest measurement) of light energy that behaves as both a wave and a particle, has neither mass nor electrical charge, and travels at the speed of light; characterized in the electromagnetic spectrum by wavelength or frequency. *See also* electromagnetic spectrum.

physical distance [MATHEMATICS] Distance measured in standard units such as feet or meters. Also known as ground distance. *See also* unit of measure.

physical geography [GEOGRAPHY] The field of geography concerned with the natural features of the earth's surface. *See also* geography.

physical network [NETWORK ANALYSIS] One of the two parts of a network system; the actual feature classes that participate in a network system. *See also* network.

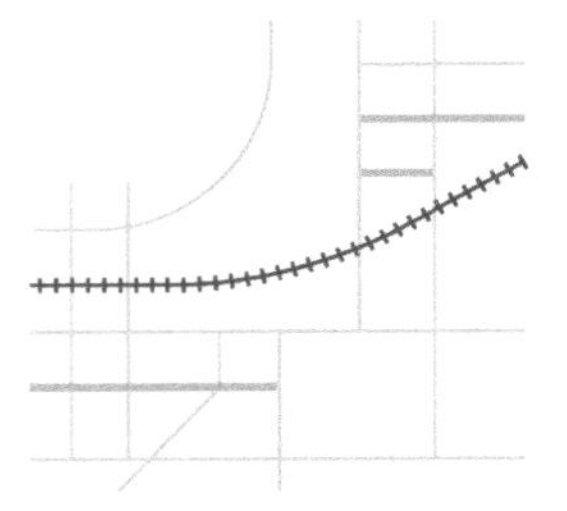

physical network

physical sensor model [REMOTE SENSING] A mathematical model based on the physical aspects and precise dimensions of a sensor and its environment that provides a ground-to-image transformation for points, has adjustable parameters, and allows for covariance terms that could propagate errors. Also called rigorous sensor model. *See also* sensor.

pictographic symbol [SYMBOLOGY] A symbol designed to look like a miniature version or caricature of the feature it represents. *See also* mimetic symbol.

picture fill [GRAPHICS (COMPUTING)] A type of fill pattern created by continuous tiling of either a .bmp (raster image) or an .emf (vector graphic) file. *See also* fill, tiling.

pie chart [STATISTICS] A chart shaped like a circle cut into wedges from a center point that represents percentage values as proportionally sized "slices." Pie charts are used to represent the relationship between parts and the whole. *See also* chart, visualization.

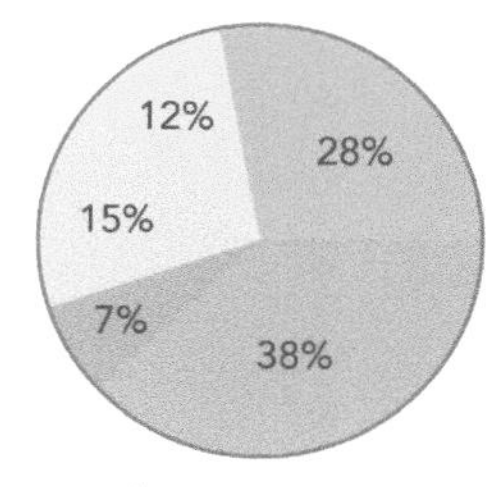

pie chart

pit

1. [GEOGRAPHY] A depression in the earth's surface.

pit 1

2. [MODELING] In modeling, a point on a surface around which all slopes are positive.

See also peak.

pixel

1. [DATA MODELS] An abbreviation of *picture element.* The smallest unit of information in a digital image or raster map, usually square or rectangular. Often used synonymously with cell.
2. [REMOTE SENSING] In remote sensing, the fundamental unit of data collection; represented as a cell in an array of data values.
3. [GRAPHICS (COMPUTING)] The smallest element of a display device, such as a video monitor, that can be independently assigned attributes, such as color and intensity.

See also cell, digital number, edge detection, histogram equalization.

pixel density [MAP DESIGN] The number of pixels per inch or per centimeter on a map or image. *See also* pixel.

pixel size [ESRI SOFTWARE] The dimensions on the ground of a single pixel in a raster, measured in map units. Pixel size is often used synonymously with cell size. *See also* measurement, pixel.

pixel space [GRAPHICS (COMPUTING)] The x,y coordinate space defined by the number of pixels in the display area, such as 1024 × 768 or 1600 × 1200. Each pixel provides a single unit of color on the screen at a given time. The more pixels in the display area, the smaller each pixel is. Since a pixel is a piece of information, a configuration with more pixels can fit more information into a given display area. *See also* pixel, image space.

pixel type *See* data type.

P

pixel value [PHOTOGRAMMETRY] A digital number (DN) stored in one cell (picture element) of an image or raster file that represents a property (such as reflectance value, elevation, temperature, and so on) for that pixel within the image. *See also* pixel.

place [GOVERNMENT] In census geography, any incorporated or unincorporated city, town, or community. *See also* Find Places, location, place-name alias.

place-name alias [GEOCODING] The formal or common name of a location, such as the name of a school, hospital, or other landmark. For example, "Memorial Hospital" is the place name for the address "893 Memorial Drive." In geocoding, the address locator can be set to accommodate the use of place-name aliases in place of their addresses for matching. *See also* identifier, location, place, secondary reference data.

plan [SURVEYING] In surveying, a high-level organization of parcels; a survey document containing data from a recorded subdivision survey plan or from a legal description. Often, many parcels are defined in one plan. Each parcel contains a reference to a plan. *See also* parcel.

plan structure [CADASTRAL AND LAND RECORDS] In survey analysis, the property of a subdivision plan that can be inferred from the adjacency of lines and line dimensions; for example, road frontage lines of adjacent parcels usually run tangent to each other, and this property of tangency can be enforced by the system. *See also* plan.

planar coordinate system [COORDINATE SYSTEMS] A two-dimensional measurement system that locates features on a plane based on their distance from an origin (0,0) along two perpendicular axes. *See also* coordinate system.

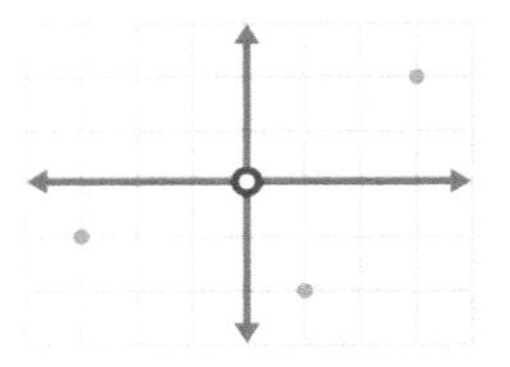

planar coordinate system

planar enforcement [DATA CONVERSION] A set of rules used to define a consistent method of building point, line, and polygon features from spaghetti-digitized data. For example, planar enforcement includes rules that polygons of differing soil types cannot overlap and that lines must be split at intersections. *See also* spaghetti data.

planar measurement [EUCLIDEAN GEOMETRY] The quantity associated with length or area within a two-dimensional plane. Planar measurements are limited to objects or shapes that can be represented in two dimensions; they do not account for depth or volume. When using geographic coordinates, planar measurements must cut through the center of the curved surface of the earth, which diminishes accuracy. *See also* geodetic measurement.

planar projection [MAP PROJECTIONS] A map projection made by projecting points from a spheroid or sphere onto a tangent or secant plane. Also called an azimuthal or zenithal projection. *See also* azimuthal projection.

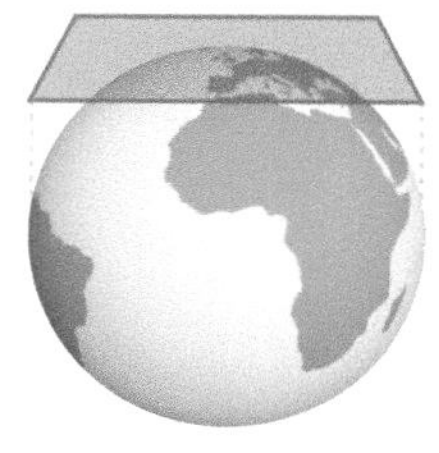

planar projection

planar projection family [MAP PROJECTIONS] A map projection family based on the use of a plane as the developable surface. *See also* developable surface, map projection families.

planarize [DATA EDITING] The process of creating multiple line features by splitting longer features at the places where they intersect other line features. This process is often applied to nontopological line work that has been spaghetti digitized or imported from a CAD drawing. *See also* spaghetti data.

plane survey [SURVEYING] A survey of a small area that does not take the curvature of the earth's surface into account. *See also* geodetic survey, surveying.

planform curvature [CARTOGRAPHY] The amount that a surface deviates from being linear perpendicular to the direction of the maximum slope. Often called plan curvature. *See also* curvature, slope.

planimeter [AERIAL PHOTOGRAPHY] A mechanical instrument used for area measurement on maps and aerial photographs.

planimetric [AERIAL PHOTOGRAPHY] Two-dimensional; showing no relief. *See also* remote sensing.

planimetric base [AERIAL PHOTOGRAPHY] A two-dimensional map that serves as a guide for contour mapping, usually prepared from aerial photographs. *See also* aerial photograph, contour.

planimetric base

planimetric coordinate method [SURVEYING] A method of area measurement based on finding the areas of trapezoids formed by drawing horizontal lines from the boundary points to the vertical y-axis. *See also* measurement, y-axis.

planimetric map [CARTOGRAPHY] A map that displays only the x,y locations of features, representing the horizontal positions and not the vertical (height or depth) dimensions. *See also* topographic map.

P

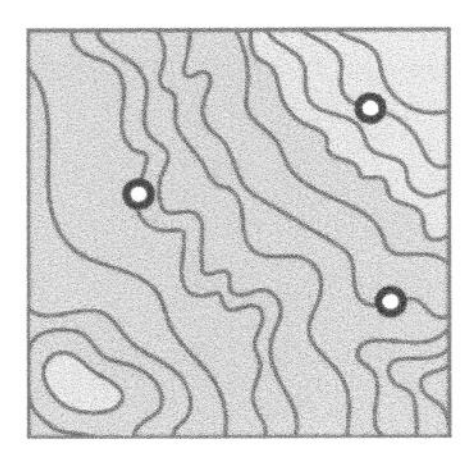

planimetric map

planimetric shift [AERIAL PHOTOGRAPHY] Deviations in the horizontal positions of features in an aerial photograph caused by differences in elevation. Planimetric shift causes changes in scale throughout a photograph. *See also* aerial photograph, elevation, scale.

planimetric-perspective map [MAP DESIGN] A map that gives an oblique view of the mapped area with no elevation information. *See also* elevation, oblique-perspective map.

plat [SURVEYING] A survey diagram, drawn to scale, of the legal boundaries and divisions of a tract of land. *See also* lot, surveying.

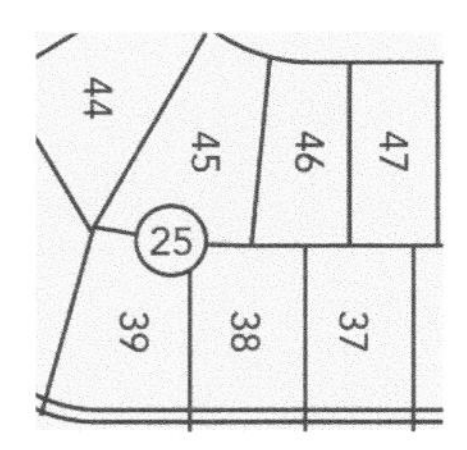

plat

plate carrée projection *See* equirectangular projection.

platform [REMOTE SENSING] The vessel, craft, or instrument used for remote sensing data gathering, such as a drone, aircraft, satellite, submersible, or other vehicle. *See also* sensor.

platted lot [CADASTRAL AND LAND RECORDS] A parcel identified by lot number in a platted subdivision. *See also* lot, parcel, plat, subdivision.

platted subdivision [CADASTRAL AND LAND RECORDS] A subdivision mapped for public record to show legal descriptions of the subdivided area. *See also* lot, plat, platted lot, subdivision.

PLSS [SURVEYING] Acronym for *Public Land Survey System.* The description of the location of land in the United States using a survey system established by the U.S. government in 1785. The system is based on the concept of a township, a square parcel of land measuring 6 miles on each side. The township's position is described as a number of 6-mile units east of a north–south line (called the meridian) and north or south of an east–west line (called the baseline). Each township is divided into 36 sections, each of which is 1 square mile. A section is divided into quarters equal to 160 acres. The quarter section may be further divided into four 40-acre parcels. The PLSS is also known as the rectangular survey, the Township and Range System, or the USPLSS. *See also* DLS, parcel, surveying.

plumb line [SURVEYING] A line that corresponds to the direction of gravity at a point on the earth's surface; the line along which an object will fall when dropped.

PNG [DATA STRUCTURES] Acronym for *Portable Network Graphics.* A digital image format that uses lossless compression to produce raster files with an RGB/RGBA color schema. Designed to replace the Graphics Interchange Format (GIF). *See also* GIF, lossless compression.

PNT [GPS] Acronym for *positioning, navigation, and timing.* Location information gathered by satellite navigation constellations about objects on or above the ground, used for routing, location-based services, and spatial analytics. Positioning provides the geographic location of an object; navigation forecasts the object's location for use with routing; and timing synchronizes position and navigation to improve travel time calculations. *See also* GPS, satellite constellation.

point [ESRI SOFTWARE, SYMBOLOGY] A geometric element defined by a pair of x,y coordinates. *See also* line, point feature, polygon, shape.

point cloud [PHOTOGRAMMETRY] A (typically) large collection of x,y,z coordinates in three-dimensional space representing the real-world surface dimensions of objects and the ground. Point clouds are collected photogrammetrically using multiple overlapping images taken from different viewpoints or with a three-dimensional scanning device, such as a lidar. *See also* lidar, real world.

point cloud scene layer [3D ANALYSIS] A type of data layer that represents a collection of points in 3D space. Each point in the cloud has spatial coordinates (x,y,z) and often additional attributes such as color, intensity, and classification. Point cloud scene layers are commonly used to provide highly detailed renditions of the 3D surface of objects, terrain, and other features. *See also* point cloud.

point digitizing *See* point mode digitizing.

point event [LINEAR REFERENCING] In linear referencing, a feature that occurs at a precise point location along a route and uses a single measure value. Examples include accident locations along highways, signals along rail lines, bus stops along bus routes, and pumping stations along pipelines. *See also* line event, linear referencing.

point feature [MAP DESIGN] A map feature that has neither length nor area at a given scale, such as a city on a world map or a building on a city map. *See also* feature.

point feature map [MAP DESIGN] A map that has symbols showing point features at specific locations. A point can be a one-, two-, or three-dimensional feature symbolized as a point for mapping purposes. *See also* point feature.

point mode digitizing [DATA CAPTURE] A method of digitizing in which the digitizer selects particular points, or vertices, to encode. *See also* stream mode digitizing.

point of beginning
1. [CADASTRAL AND LAND RECORDS] An established point from which irregular parcels are defined using a connected path around the parcel's boundary and noting landmarks along the way.
2. [SURVEYING] The first point surveyed.

See also parcel, path.

point of interest [CARTOGRAPHY] Also known by the acronym *POI*. Any place or feature that is of potential interest to visitors. For example, a point of interest may be a business, a government building such as a post office, a landmark such as the Eiffel Tower, or a natural feature such as a lake or a trailhead. *See also* place.

point of tangency [CARTOGRAPHY] The point at which a tangent-case developable surface touches the generating globe. This point on the map is true in scale to the equivalent point on the spherical or elliptical approximation of the earth. *See also* developable surface.

point sample [DATA CAPTURE] A sampling method in which data is collected at point locations.

point size [GRAPHICS (MAP DISPLAY)] A unit of measure for fonts, nearly equal to 1/72 of an inch. *See also* font, text element.

point thinning [3D GIS] The act of reducing point data in a dataset. Point thinning reduces the number of point measurements needed to represent a surface for a given area. *See also* point, weeding.

point-in-polygon overlay [SPATIAL ANALYSIS] A spatial operation in which points from one feature dataset are overlaid on the polygons of another to determine which points are contained within the polygons. *See also* line-on-point overlay, line-on-line overlay.

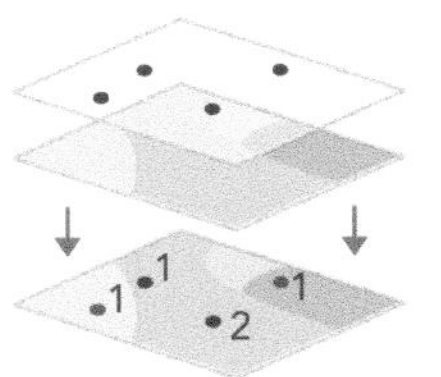

point-in-polygon overlay

point-on-line overlay *See* line-on-point overlay.

point-to-point correspondence [MAP PROJECTIONS] A map projection that transforms all real-world coordinates to corresponding points on the map. *See also* projection, real world.

polar aspect [MAP PROJECTIONS] A planar projection where the point of tangency is located at (or the line or lines of tangency encircle) either the north or south pole. *See also* planar projection, point of tangency.

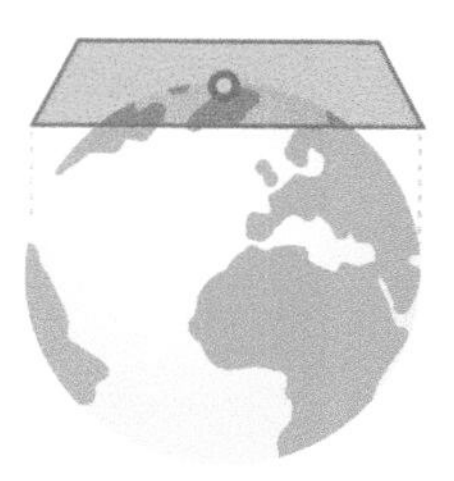

polar aspect

polar axis [CARTOGRAPHY] The axis from the center of the earth to the poles. *See also* axis.

polar flattening *See* flattening.

polar orbit [REMOTE SENSING] A satellite orbit with an inclination near 90 degrees that passes over each polar region. *See also* remote sensing.

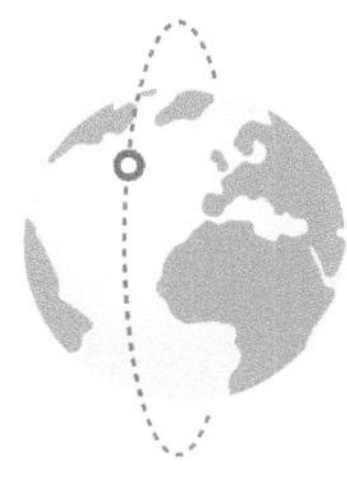

polar orbit

polar planimeter [AERIAL PHOTOGRAPHY] A planimeter attached to a weighted base by an arm (called a pole arm) that allows movement with a circular area. *See also* counting dial, planimeter, vernier scale.

polar radius [GEODESY] The distance from the earth's geometric center to either pole. *See also* geodesy, semiminor axis.

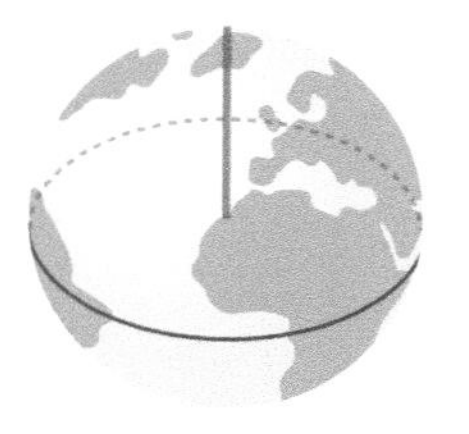

polar radius

polygon

1. [DATA MODELS] On a map, a closed shape defined by a connected sequence of x,y coordinate pairs, where the first and last coordinate pair are the same and all other pairs are unique.

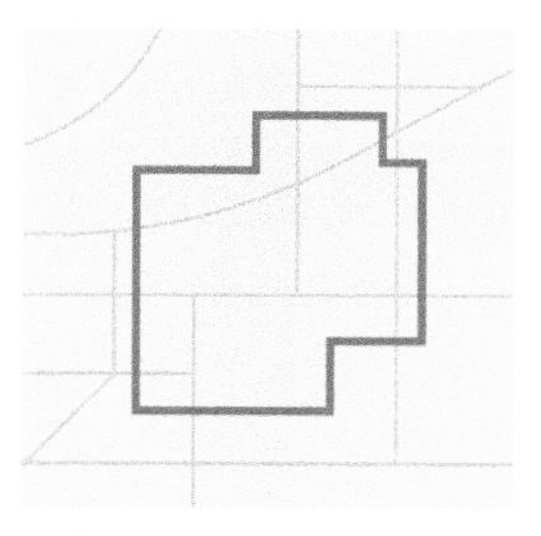

polygon 1

2. [ESRI SOFTWARE, SYMBOLOGY] A shape defined by one or more rings, where a ring is a path that starts and ends at the same point. If a polygon has more than one ring, the rings may be separate from one another or they may nest inside one another, but they may not overlap.

See also polygon feature, ring, x,y coordinates.

polygon feature

1. [DATA MODELS] A map feature that bounds an area at a given scale, such

as a country on a world map or a district on a city map.

2. [ESRI SOFTWARE] A digital map feature that represents a place or thing that has area at a given scale. A polygon feature may have one or more parts. For example, a building footprint is typically a polygon feature with one part. If the building has a detached unit, it might be represented as a multipart feature with discontinuous parts. If the detached unit is in an interior courtyard, the building might be represented as a multipart feature with nested parts. A multipart polygon feature is associated with a single record in an attribute table.

See also equal-area classification, feature.

polygon overlay [DATA EDITING] The process of superimposing two or more geographic polygon layers and their attributes to produce a new polygon layer. *See also* overlay.

polygon-arc topology [DATA MODELS] In a polygon coverage, the list of topologically connected arcs that define the boundary of a polygon feature and the label point that links it to an attribute record in the coverage point attribute table. *See also* arc, coverage, topology.

polyhedron [EUCLIDEAN GEOMETRY] A three-dimensional object or volume defined by several plane faces or polygons. *See also* 3D shape.

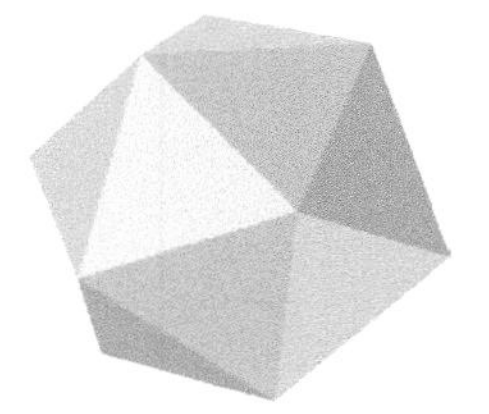

polyhedron

polyline [SYMBOLOGY, ESRI SOFTWARE] A shape defined by one or more paths, in which a path is a series of connected segments. If a polyline has more than one path (a multipart polyline), the paths may either branch or be discontinuous. *See also* line, path, polyline feature.

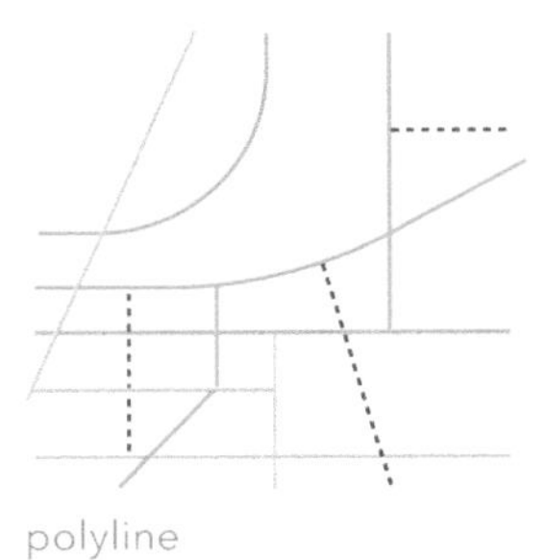

polyline

polyline feature [SYMBOLOGY] A digital map feature representing a place or thing that has length but not area at a given scale. A polyline feature may have one or more parts. For example, a stream is typically a polyline feature with one part; however, if it diverges, goes underground and reemerges, or otherwise has disrupted visibility, it might be represented as a multipart polyline feature with branching or discontinuous parts. A multipart polyline feature is associated with a single record in an attribute table. *See also* feature, line feature.

population count [GOVERNMENT] The primary function of a census. A population count can refer to both the method and the product of systematically recording information about regional inhabitants. GIS typically uses population counts to describe people, but these can also refer to a count of any type of species. *See also* data collection unit, population density.

population density [GOVERNMENT] The concentration of a species within a region. Density is calculated by dividing regional population count by total regional area, such as total square miles or kilometers. *See also* data collection unit, density, population count.

pop-up [VISUALIZATION] A read-only display of feature attributes used with interactive maps. *See also* interactive map, web map.

position [SURVEYING] The latitude, longitude, and altitude (x,y,z coordinates) of a point, often accompanied by an estimate of error. Position may refer to an object's orientation (facing east, for example) without referring to its location. *See also* x,y,z coordinates.

positional accuracy [ACCURACY] A measure of the overall correctness of the reported location of an object or feature compared to its true (or accepted) location. Often determined by comparing the horizontal and vertical position provided by a GPS receiver for an object to its actual ground coordinates. Positional accuracy is typically stated as a statistical estimate.

positional displacement [PHOTOGRAMMETRY] The condition when a feature is offset from its true position on an oblique-perspective map. Positional displacement occurs in direct proportion to the feature's symbol height. *See also* oblique-perspective map, radial displacement, relief displacement.

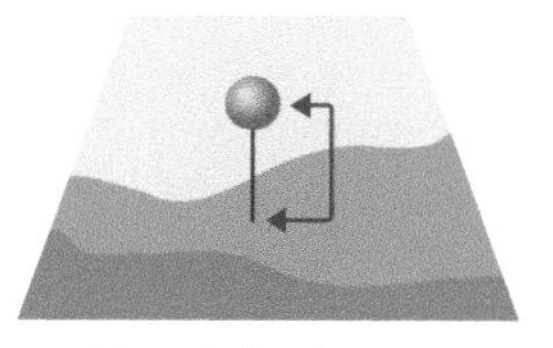

positional displacement

positional error [ACCURACY] A measurable discrepancy between the coordinates of a feature on the map and the feature's actual location on the earth. *See also* error, positional accuracy.

positional reference system [PHYSICS] A method of pinpointing the position of objects in space using geometry. *See also* geometry.

positioning, navigation, and timing *See* PNT.

postal code [GEOCODING] A series of letters or numbers, or both, in a specific format, used by the postal service of a country to divide geographic areas into zones to simplify delivery of mail. *See also* ZIP code.

P

precise code *See* P-code.

precision

1. [DATA QUALITY] The closeness of a repeated set of observations of the same quantity to one another. Precision is a measure of control over random error. For example, an assessment of the quality of a surveyor's work is based in part on the precision of their measured values.
2. [DATA MANAGEMENT] The number of significant digits used to store numbers, particularly coordinate values. Precision is important for accurate feature representation, analysis, and mapping.
3. [STATISTICS] A statistical measure of repeatability, usually expressed as the variance of repeated measures about the mean.

See also single precision, double precision, dataset precision.

P

precision code *See* P-code.

prediction [SPATIAL STATISTICS (USE FOR GEOSTATISTICS)] In spatial modeling, the process of forming a statistic from observed data to assign values to random variables at locations where data has not been collected. *See also* estimation.

prediction standard error [SPATIAL STATISTICS (USE FOR GEOSTATISTICS)] A value quantifying the uncertainty of a prediction; mathematically, the square root of the prediction variance. (The prediction variance is the variation associated with the difference between the true and predicted value.) As a rule, 95 percent of the time the true value will lie within the predicted value plus or minus two times the prediction standard error if data is normally distributed. *See also* kriging.

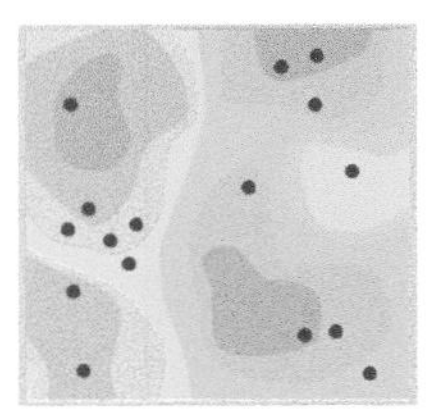

prediction standard error

primary colors [GRAPHICS (COMPUTING)] The individual components in a color system from which all other colors are derived. In the additive color system, the primary colors are red, green, and blue; in the subtractive color system, the primary colors are cyan, yellow, and magenta; in traditional pigments, they are red, blue, and yellow. *See also* RGB, CMYK.

primary key [DATABASE STRUCTURES] An attribute or set of attributes in a database that uniquely identifies each record. A primary key allows no duplicate values and cannot be null. *See also* key.

primary reference data [GEOLOCATING] In geocoding, the most basic reference material used in an address locator, usually consisting of the geometry of features in a region and an associated address attribute table. *See also* address, locator.

primary table [DATABASE STRUCTURES] In geocoding, the attribute table associated with the primary reference data. Based on the address locator style selected, certain address elements must be present in the primary table. *See also* locator, primary reference data.

prime meridian [COORDINATE SYSTEMS]
1. The zero meridian (0°) used as the reference from which longitude east and west is measured.
2. In a coordinate system, any line of longitude designated as 0 degrees east and west, to which all other meridians are referenced. The Greenwich meridian is internationally recognized as the prime meridian for most official purposes, such as civil timekeeping.

See also Greenwich meridian.

prime vertical [GEODESY] In astronomy and geodesy, the vertical circle that passes through an observer's zenith and through the east and west points of the horizon. *See also* geodesy, zenith.

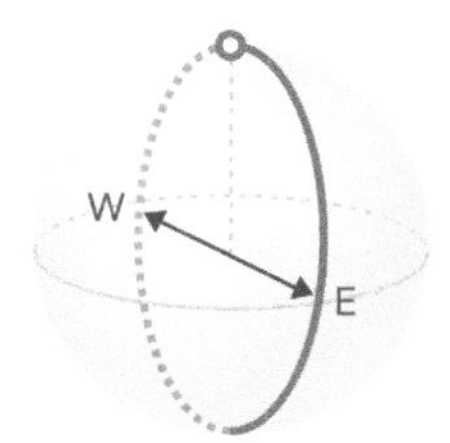

prime vertical

principal component analysis
1. [DATA ANALYSIS] In data analysis and predictive modeling, a data transformation method that reduces the dimensions of large datasets by transforming large sets of variables into a smaller one that still contains most of the information from the larger sets. This method compresses data by eliminating redundancy and assists in separating features within the data.
2. [REMOTE SENSING] In remote sensing, a data transformation method that rotates the axes of the input bands to a new multivariate attribute space in which the axes are not correlated. The first component accounts for the greatest variability in the data, the second component accounts for the next largest amount of variability, and so on.

See also band, geographic transformation.

principal meridian [SURVEYING] A meridian determined by government land surveyors that intersects a surveyed baseline to establish an initial point in the U.S. Public Land Survey System and Canada's Dominion Land Survey. *See also* DLS, initial point, PLSS.

principal point [REMOTE SENSING] In remote sensing, the location where the optical principal axis intersects the focal plane. *See also* fiducial mark.

principal scale [MAP PROJECTIONS] The scale of the generating globe used to make a map projection. *See also* actual scale, generating globe, projection, scale factor.

PRJ [ESRI SOFTWARE] A file type associated with a coverage, GRID, or TIN; used with shapefiles. The PRJ file contains the coordinate system information for the data. PRJ may be used to refer to the coordinate system even if the information is not stored in a PRJ file. For example, "The PRJ of the shapefile is WGS 1984 UTM zone 15 north." *See also* coordinate system, TIN.

PRN [GPS]
1. Acronym for *pseudo-random noise.* A signal carrying a code that can be reproduced exactly but appears to be randomly distributed like noise. Each Navstar satellite has a unique PRN code.
2. Acronym for *pseudo-random number.* A number representing a unique GPS satellite ID or code.

See also GPS, Navstar, noise, PRN code, random noise, signal.

P

PRN code [GPS] PRN is an acronym for *pseudo-random noise.* A repeating radio signal broadcast by each GPS satellite and generated by each GPS receiver. In a given cycle, the satellite and the receiver start generating their codes at the same moment, and the receiver measures how much later the satellite's broadcast reaches it. By multiplying that time by the speed of radio waves, the receiver can compute the distance between the satellite's antenna and its own. *See also* civilian code, P-code.

probability [STATISTICS] A measure of the likelihood that a particular outcome, such as a spatial pattern or event, will occur given a set of possible outcomes. Probability values range from 0 for impossible outcomes to 1 for completely certain outcomes. The probability that a tossed coin will land heads-up, for example, is 0.5, since landing heads-up is one of two possible outcomes. *See also* Bayesian statistics, correlation, F test, Monte Carlo method.

probability map [STATISTICS] A surface that gives the probability that the variable of interest is above or below a specified threshold value. *See also* probability.

process [MODELING] In geoprocessing, a tool and its parameter values. One process, or multiple connected processes, creates a model. *See also* geoprocessing.

processing error [ACCURACY] A subtle error introduced as data is processed to make a map. Detection requires knowledge of the data and the methods used to create the map; includes numerical error and topological error. *See also* error, numerical error, topological error.

processing template [DIGITAL IMAGE PROCESSING] A stored image processing function chain that can be repeatedly applied to multiple, similar image datasets. Typically used to display the processed results as you pan and zoom the imagery to generate on-the-fly information layers; the processed results can also be explicitly saved as a file. Also known as a raster function template (RFT). *See also* function chain.

profile [CARTOGRAPHY] A diagram that shows the change in elevation of a surface along a path on the ground. Also known as an elevation profile, and in architecture as an elevation drawing. *See also* elevation drawing.

profile curvature [CARTOGRAPHY] The amount that a surface deviates from being linear parallel to the direction of the maximum slope. *See also* curvature, maximum slope.

profile graph [CARTOGRAPHY] A graph of the elevation of a surface along a specified line. *See also* chart, elevation.

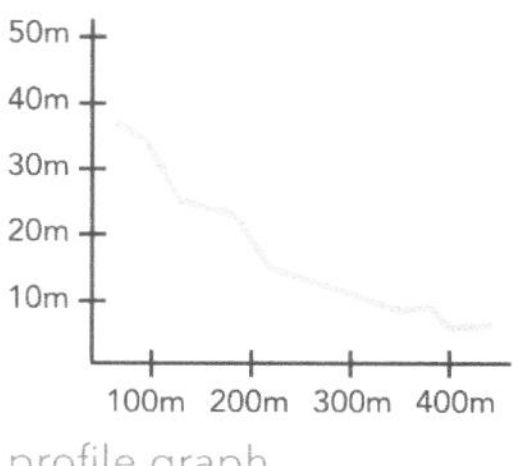

profile graph

profile line [CARTOGRAPHY] A line on a map along which a profile is constructed. *See also* irregular line profile.

projected coordinate system [COORDINATE SYSTEMS] A two-dimensional Cartesian coordinate system that uses x, y, and z coordinates to represent point, line, and area feature locations in two or three dimensions. A projected coordinate system is defined by a geographic coordinate system, a map projection, any parameters needed by the map projection, and a linear unit of measure. *See also* geographic coordinate system, projection.

projected coordinates [COORDINATE SYSTEMS] A measurement of locations on the earth's surface expressed in a two-dimensional system that locates features based on their distance from an origin (0,0) along two axes, a horizontal x-axis representing east–west and a vertical y-axis representing north–south. Projected coordinates are converted from latitude and longitude to x,y coordinates using a map projection. *See also* geographic coordinates.

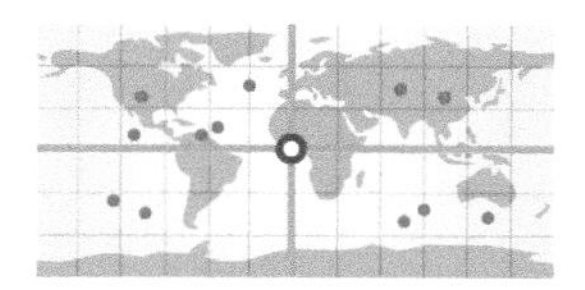
projected coordinates

projection [MAP PROJECTIONS]

1. A geometric transformation of the earth's spherical or ellipsoidal surface onto a flat map surface—also known as a developable surface. A projection generally requires a systematic mathematical transformation of the earth's graticule of lines of longitude and latitude onto a plane. Every map projection distorts distance, area, shape, direction, or some combination thereof.

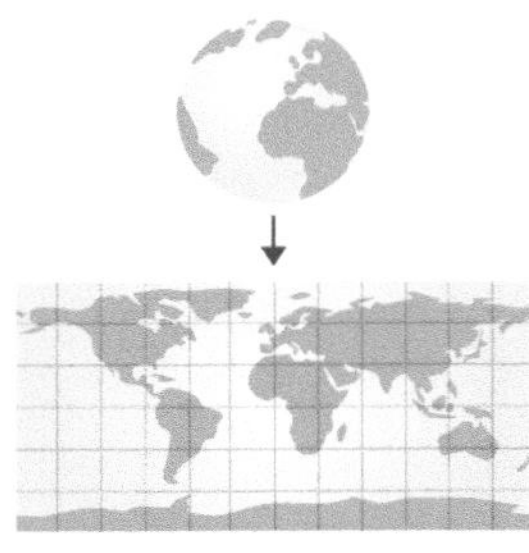

projection 1

2. The conversion of a spherical surface (a globe) to a flat map surface using one of many mathematical models.

See also equal-area projection, equidistant projection, projection transformation, secant projection, stereographic projection.

projection transformation [DATA CONVERSION] A mathematical conversion of a map from one coordinate system to a projected coordinate system. For example, from a geographic coordinate system (latitude and longitude) to a projected coordinate system (such as UTM) or from one projected coordinate system to another. Generally used to integrate maps with differing coordinate systems into a GIS. *See also* projected coordinate system.

projective transformation [DATA CONVERSION] A transformation used only to transform coordinates digitized directly from high-altitude aerial photographs of relatively flat terrain, assuming there is no systematic distortion in the photographs. *See also* transformation.

prolate ellipsoid [MATHEMATICS] An ellipsoid created by rotating an ellipse around its major axis. *See also* ellipsoid, oblate ellipsoid, spheroid.

propaganda map [MAP DESIGN] A map of persuasion designed to distort or misrepresent information. *See also* persuasion map.

proportional symbol [SYMBOLOGY] A map graphic that changes in size correlated to the data value it represents. For example, on a weather map, proportional symbol snowflakes may indicate areas with higher or lower snowfall. *See also* symbol.

protected code *See* P-code.

protractor [CARTOGRAPHY] A measuring device that divides a circle into equal angular intervals, usually degrees.

proximity [PHYSICS] The nearness of features or locations in space. *See also* buffer, proximity analysis, proximity query.

proximity analysis [SPATIAL ANALYSIS] A type of analysis in which geographic features (points, lines, polygons, or raster cells) are selected based on their distance from other features or cells. *See also* buffer, distance.

proximity query [SPATIAL ANALYSIS] A form of spatial query in which geographic features within a specified distance of a particular feature are selected. *See also* distance, spatial query.

proxy object [PROGRAMMING] A local representation of a remote object, supporting the same interfaces as the remote object. All interaction with the remote object from the local process is mediated through the proxy object. A local object makes calls on the members of a proxy object as if it were working directly with the remote object. *See also* object.

pseudo node [ESRI SOFTWARE] In a geodatabase topology, a temporary feature marking the location where an edge has been split during an edit session. This type of pseudo node becomes a vertex when the edit is saved. *See also* edge, vertex.

pseudo-contiguous cartogram [MAP DESIGN] A cartogram on which transformed data collection areas share common boundaries, but the boundaries are not the same as those found on the earth or represented on a conventional map. *See also* boundary, cartogram, data collection unit.

pseudo-cylindrical map projection [MAP PROJECTIONS] A map projection that is similar to a cylindrical projection in that the parallels are horizontal lines, and the meridians are equally spaced; the difference is that all meridians except the vertical-line central meridian are curved instead of straight. *See also* cylindrical projection.

pseudo-random noise *See* PRN.

pseudo-random noise code *See* PRN code.

pseudo-random number *See* PRN.

Public Land Survey System *See* PLSS.

public participation [ETHICS] The active involvement of stakeholders outside an organization in the decision-making or planning processes of that organization. Public participation in GIS processes may include making GIS tools and data accessible, at an appropriate technical level, to stakeholders, or it may result in knowledge gained from stakeholders who are incorporated into GIS analyses.

push broom scanner *See* along-track scanner.

p-value [STATISTICS] A probability resulting from a statistical test of the coefficient associated with each independent variable in a regression model. The null hypothesis for this statistical test states that the coefficient is not significantly different from zero. Small p-values reflect small probabilities. They suggest that the coefficient is significantly different from zero, and consequently, that the associated explanatory variable is helping to model or predict the dependent variable. Variables with coefficients near zero do not help predict or model the dependent variable; they are almost always removed from the regression equation (unless there are strong theoretical reasons to keep them). *See also* dependent variable, independent variable, null hypothesis, regression coefficient.

pyramid

1. [DATA STRUCTURES] In raster datasets, a method of multiscale image processing that copies the original data in decreasing levels of resolution to enhance performance. The coarsest level of resolution is used to quickly draw the entire dataset; subsequent more detailed layers are drawn at fast speeds, since fewer pixels are required.

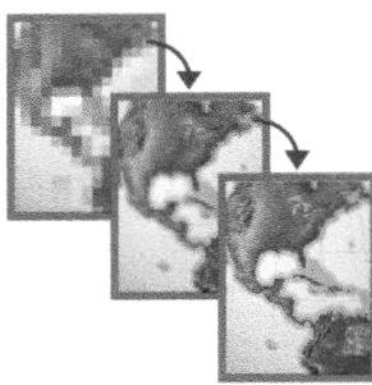

pyramid 1

2. [DIGITAL IMAGE PROCESSING] In image processing, a reduced-resolution dataset stored with imagery that is used to read and display imagery at lower resolutions as a means to increase display performance.

See also raster, resolution, terrain dataset pyramid.

QQ plot [STATISTICS] A scatter chart in which the quantiles of two distributions are plotted against each other. *See also* quantile, scatter chart.

quad sheet *See* quadrangle.

quadrangle [CARTOGRAPHY] A rectangular map bounded by increments of latitude and longitude (quadrilaterals). Often a map sheet in either the 7.5-minute or 15-minute series published by the U.S. Geological Survey. Often known as topo sheet or shortened to quad. Each map in the series is usually named after a local physiographic feature in the mapped series, such as the Ranger Creek, Texas, quad map. *See also* map sheet, orthophotoquad, U.S. Geological Survey.

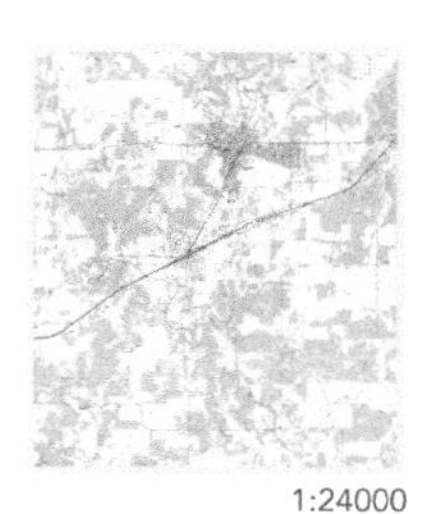

quadrangle

quadrant

1. [COORDINATE SYSTEMS] In a rectangular coordinate system, any of the quarters formed by the central intersection of x- and y-axes that divide a plane into four equal parts.

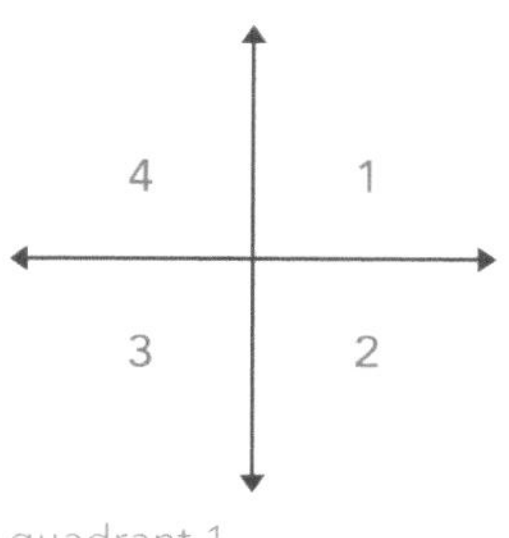

quadrant 1

2. [EUCLIDEAN GEOMETRY] One quarter of a circle measured from the center, having an arc of 90 degrees.
3. [COORDINATE SYSTEMS] The four divisions of the Cartesian coordinate system, formed by the intersection of the two axes, labeled counterclockwise starting from the upper-right quadrant by the Roman numerals I, II, III, and IV. Because the signs for both coordinates are positive in quadrant I, that is the quadrant used for map grids.

See also axis, Cartesian coordinate system, coordinate system, quadtree.

quadrat [SPATIAL STATISTICS (USE FOR GEOSTATISTICS)] In spatial sampling, one of a set of identically sized areas—or data collection units, often rectangular—within which the members of a population are counted.

Q

The size, number, and location of quadrats within the data collection unit are chosen by the sampler. Population counts in each quadrat are compared to determine distribution patterns. *See also* quadrat analysis.

quadrat analysis [SPATIAL STATISTICS (USE FOR GEOSTATISTICS)] A comparison of the statistically expected and actual counts of objects within data collection units (quadrat spatial sampling areas) to test for distribution patterns such as randomness and clustering. *See also* quadrat.

quadrilateral [CARTOGRAPHY] An area bounded by increments of latitude and longitude, although the increments for latitude need not be the same as for longitude (for example, 1 degree × 2 degree). *See also* latitude, longitude.

quadtree [DATA STRUCTURES] A method for encoding raster data that reduces storage requirements and improves access speeds by storing values only for homogeneous regions rather than for every pixel. The raster is recursively subdivided into quadrants until all regions are homogeneous or until some specified level has been reached. *See also* raster.

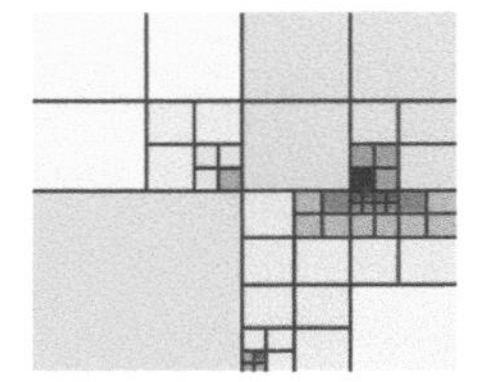

quadtree

qualitative analysis [DATA QUALITY] A systematic examination of a problem or complex entity that does not involve measurements and numbers. *See also* analysis, qualitative data.

qualitative change map [MAP DESIGN] A map that shows change in a nominal-level attribute over time or change in the location of a feature over time. *See also* qualitative analysis.

qualitative data [DATA STRUCTURES] Data classified or shown by category, rather than by amount or rank, such as soil by type or animals by species. *See also* quantitative data.

quality assurance [QUALITY ASSURANCE] A process used to verify the quality of a product after its production. *See also* quality control.

quality control [QUALITY CONTROL] A process used during production of a product to ensure its quality. *See also* quality assurance.

quantile [STATISTICS] In a data distribution, a value representing a class break, where classes contain approximately equal numbers of observations. The p-th quantile, where p is between 0 and 1, is that value that has a proportion p of the data below the value. For theoretical distributions, the p-th quantile is the value that has p probability below the value. *See also* class intervals, distribution.

quantile classification [DATA STRUCTURES] A data classification method that distributes a set of values into groups that contain an equal number of values. *See also* classification.

quantile intervals [DATA MANAGEMENT] A procedure used to assign class intervals to a numeric distribution so that the numbers of features in each class are as close as possible to equal. *See also* class intervals.

quantitative analysis [DATA ANALYSIS] A systematic examination of a problem or complex entity that involves measurements and numbers. *See also* quantitative data.

quantitative change map [MAP DESIGN] A map that shows the increase or decrease in an attribute value over a specified period. *See also* quantitative data.

quantitative data [DATA STRUCTURES] Data grouped or shown by measurements of number, magnitude, intensity, or amount, such as population per unit area. *See also* qualitative data.

quantitative geography [GEOGRAPHY] The application of mathematical and statistical concepts and methods to the study of geography. *See also* geography, geoinformatics.

quantitative thematic map [MAP DESIGN] A map that shows a single theme or a few related themes of quantitative information. *See also* quantitative data.

quarter point [NAVIGATION] One of the 16 compass points indicating north by east, northeast by north, northeast by east, and so on. *See also* cardinal point, compass, compass point.

quarter section [SURVEYING] One-half of one-half of a square-mile section (160 acres) in the U.S. Public Land Survey System and Canada's Dominion Land Survey. *See also* DLS, PLSS, quarter-quarter section, section.

quarter-quarter section [SURVEYING] One-quarter of one-quarter of a section, or one-16th of a section (40 acres), in the U.S. Public Land Survey System and Canada's Dominion Land Survey. *See also* DLS, PLSS, quarter section, section.

quartile [DATA MANAGEMENT] One class in a four-class data grouping using quantile intervals. *See also* quantile intervals, quintile.

Quasi-Zenith Satellite System *See* QZSS.

quintile [DATA MANAGEMENT] One class in a five-class data grouping using quantile intervals. *See also* quantile intervals, quartile.

QZSS [GPS] Acronym for *Quasi-Zenith Satellite System*. A regional navigation satellite system (RNSS) and a satellite-based augmentation system (SBAS) compatible with the United States' GPS. Used for positioning, navigation, and timing (PNT) in the Asia-Oceania region with a focus on Japan. Developed and operated by the Japanese government. *See also* GPS augmentation, PNT, satellite constellation.

radar [PHYSICS] An abbreviation of *radio detection and ranging*. An active sensor system that detects surface features on the earth by bouncing polarized radio waves off them and measuring the energy, including distance, direction, and polarization, that is reflected back. *See also* lidar, sonar.

radar

radar altimeter [PHYSICS] An instrument that determines elevation, usually from mean sea level, by measuring the amount of time an electromagnetic pulse takes to travel from an aircraft to the ground and back again. *See also* electromagnetic radiation, elevation.

R

radar interferometry [REMOTE SENSING] The analysis of interferograms that have been created by IFSAR, or artificially. Radar interferometry involves the comparison of two or more images of the same area taken from different positions and calibrated with surveyed ground points to generate 3D digital elevation models (DEMs), or models demonstrating slight movements of surface features. *See also* digital elevation model, IFSAR, interferogram.

radial displacement [PHOTOGRAMMETRY] In imagery, the apparent distortion of the top of a 3D or vertical photograph in relation to the nadir point; specifically related to vertical images collected from a pinhole sensor where there is apparent displacement of objects that have elevation related to their base. The direction of displacement is radial from the principal point on a true vertical, or from the isocenter on a vertical image distorted by tip or tilt. Often referred to as building lean. *See also* positional displacement, relief displacement.

radial flow map [MAP DESIGN] A flow map that has a spokelike pattern because the features and places are mapped as nodes (line endpoints), with one place being a common origin or destination node. *See also* destination, flow map, node, origin.

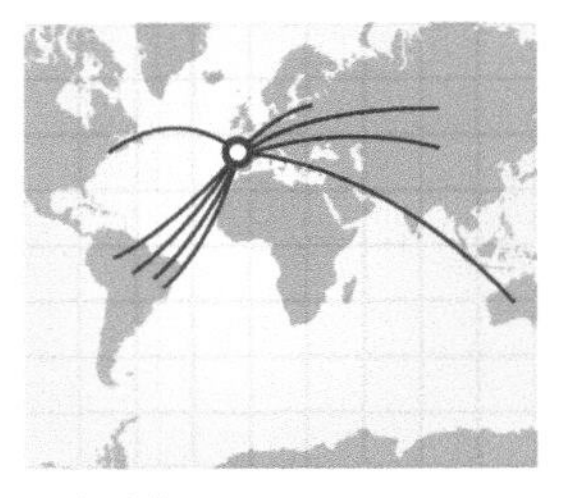

radial flow map

radial lens distortion

1. [PHOTOGRAMMETRY] In imagery, a deviation that arises from light rays appearing to bend more through the edges of a camera lens than they do through its optical center.
2. [DIGITAL IMAGE PROCESSING] The distance that a pixel deviates from its ideal location in a radial direction, away from the principal point, preventing the preservation of straight lines being projected between the object world and the image being formed.

radian [EUCLIDEAN GEOMETRY] The angle subtended by an arc of a circle that is the same length as the radius of the circle, approximately 57 degrees, 17 minutes, and 44.6 seconds. There are 2π radians in one complete rotation. *See also* degree, gradian, steradian.

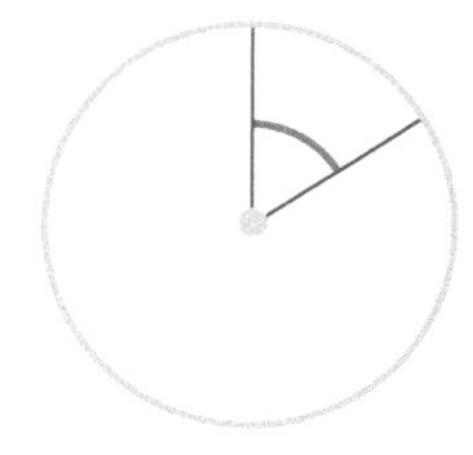

radian

radiance [PHOTOGRAMMETRY] Radiation that is reflected from the surface of a feature, bounced into the optical path from surrounding areas, and reflected from clouds above the area of a pixel. It is a measurement of the flux density of radiant energy per unit solid angle and per unit projected area of radiating surface, measured in watts per steradian per square meter (W·sr–1·m–2), as a function of wavelength. *See also* radiation.

radiation [PHYSICS] The emission and propagation of energy through space in the form of waves. Electromagnetic energy and sound are examples of radiation. *See also* albedo, reflectance.

radiometer [PHYSICS] An instrument that measures the intensity of radiation in a particular band of wavelengths in the electromagnetic spectrum, such as infrared or microwave. *See also* bolometer, photometer, spectrometer, wavelength.

radiometric correction [REMOTE SENSING] Procedures that correct or calibrate aberrations in data values due to specific distortions from such things as atmosphere effects (such as haze) or instrumentation errors (such as striping) in remotely sensed data. *See also* distortion, radiometric resolution.

radiometric resolution [REMOTE SENSING] The sensitivity of a sensor to small fluctuations in the amount of energy in the electromagnetic spectrum. There are three aspects to this characteristic:

1. The sensor's bandwidth sensitivity (full width at half maximum, or FWHM) determines the range of wavelengths in the specific image band of the electromagnetic spectrum that the sensor can detect.
2. Typically, sensors have 8-, 11-, 12-, 14-, or 16-bit depth per band. The

higher the bit depth, the higher the sensor's potential radiometric resolution.

3. The sensor's signal-to-noise ratio (S/N) describes the amount of signal resolution information within the band. For example, two sensors can each return an image containing 8 bits of information (256 possible gray levels), but one may have 60 S/N and the other 120 S/N, or double the amount of signal resolution information.

See also band, electromagnetic spectrum, radiometric correction, sensor.

radius [EUCLIDEAN GEOMETRY] The distance from the center to a point on the outer edge of a circle, circular curve, or sphere. *See also* circle, semimajor axis, sphere.

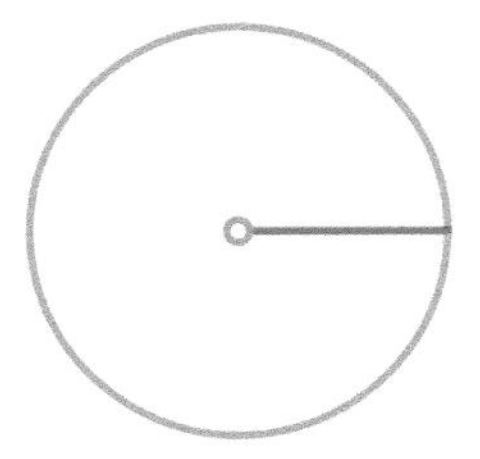

radius

radius of curvature [MATHEMATICS] The measure of how curved a surface is. On an oblate ellipsoid, the radius of curvature is greatest at the pole and smallest at the equator; on a sphere, it is equal everywhere. *See also* oblate ellipsoid, radius, sphere.

railroad map [MAP DESIGN] A map of a railroad network.

raised-relief globe [MAP DESIGN] A globe on which the terrain surface is raised up from the base elevation and exaggerated to give a more realistic impression of relief. *See also* relief, terrain model.

Ramer-Douglas-Peucker algorithm [DATA EDITING] A line simplification algorithm that reduces the number of vertices in a polyline or polygon while preserving the overall shape and minimizing the loss of important details. Developed independently by Swiss scientist Urs Ramer in 1972 and Canadian geographers David H. Douglas and Thomas K. Peucker in 1973. Also known as the Douglas-Peucker algorithm.

random error [UNCERTAINTY] A measured, observed, calculated, or interpreted value (or result) that differs from the expected or true value but cannot be attributed to a known (systematic) source or the source of which can vary based on the measurement methodology. *See also* error.

random noise [SPATIAL STATISTICS (USE FOR GEOSTATISTICS)] In a spatial model, variation in the value of a variable that cannot be described by a mathematical function and is not spatially correlated: It includes measurement error and microscale variation (variation at a finer scale than that at which the data has been sampled). Random noise is one of the three main components—along with drift and spatially correlated variation—that contribute to the change in value of a variable over a surface. In a semivariogram, random noise is represented by the nugget. Random noise is sometimes called white noise. *See also* correlation, drift, kriging, nugget.

random sample [DATA CAPTURE] A sampling method in which all locations or all features in a population are given an equal probability of being selected for the sample. *See also* sampling method.

random spatial arrangement [DATA MODELS] An arrangement that occurs when there is no apparent order or pattern in the spacing between features on a map. *See also* pattern, spatial data.

range

1. [SPATIAL STATISTICS (USE FOR GEOSTATISTICS)] A parameter of a variogram or semivariogram model that represents a distance beyond which there is little or no autocorrelation among variables.
2. [MATHEMATICS] The minimum and maximum values a set of numbers can have.
3. [BIOLOGY] The area where a species is found naturally. Also known as species range.

See also autocorrelation, variogram.

range domain [DATA STRUCTURES] A type of attribute domain that defines the range of permissible values for a numeric attribute. For example, the permissible range of values for a pipe diameter could be between 1 and 32 inches. *See also* domain.

range grading [MAP DESIGN] A map portrayal of a distribution of data using a limited number of quantitative map symbols, each representing a different range of data values. *See also* quantitative data.

ranging [SURVEYING] Finding the distance from one location to another.

rank [ACCURACY] A method of assigning an accuracy value to feature classes. Rank helps avoid having vertices from a feature class that were collected with a high level of accuracy being snapped to vertices from a less accurate feature class. *See also* snapping.

raster [DATA MODELS] A fundamental data structure consisting of a matrix of equally sized cells, or pixels, arranged in rows and columns and composed of single or multiple bands. Each cell represents a location on the earth's surface and contains a numeric value that represents a particular attribute or phenomenon, such as temperature at a particular height or depth, elevation, or image brightness value. Groups of cells that share the same cell value represent the same geographic feature. *See also* vector, lattice.

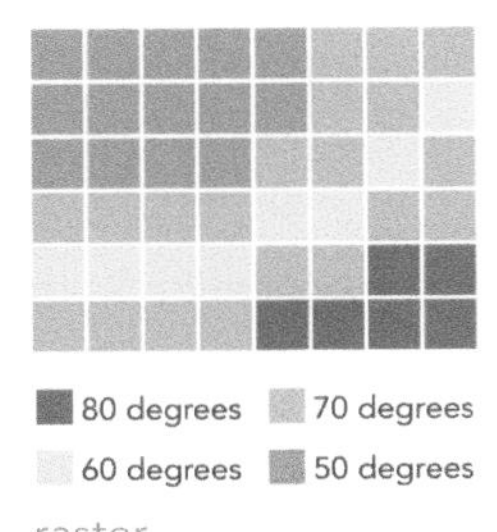

raster

raster band *See* raster dataset band.

raster catalog [ESRI SOFTWARE] A collection of raster datasets defined in a table of any format, in which the records define the individual raster datasets that are included in the catalog. Raster catalogs can be used to display adjacent or overlapping raster datasets without having to mosaic them together into one large file. *See also* raster.

raster cell *See* cell.

raster data [VISUALIZATION] Data collected by grid cells or pixels in a row and column grid or matrix. *See also* kernel.

R

raster data model [DATA MODELS] A representation of the world as a surface divided into a regular grid of cells. Raster models are useful for storing data that varies continuously, as in an aerial photograph, a satellite image, a surface of chemical concentrations, or an elevation surface. *See also* vector data model.

raster dataset [DATA MODELS] A simple spatial data model that displays an image or raster based on the image format (for example, a JPEG) on a device. Within this model, the device is unaware of geopositioning unless it is explicitly included in the basic format, such as in a GEOTIFF or JPEG2000. *See also* feature dataset, geopositioning, GEOTIFF, JPEG, raster data model.

raster dataset band [REMOTE SENSING] One layer in a raster dataset that represents data values for a specific range in the electromagnetic spectrum (such as ultraviolet, blue, green, red, and infrared), or radar, or other values derived by manipulating the original image bands. A raster dataset can contain more than one band. For example, satellite imagery commonly has multiple bands representing different wavelengths of energy from along the electromagnetic spectrum. *See also* band, layer, wavelength.

raster function chain *See* function chain.

raster intersection [DATA MODELS] Three or more lines in a raster that meet at a common point. *See also* intersection, raster.

raster layer [ESRI SOFTWARE] A layer that references a raster as its data source and a raster renderer that defines how the raster data should be rendered and any additional display properties. *See also* layer, renderer.

raster model *See* raster data model.

raster preprocessing [DATA CONVERSION] Simple raster editing that prepares images for viewing and analysis. Preprocessing includes georeferencing, clipping, positioning, resizing, enhancing, and mosaicking. *See also* data conversion, georeferencing, raster.

Raster Product Format *See* RPF.

raster slope map [MAP DESIGN] A slope zone map generated using a GIS and raster elevation data. *See also* elevation, slope map.

raster snapping *See* snapping.

raster statistics [ESRI SOFTWARE] Statistics that are calculated from the cell values of each band in a raster. The statistics that are calculated include the minimum, maximum, mean, and standard deviation cell values, and if the dataset is thematic, the number of classes. Statistics are required for some rendering and geoprocessing operations. *See also* geoprocessing.

raster tile *See* tiling.

raster tracing [DATA CONVERSION] An interactive vectorization process that involves drawing along the boundary of contiguous raster cells to create vector features. *See also* vectorization.

raster type [DATA MODELS, ESRI SOFTWARE] Metadata attached to a raster that provides identifying information, such as georeferencing, acquisition date, raster format, and sensor type. *See also* metadata.

rasterization [DATA CONVERSION] The conversion of points, lines, and polygons into cell data. *See also* vectorization.

rate [MATHEMATICS] A ratio commonly used in population statistics; rate is obtained by dividing a numerator that equals the number of occurrences of an event during a specified time by a denominator that equals the number of possible occurrences in the same period. For birth and death rates, it is expressed as the number of events per population in the time period, such as the number of births or deaths. *See also* ratio.

ratio [MATHEMATICS] A value obtained from dividing one number by another.

ratio bands [PHOTOGRAMMETRY] In spectral imagery, a derivative band generated from dividing one band by another that is used to improve visibility. For example, to remove haze related to atmospheric Rayleigh scattering.

ratio data [DATA STRUCTURES] Data classified relative to a fixed zero point on a linear scale. Mathematical operations can be used on these values with predictable and meaningful results. Examples of ratio measurements are age, distance, weight, and volume. *See also* data, scale.

R

ratioing [DIGITAL IMAGE PROCESSING] In digital image processing, enhancing the contrast between features in an image by dividing the values of pixels in one image by the values of corresponding pixels in a second image. *See also* contrast ratio, digital image processing.

raw data [DATA MANAGEMENT] Data in its most basic form that is not processed or manipulated in any manner. *See also* data.

ray tracing [3D GIS] A technique that traces imaginary rays of light from a viewer's eye to the objects in a three-dimensional scene to determine which parts of the scene should be displayed from that perspective. *See also* viewshed.

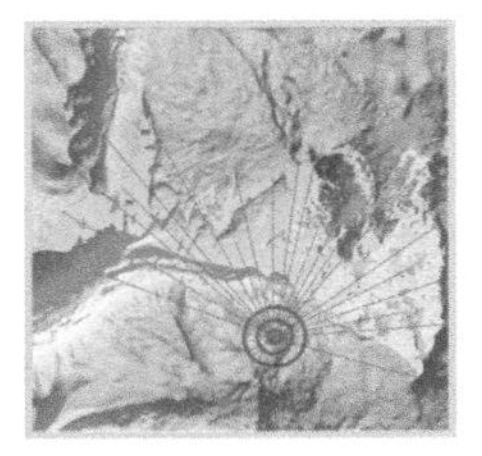

ray tracing

real time [DATA CAPTURE] Technology or data collection that takes place at the same rate, and sometimes at the same time, as it does in the real world. A common GIS example of real time is a weather feed that provides meteorological data in near real time. *See also* live feed, real-time data, real world.

real world [MODELING] Not theoretical or modeled. The objects, processes, or functions that exist or occur in reality. *See also* real time.

real-time data [DATA CAPTURE] Data that is displayed immediately as it is collected. Real-time data is often used for navigation or tracking. *See also* live feed, real time, streaming.

reclassification [SPATIAL ANALYSIS] The process of taking input cell values and replacing them with new output cell values. Reclassification is often used to simplify or change the interpretation of raster data by changing a single value to a new value or grouping ranges of values into single values—for example, assigning a value of 1 to cells that have values of 1 to 50, 2 to cells that range from 51 to 100, and so on. *See also* cell, classification, raster, simplification.

record [DATABASE STRUCTURES]

1. A set of related attribute values in a database for a single feature. For example, in an address database, the street, city, country, and postal code may compose one address record.
2. A row in a table.

FID	Species	Color
1	oak	brown
2	pine	green
3	fir	green

record 2

See also database, table, tuple.

rectangular cartogram [DATA CONVERSION] A pseudo-contiguous cartogram made by transforming data collection units into rectangles proportional in size to the magnitude being shown. *See also* data collection unit, pseudo-contiguous cartogram.

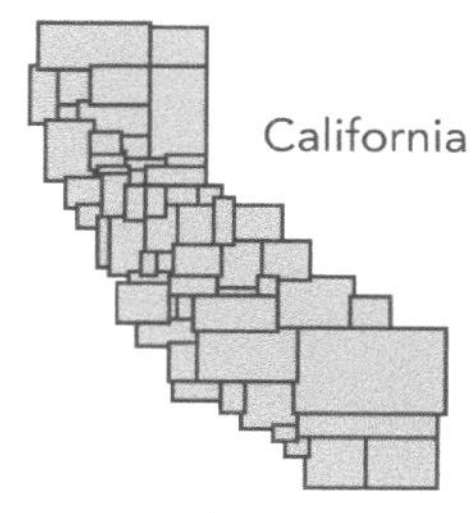

rectangular cartogram

rectangular distance table [DATA MANAGEMENT] A table in which locations are listed alphabetically along the top and side to create a row, column matrix of distances. *See also* table.

rectangular survey *See* PLSS.

rectification [DATA CONVERSION] The process of applying a mathematical transformation to an image so that the result is a planimetric image. *See also* geometric correction, orthorectification, planimetric, transformation.

rectified aerial photograph [DIGITAL IMAGE PROCESSING] An aerial photograph with geometry corrected to remove systematic scale distortions. *See also* aerial photograph, distortion, geometry.

rectilinear

1. [MATHEMATICS] Characterized by straight lines, usually parallel to orthogonal axes.

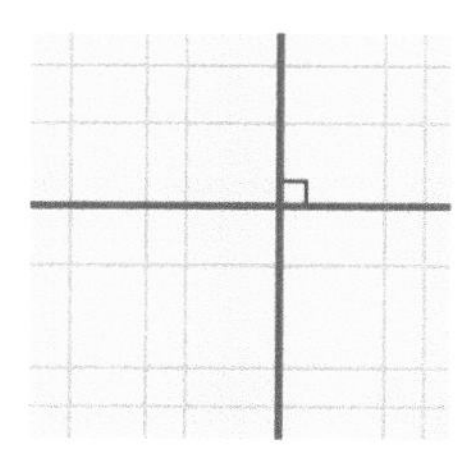
rectilinear 1

2. [MAP DESIGN] A map or image with identical horizontal and vertical scales.

See also orthogonal.

redistricting [GOVERNMENT] The process of revising the boundaries of administrative, legislative, or election districts. *See also* boundary, congressional district.

reference data

1. [GEOCODING] In geocoding, metadata containing an object's significant location information, such as street address or latitude-longitude coordinates. Reference data consists of the spatial representation of the data and its related attribute table.
2. [PHOTOGRAMMETRY] In imagery, a baseline dataset used to compare against processed data to assess the accuracy of results. Typically used for geopositional or feature classification accuracy; verifiable with ground surveying.

See also geocoding reference data, primary reference data, reference data source, secondary reference data.

reference data source [ESRI SOFTWARE] A spatial data layer that a geocoding service uses to perform address geocoding. A reference data source can be any point, line, or polygon feature class that contains the necessary address attributes. *See also* geocoding service.

reference datum [GEODESY] Any datum, plane, or surface from which other quantities are measured. *See also* datum.

reference ellipsoid [GEODESY] An ellipsoid associated with a geodetic reference system or geodetic datum. *See also* geodetic datum.

reference frame [MODELING] A materialization or realization of a reference system consisting of a set of precise coordinates of well-defined physical objects on the ground or in space. *See also* ITRF, ITRS.

R

reference grid *See* alphanumeric grid.

reference level *See* datum level.

reference line [CARTOGRAPHY] The line between an observer and a reference point (an object or known position), along which direction is measured. Also known as a baseline. *See also* direction line, grid azimuth.

reference map [MAP DESIGN] A map designed to show where geographic features are in relation to each other. *See also* map, thematic map.

reference scale [SYMBOLOGY] The scale at which symbols appear on a digital page at their true size, specified in page units. As the extent is changed, text and symbols will change scale along with the display. Without a reference scale, symbols will look the same at all map scales. *See also* scale.

reference spheroid *See* reference ellipsoid.

reference system

1. [COORDINATE SYSTEMS] A method for identifying positions on the globe. This is often constructed with a grid that either refers to the earth's latitude and longitude (graticule), or a planar equivalent that divides grid lines by a fixed length from a predefined point of origin.
2. [MODELING, STANDARDS] A collection of abstract principles, numerical standards, conventional values, models, and specifications for quantitatively describing the positions of points in space and how these positions vary over time.

See also graticule, ITRS, spherical coordinate system.

referential integrity [DATA QUALITY] A mechanism for ensuring that data remains accurate and consistent as a database changes. When changes are made to a table related to another table by a common key, the changes are automatically reflected in both tables.

reflectance

1. [PHYSICS] The proportion of incident radiant energy that is reflected by a surface. Reflectance varies according to the wavelengths of the incident radiant energy and the color and composition of the surface.
2. [PHOTOGRAMMETRY] In imagery, the ratio of light reflected from an object or that is captured by a target above the object. This ratio is wavelength-dependent, and the value's range is between 0 and 1 or can be expressed as a percentage.

See also albedo, radiation, wavelength.

reflective optics [PHOTOGRAMMETRY] A lens that uses glass to bend light and focus it onto a focal plane. *See also* refractive optics.

refractive optics [PHOTOGRAMMETRY] A lens that uses a mirror to focus light onto a focal plane. Typically used to provide lightweight, long focal length optics for sensors. *See also* reflective optics.

region

1. [GEOGRAPHY] In geography, an area usually distinguished by common cultural or physical characteristics, such as Southern California, Western Europe, or Southeast Asia.

region 1

2. [DATA MODELS] A set of contiguous cells with the same value.
3. [ESRI SOFTWARE] In the coverage data structure, a polygon feature made up of multiple polygons that may be separate, overlapping, nested, or adjacent. The polygons that compose a region are stored in a polygon feature class, whereas the region is stored in a subclass of this feature class. A region has its own attributes but no shape geometry; its shape is defined by the shape geometry of the polygons that compose it.

See also cell, cultural feature, natural feature, polygon feature.

register

1. [DATA EDITING] To align two or more maps or images so that equivalent geographic coordinates coincide.
2. [GEOLOCATING] To link map coordinates to ground control points.

See also map, tic.

registration [DATA MANAGEMENT, COORDINATE SYSTEMS] Aligning different datasets or layers to a common coordinate system or reference framework. *See also* rescale, rotate, translate.

regression [STATISTICS] A statistical method for evaluating the relationship between a single dependent variable and one or more independent variables thought to influence the dependent variable. Regression is used to predict the value of the dependent variable or to determine whether an independent variable in fact influences the dependent variable, and to what extent. *See also* dependent variable, independent variable, regression equation.

regression coefficient [STATISTICS] A value associated with each independent variable in a regression equation, representing the strength and type of relationship the independent variable has to the dependent variable. For example, fire frequency might be modeled as a function of solar radiation, vegetation, precipitation, and aspect.

A positive relationship between fire frequency and solar radiation is likely (more sun increases the frequency of fire incidents). When the relationship is positive, the sign for the associated coefficient is also positive. A negative relationship between fire frequency and precipitation is also likely (places with more rain have fewer fires). Coefficients for negative relationships have negative signs. If the relationship is strong, the absolute value of the coefficient is large. Weak relationships are associated with coefficients near zero.

See also dependent variable, independent variable, regression equation.

regression equation [STATISTICS] The mathematical formula applied to independent variables to predict the dependent variable being modeled. The notation in regression equations is always Y for the dependent variable and X for the independent variables. Each independent variable is associated with a regression coefficient describing the strength and sign of that variable's relationship to the dependent variable. A regression equation might look like this (where b represents a regression coefficient):

$$Y = b_0 + b_1X_1 + b_2X_2 + \dots b_nX_n.$$

See also dependent variable, independent variable, regression coefficient.

regular spatial arrangement [DATA MANAGEMENT] A spatial configuration with equal spacing between geographic features. *See also* spatial configuration.

reject processing [GEOCODING] Handling unmatched addresses through fine-tuning the geocoding process. After a table of addresses is matched the first time, unmatched addresses can be reviewed or edited. Reject processing attempts to find possible matches by correcting errors or adjusting search criteria for the addresses that fail the first time. *See also* address matching, geocoding.

relate [DATABASE STRUCTURES] An operation that establishes a temporary connection between records in two tables using a key common to both. *See also* joining, key.

relational database [DATABASE STRUCTURES] A data structure in which collections of tables are logically associated with each other by shared fields. *See also* joining, primary key, relate.

relational join *See* joining.

relational operator [SPATIAL ANALYSIS] An expression used to compare values associated with data: greater than, less than, maximum, minimum, contains, and so on. *See also* operator.

relative accuracy [ACCURACY] A measure of positional consistency between a data point and other, near data points. Relative accuracy compares the scaled distance of objects on a map with the same measured distance on the ground. *See also* absolute accuracy.

relative bearing [NAVIGATION] A bearing measured relative to a vessel or aircraft's heading. *See also* bearing.

relative coordinates [MAP PROJECTIONS] Coordinates identifying the position of a point with respect to another point. *See also* absolute coordinates.

relative direction [WAYFINDING] Direction relative to your own body. The most common relative directions are left, right, forward, backward, up, and down. Also called egocentric direction.

relative relief [GEOGRAPHY] The local range (absolute difference) between high and low elevations. *See also* elevation, range, relief.

relative relief mapping method [MAP DESIGN] A mapping method designed to give a general impression of the relative heights of different landform features. *See also* mapping methods, relative relief.

relative steepness [GEOGRAPHY] Steepness of a hillside in relation to the steepness of surrounding hillsides.

reliability diagram [CARTOGRAPHY] A map element that contains a simplified view of the sources used to compile a map. *See also* map element.

relief [CARTOGRAPHY] Elevations and depressions of the earth's surface, including those of the ocean floor. Relief can be represented on maps by contours, shading, hypsometric tints, digital terrain modeling, or spot elevations. *See also* contour, shading, hypsometric tinting, digital terrain model, spot elevation.

relief

relief displacement [PHOTOGRAMMETRY] The radial distance between where an object appears in an image to where it should be located according to a planimetric coordinate system. This dislocation of ground features in an image with relief displacement is due to terrain relief. Relief displacement is corrected by imagery orthorectification. *See also* positional displacement, radial displacement.

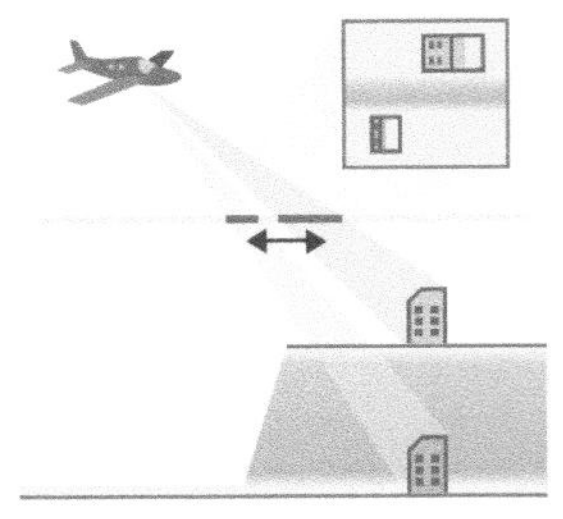

relief displacement

relief portrayal [MAP DESIGN] A mapping technique used to portray the terrain surface; a deliberate attempt on the part of the mapmaker to give a sense of the three-dimensional nature of the surface. *See also* relief, terrain surface.

relief reversal [GRAPHICS (MAP DISPLAY)] The undesirable effect that occurs in map design when hills look like valleys and valley bottoms look like ridgetops. This effect is most pronounced when illumination (by the sun or an imaginary light source) is at an oblique angle, such as at the lower right.

relief shading *See* hillshading.

rematching [GEOCODING] The process of regeocoding a feature or features in a geocoded feature class.

remote sensing [REMOTE SENSING] Collecting and interpreting information about the surface and environment of the earth or other celestial bodies from a distance. Remote sensing gathers data by sensing radiation that is naturally emitted or reflected by the object's surface or from the atmosphere, or by sensing signals transmitted from a device and reflected back to it. Examples of remote sensing methods include aerial photography, radar, and satellite imagery. *See also* aerial photograph, radar, satellite imagery, sensor.

remote sensing imagery [REMOTE SENSING] Imagery acquired from satellites, aircraft, and drones, including panchromatic, radar, microwave, color, and multispectral imagery. *See also* multispectral image, pixel, satellite imagery, sensor.

remote sensing imagery

renderer [ESRI SOFTWARE] A mechanism that defines how data appears when displayed. For example, the hillshade renderer for raster data calculates and applies shading based

R

on existing data values for slope and aspect. *See also* data, raster.

rendering [GRAPHICS (COMPUTING)] The process of drawing to a display; the conversion of the geometry, coloring, texturing, lighting, and other characteristics of an object into a display image. *See also* raster statistics, streaming, stretch.

replacement sensor model [REMOTE SENSING] In remote sensing and photogrammetry, a nonphysical sensor model with a ground-to-image transformation, adjustable parameters, and covariance terms. Computationally equivalent to their physical counterparts, these models mask the physical design of the sensor and provide a mechanism to build a multisensory bundle adjustment. *See also* RPC, sensor.

representation

1. [CARTOGRAPHY] A method of illustrating data so it can be viewed and understood. In cartography, representation is used to depict likenesses of real-world features in such a way that the depictions symbolize or correspond to the real features. Representation is used to present information in a format that is viewable, storable, and transferable.
2. [SYMBOLOGY] A visual likeness or depiction of an entity that acts as a substitute for the actual entity.
3. [ESRI SOFTWARE] Extra information added to a feature or feature class that defines the rules and overrides for display on a map.

See also real world, symbology, visualization.

representative fraction [CARTOGRAPHY] The ratio of a distance on a map to the equivalent distance measured in the same units on the ground. A scale of 1:50,000 means that one inch on the map equals 50,000 inches on the ground. *See also* scale.

reprojection [COORDINATE SYSTEMS] The process of mathematically converting a map or raster from one projected coordinate system to another. Typically used to integrate maps from two or more projected coordinate systems within a GIS. *See also* coordinate system, geographic transformation.

resampling [MATHEMATICS] The process of interpolating new cell values when transforming rasters to a new coordinate space or cell size. *See also* bilinear interpolation, cubic convolution, majority resampling, nearest neighbor resampling.

rescale [DATA MANAGEMENT] Adjusting the dimensions or resolution of a dataset while maintaining its spatial relationships and proportions, such as enlarging or reducing the size of a map or image. *See also* registration, rotate, scale, translate.

resection [PHOTOGRAMMETRY] A computational technique that uses ground reference points with known coordinates to locate the exterior orientation (camera station, coordinates, and attitudes) of an image. Uses three or more image rays, ensuring that for each ray, the image point, sensor's perspective center, and ground point are all collinear, providing parameters to correlate the image coordinates to ground coordinates with the use of a sensor model. *See also* camera station, coordinates.

resection method [WAYFINDING] A technique for locating position that involves plotting lines that cross, or intersect, at the observer's position. *See also* compass method, inspection method.

residual [STATISTICS] In a regression model, the difference between the observed y-value and the predicted y-value; the unexplained portion of the dependent variable. Predicted values rarely match observed values exactly. The residual is one measure of model fit. Large residuals indicate poor model fit. *See also* dependent variable, independent variable, regression equation, trend, trend surface analysis.

resolution

1. [CARTOGRAPHY] The detail with which a map depicts the location and shape of geographic features. The larger the map scale, the higher the possible resolution. As scale decreases, resolution diminishes and feature boundaries must be smoothed, simplified, or not shown at all; for example, small areas may have to be represented as points.

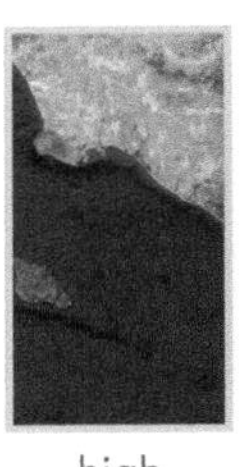

resolution 1

2. [GRAPHICS (MAP DISPLAY)] The dimension of each cell or pixel in raster data that represents the area covered on the ground. Often used interchangeably with spatial resolution.
3. [GRAPHICS (COMPUTING)] The smallest spacing between two display elements, expressed as dots per inch, pixels per line, or lines per millimeter.
4. [ESRI SOFTWARE] The smallest allowable separation between two coordinate values in a feature class. A spatial reference can include x, y, z, and m resolution values. The inverse of a resolution value was previously known as a precision or scale value.

See also cell, dpi, level of detail, scale, spatial resolution.

resolution merging *See* pan sharpening.

resolving power [REMOTE SENSING] The amount of detail a sensor can capture in each image, which is determined by a combination of atmospheric conditions, a sensor's range to the ground, the quality of the sensor's optics, and the resolution of the focal plane. *See also* sensor.

restriction [NETWORK ANALYSIS] A Boolean network element attribute used for limiting traversal through a network dataset. "One-way street," "no trucks allowed," and "buses only" are examples of restrictions. *See also* cost, descriptor, hierarchy, network attribute.

reverse geocoding [GEOCODING] The process of finding a street address from a point on a map. *See also* geocoding.

RF *See* representative fraction.

RGB

1. [GRAPHICS (COMPUTING)] Acronym for *red, green, blue.* A color model that uses red, green, and blue, the primary additive colors used to display images on a monitor. RGB colors are produced by emitting light, rather than by absorbing it as is the case with ink on paper. Adding 100 percent of all three colors results in white.

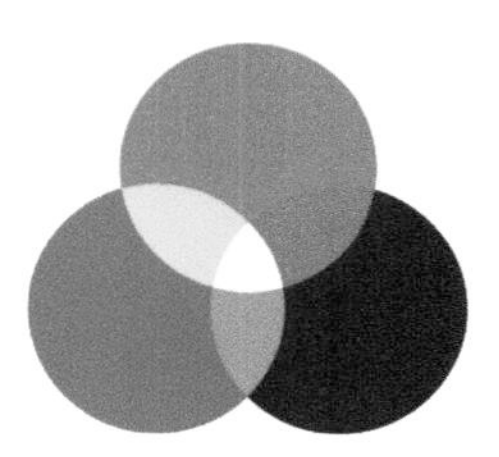

RGB 1

2. [DIGITAL IMAGE PROCESSING] The three components used to specify a 24-bit (true) color.

See also HSV.

rhumb line [GEODESY] A complex curve on the earth's surface that crosses every meridian at the same oblique angle. A rhumb line path follows a single compass bearing; it is a straight line on a Mercator projection, or a logarithmic spiral on a polar projection. A rhumb line is not the shortest distance between two points on a sphere. *See also* great circle, sphere.

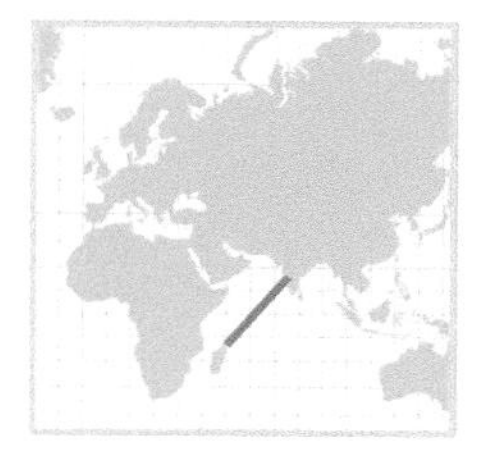

rhumb line

ring [ESRI SOFTWARE] A geometric element from which polygons are constructed. A ring is any closed path (one that begins and ends at the same point). *See also* path.

ring study [ANALYSIS/GEOPROCESSING, BUSINESS] The simplest and most widely used type of market-area analysis, in which a circle is generated around an area on a map. The underlying demographics are then extracted from the area delineated by the circle. Generally, a ring study is used to generate a rough visualization of the market area around a point. *See also* market area.

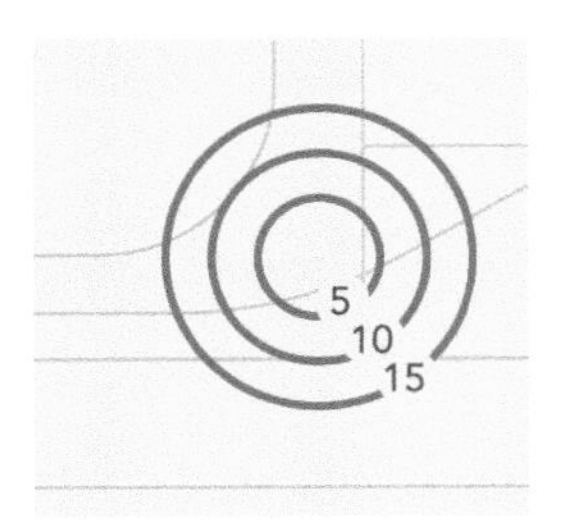

ring study

rise [MAP DESIGN] The elevation difference between two locations used in a "rise over run" slope determination. *See also* run, slope, slope ratio.

river addressing [LINEAR REFERENCING] In hydrology applications, another name for linear referencing. River addressing allows objects such as gauging stations to be located by their relative positions along a line feature. *See also* linear referencing.

RMS error [GEOREFERENCING] RMS is an acronym for *root mean square*. A measure of how well georeferenced points align with real-world geography in spatial data calculations. A low RMS error indicates generally good alignment; a high RMS error indicates a higher probability of stretching or shrinking. *See also* georeferencing, real world, rubber sheeting.

RMSE *See* RMS error.

road and recreation atlas [NAVIGATION] A book for travel and trip planning that contains a series of maps covering large regions. *See also* atlas.

road map [NAVIGATION] A map designed to aid road navigation but with many general purposes.

roamer [NAVIGATION] A transparent gauge that represents easting and northing distances at a given map scale, used to locate positions on a map. *See also* easting, northing.

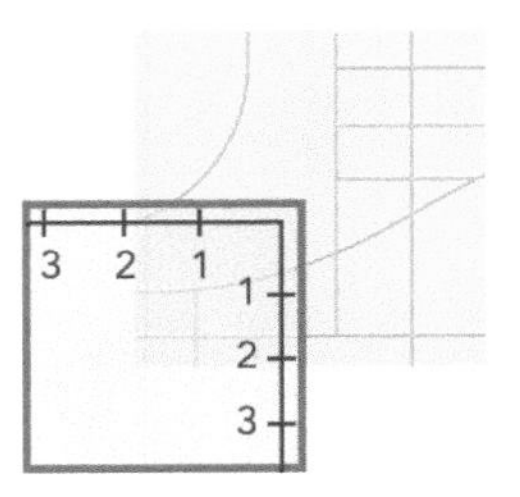

roamer

Robinson projection [MAP PROJECTIONS] A pseudo-cylindrical map projection that is neither equal area nor conformal. Named for U.S. geographer and cartographer Arthur H. Robinson (1915–2004). *See also* projection.

root mean square error *See* RMS error.

rotate [DATA MANAGEMENT, COORDINATE SYSTEMS] The movement of a dataset around a specific point or axis to change its orientation. *See also* registration, rescale, translate.

route

1. [LINEAR REFERENCING] Any line feature, such as a street, highway, river, or pipe, that has a unique identifier.
2. [NETWORK ANALYSIS] A path through a network.

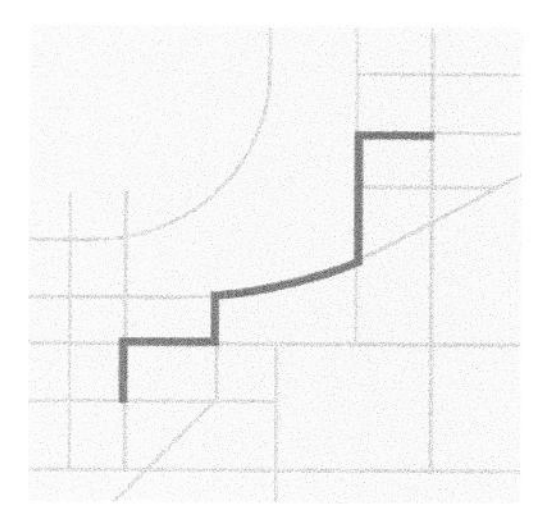

route 2

3. [NETWORK ANALYSIS, ESRI SOFTWARE] A path through a network that visits a set of specified network locations. In vehicle routing problem (VRP) analysis, a route may also refer to a vehicle and its associated properties and constraints.

See also network, path, stop.

route analysis [NETWORK ANALYSIS] A type of network analysis that determines the best route from one network location to one or more other locations. It can also calculate the quickest or shortest route depending on the impedance chosen. The order of the stops may be determined by the user. For example, if the impedance is time, the best route is the quickest route. *See also* linear referencing, network analysis.

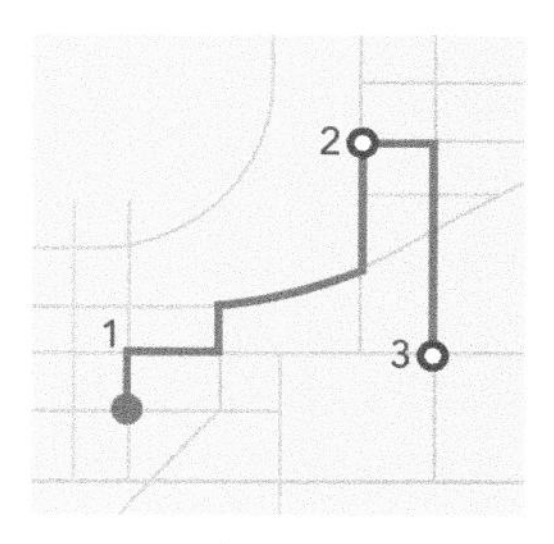

route analysis

route event [LINEAR REFERENCING] In linear referencing, linear, continuous, or point features occurring along a base route system. *See also* event, route event table.

route event source [LINEAR REFERENCING, ESRI SOFTWARE] In linear referencing, a component that stores and manages information about events along a route. Events can include incidents, accidents, or points of interest. Events are typically stored in an event table that can be used as a dynamic feature class; each table row is a feature that can be calculated as needed. *See also* dynamic feature class, dynamic segmentation, route event table.

route event table [LINEAR REFERENCING, ESRI SOFTWARE] In linear referencing, a table that stores route locations and their attributes. A route event table, at a minimum, consists of a route identifier field and a measure location field (point events) or fields (line events). *See also* event table, route event, route event source.

route feature class *See* route reference.

route identifier [LINEAR REFERENCING] In linear referencing, a numeric or character value used to identify a route. *See also* identifier, linear referencing.

route location [LINEAR REFERENCING] In linear referencing, a discrete location on a route (a point) or a portion of a route (a line). A point route location uses a single measure value to describe the location. A line route location uses both a from- and to-measure value to describe the location. *See also* linear referencing, point.

route measure [LINEAR REFERENCING] In linear referencing, a value stored along a linear feature that represents a location relative to the beginning of the feature, or some point along it, rather than as an x,y coordinate. Measures are used to map events such as distance, time, or addresses along line features. *See also* dynamic segmentation, hatches, linear referencing, m-value, route.

route measure anomalies [LINEAR REFERENCING] In linear referencing, route measure values that do not adhere to the expected behavior. *See also* route.

route reference [LINEAR REFERENCING] In linear referencing, a collection of routes with a common system of measurement stored in a single feature class (for example, a set of all highways in a county). *See also* linear referencing, route.

route renewal [NETWORK ANALYSIS] In network analysis, an object used in vehicle routing problem (VRP) analysis. A route renewal object specifies a depot that can be used by a particular route to load/unload the vehicle along the route as necessary so that the capacity is reset and the route can service more orders. *See also* depot, vehicle routing problem.

route service [INTERNET] A type of web service that determines driving directions between a set of route stops. *See also* linear referencing, network, route.

route zone [NETWORK ANALYSIS, ESRI SOFTWARE] A feature used in vehicle routing problem (VRP) analysis. A route zone has a polygon geometry and can be used to define the area of coverage or available service for a specified route. *See also* linear referencing, vehicle routing problem.

routing service *See* route service.

rover [GPS] A portable GPS receiver used to collect data in the field. The rover's position can be calculated relative to a second, stationary GPS receiver. *See also* GPS.

roving window *See* filter.

row [DATA MODELS] A horizontal group of cells in a raster, or pixels in an image. *See also* cell, pixel, raster.

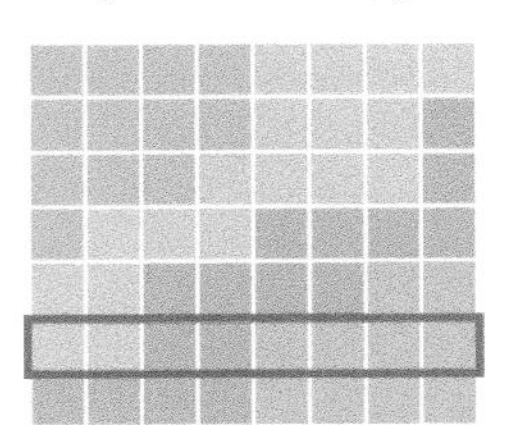

row

row standardization [SPATIAL STATISTICS (USE FOR GEOSTATISTICS)] A technique for adjusting the weights in a spatial weights matrix. When weights are row standardized, each weight is

R

divided by its row sum. The row sum is the sum of weights for a feature's neighbors. *See also* spatial weights matrix, weight.

row, column [PHOTOGRAMMETRY] A method of referencing the location of a pixel in a raster or image space. *See also* grid, pixel, raster.

RPC [REMOTE SENSING] Acronym for *rational polynomial coefficients.* In remote sensing and photogrammetry, typically 3rd–5th order polynomials that are used in ratios to provide a ground-to-image transformation. RPCs are generated from physical sensor models and can be accurate to subpixel levels, provided that the input to the generator software is accurate to that granularity. They are stable for long focal length, narrow field of view sensors, such as space-borne sensors. Because they lack a covariance model, RPCs are not considered to be full replacement sensor models. *See also* transformation.

RPF [DATA STRUCTURES] Acronym for *raster product format.* A data format composed of rectangular pixel arrays (compressed or uncompressed), produced by the National Geospatial-Intelligence Agency and U.S. allies for military applications. *See also* raster.

R-squared [STATISTICS] A statistic computed by the regression equation to quantify model performance. The value of R-squared ranges from 0 to 100 percent. If a model fits the observed dependent variable values perfectly, the R-squared value is 1.0, although this is highly unlikely. An R-squared value like 0.49, for example, is far more likely, and means that the model explains 49 percent of the variation in the dependent variable. *See also* dependent variable, independent variable.

rubber banding *See* rubber sheeting.

rubber sheeting [DATA EDITING] A procedure for adjusting the coordinates of all points in a dataset to accurately match identifiable points in the dataset to known map locations. The dataset may be composed of image (raster) or feature (vector) data. Rubber sheeting moves not only the identifiable points to new x,y coordinates but also the neighboring data, as if the map or image was printed on a rubber sheet. Rubber sheeting preserves the interconnectivity between points and objects through stretching, shrinking, or reorienting their interconnecting lines (in a feature dataset), or their neighboring pixels (in an image). *See also* edgematching.

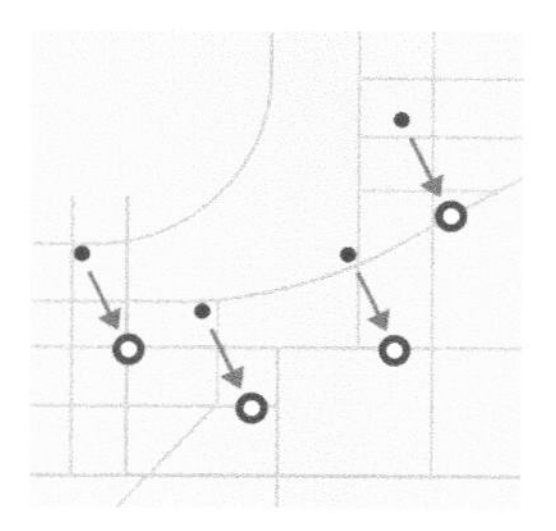

rubber sheeting

run [MAP DESIGN] The horizontal ground distance component used in a "rise over run" slope determination. *See also* rise, slope, slope ratio.

R

S/A *See* selective availability.

sampling

1. [DATA CAPTURE] The collection of data for only a portion of the region or population used to make inferences about the characteristics of the population from which the portion (sample) was drawn.
2. [REMOTE SENSING] In imagery, the capture and recording of a discrete, spatial, electromagnetic signal with a sensor. For electro-optical sensors, sampling rate is determined by the optics and spacing of photosensitive cells on the focal plane, resulting in a point spread function and other source image characteristics. If an image is geometrically processed further, it is considered resampled.

See also sampling method, sensor.

sampling method [DATA CAPTURE] The means used to obtain a sample from a spatial setting or from the total population. *See also* sampling.

sampling scheme [REMOTE SENSING] In remote sensing, a methodology that defines the rules for obtaining sampling analysis data to ensure accuracy, precision, and lack of bias. A reference dataset and derived dataset are sampled according to a random sampling scheme—such as stratified random, measure accuracy, and error. *See also* derived data, reference data, sampling.

satellite constellation

1. [REMOTE SENSING] A group of artificial satellites working together as a system.

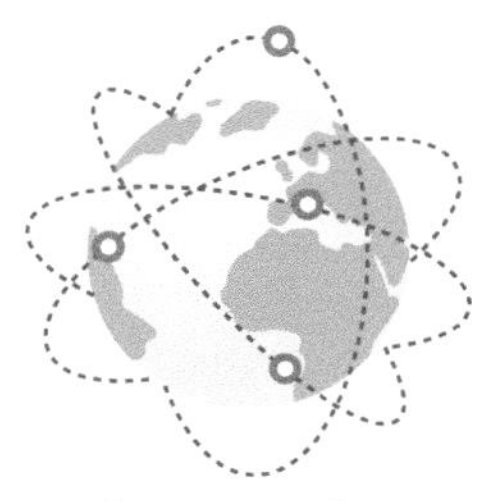

satellite constellation 1

2. [GPS] All the satellites that compose a GPS system.

See also data message, GNSS.

satellite imagery [SATELLITE IMAGING] Data collected by sensors on satellite platforms, typically of the earth. Satellite imagery often consists of multiple bands; sensors can be both active and passive. Satellite imagery metadata includes characteristics such as sensor position and pointing geometry, sensor gain and bias, acquisition date and time, and associated data. *See also* band, data message, GPS, sensor.

satellite imagery map [MAP DESIGN] A conventional map that uses satellite imagery as the base. *See also* satellite imagery.

satellite navigation [GPS] A satellite constellation that provides independent geopositioning data. China's BeiDou Navigation Satellite System, the European Union's Galileo, Russia's GLONASS, and the United States' GPS are a combined global navigation satellite system (GNSS). India's NavIC is a regional navigation satellite system (RNSS). Japan's QZSS is both an RNSS and a satellite-based augmentation system (SBAS), designed to enhance the accuracy of GNSS. *See also* BeiDou Navigation Satellite System, Galileo, geopositioning, GLONASS, GPS, GPS augmentation, NavIC, QZSS, satellite constellation.

saturation

1. [GRAPHICS (MAP DISPLAY)] The intensity or purity of a color; the perceived amount of white in a hue relative to its brightness, or how free it is of gray of the same value.

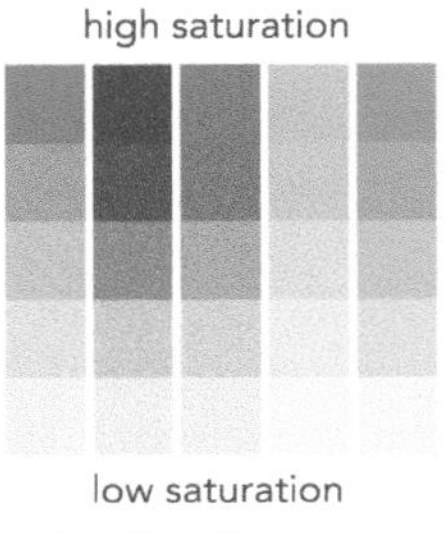

saturation 1

2. [PHYSICS] The point at which energy flux exceeds the sensitivity range of a detector.
3. [METEOROLOGY] The condition that exists when air holds the maximum possible amount of water vapor for its temperature and atmospheric pressure; when the relative humidity is 100 percent.

See also chroma, hue, intensity, value.

SBET [GEOREFERENCING] Acronym for *smoothed best estimate of trajectory*. In direct georeferencing, a process of adjusting trajectory using Global Navigation Satellite Systems (GNSS) and inertial data with a Kalman filter, running it forward and reversed in time and then averaging the result. *See also* GNSS.

scalable [COMPUTING] The ability to grow in size or complexity without showing negative effects.

scalable vector graphics *See* SVG.

scale

1. [CARTOGRAPHY] A ratio between the distance shown on a map and the corresponding distance on the ground. For example, a map scale of 1:100,000 means one unit of measure on the map represents 100,000 of those same units on the earth.
2. [DATA QUALITY] In mathematical or double precision, the number of digits to the right of a decimal point in a number. For example, the number 567.89 has a scale of 2.

See also large scale, representative fraction, scale bar, small scale.

scale bar [SYMBOLOGY] A map element—typically a line marked like a ruler—used to graphically represent proportional distance on the earth. *See also* map element, scale, tick marks.

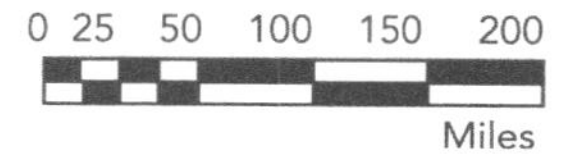

scale bar

scale bar extension [SYMBOLOGY] An extension to the left of the zero point on the scale bar that allows the determination of distance not only in whole units, but also in fractions of units, such as tenths of a mile or kilometer. *See also* map element, scale bar.

scale conversion [SYMBOLOGY] The conversion of map scale between a verbal scale, representative fraction, and scale bar. *See also* representative fraction, scale, scale bar, verbal scale.

scale factor

1. [CARTOGRAPHY] The reciprocal of the ratio used to specify scale on a map. For example, if the scale of a map is given as 1:50,000, the scale factor is 50,000.
2. [COORDINATE SYSTEMS] In a coordinate system, a value (usually less than one) that converts a tangent projection to a secant projection, represented by "k0" or "k." If a projected coordinate system doesn't support a scale factor, the standard lines of the projection have a scale factor of 1.0. Other points on the map have scale factors greater or less than 1.0. If a projected coordinate system supports a scale factor, the defining parameters no longer have a scale factor of 1.0.
3. [CARTOGRAPHY] The relation between the denominators of the representative fractions for the actual and principal scales at particular locations on the map.

See also actual scale, principal scale, representative fraction, secant projection, tangent projection.

scale range [MAP DISPLAY] The scales at which a layer is visible on a web map. Scale ranges are commonly used to prevent detailed layers from displaying at small scales (zoomed out) and to prevent general layers from displaying at large scales (zoomed in). *See also* range, representation, scale, web map.

scale variation [MAP PROJECTIONS] Variation in scale from place to place on a map that is a result of the projection process. No map projection maintains correct map scale throughout. *See also* scale, scale factor.

scanner [DATA CAPTURE] A device that captures a print or hard-copy image, such as a text document or map, and records the information in digital format.

scanning [DATA CAPTURE] The process of capturing data from hard-copy maps or images in digital format using a device called a scanner. *See also* scanner.

scatter chart [STATISTICS] A chart or graph in which the relation between two variables is shown with increasing values scaled along perpendicular x- and y-axes. Scatter charts are frequently used in analysis to find data trends. Commonly known as a scatterplot. *See also* chart, QQ plot, trend.

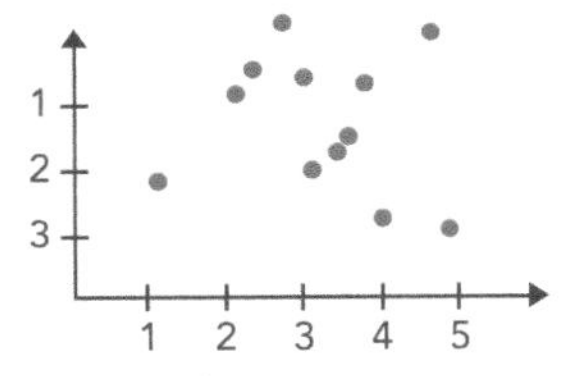

scatter chart

scatterplot *See* scatter chart.

scene [3D ANALYSIS] In 3D analysis, a method of viewing 3D data in perspective. *See also* 3D scene, camera.

schema [COMPUTING] The structure or design of a database or database object, such as a table, view, index, stored procedure, or trigger. In a relational database, the schema defines the tables, the fields in each table, the relationships between fields and tables, and the grouping of objects within the database. Schemas are generally documented in a data dictionary. A database schema provides a logical classification of database objects. *See also* data dictionary.

S-code *See* civilian code.

scrubbing [QUALITY CONTROL] Improving the appearance of data by closing open polygons, fixing overshoots and undershoots, refining thick lines, and so on. *See also* data conversion, overshoot, undershoot.

SDC feature class [ESRI SOFTWARE] A highly compressed, read-only data structure that can store spatial geometry (points, lines, and polygons) and attribute data. The SDC structure supports geocoding, routing, and most spatial operations. *See also* feature class.

SDCM [GPS] Acronym for the *System for Differential Corrections and Monitoring.* A regional satellite-based augmentation system (SBAS) operated by the Russian space agency, Roscosmos, that monitors and provides correction data to improve the precision of GLONASS over Russian airspace. *See also* GLONASS, GPS augmentation.

SDI [DATA SHARING] Acronym for *spatial data infrastructure.* A framework of technologies, policies, standards, and human resources necessary to acquire, process, store, distribute, and improve the use of geospatial data across multiple public and private organizations. *See also* global spatial data infrastructure, National Spatial Data Infrastructure.

SDK [PROGRAMMING] Acronym for *software development kit.* A set of code libraries and related tools used to develop platform-specific applications.

See also application programming interface.

SDTS [STANDARDS] Acronym for *Spatial Data Transfer Standard.* A data exchange protocol developed by the American National Standards Institute used for the archiving and transfer of spatial data (including metadata) between computer systems. All U.S. federal agencies are required to make digital map data available in SDTS format upon request. *See also* American National Standards Institute, data transfer.

seamline [PHOTOGRAMMETRY] In imagery, a polygon or polyline defining the boundary between adjoining images (raster datasets) in an image mosaic. Overlapping raster datasets can be blended and color balanced along the seamline by a specified width. *See also* image service, mosaic, raster dataset.

search [GEOLOCATING] To query a map for a specified location; to query a set of records based on a specific keyword or parameters. *See also* geolocation, keyword.

search radius [ANALYSIS/GEOPROCESSING, ESRI SOFTWARE] The maximum distance in coverage units a feature can be from the current point for consideration as the closest feature. The default is the width or height of the near coverage BND (boundary) divided by 100, whichever is larger.

search tolerance [NETWORK ANALYSIS] In network analysis, the threshold distance used to find the closest network element to a network location. *See also* tolerance.

searching neighborhood [SPATIAL STATISTICS (USE FOR GEOSTATISTICS)] In spatial interpolation, a polygon that forms a subset of data around the prediction location. Only data within the searching neighborhood is used for interpolation. *See also* interpolation.

secant [MATHEMATICS] A straight line that intersects a curve or surface at two or more points. *See also* line.

secant

secant projection [MAP PROJECTIONS] A projection whose surface intersects the surface of a globe either at a circle of tangency, for planar projections, or along two lines of tangency, for conic or cylindrical projections. At the lines of intersection, the projection is free from distortion. *See also* tangent projection.

secant projection

second

1. [GEODESY] An angle equal to one sixtieth of a minute of latitude or longitude.
2. [MATHEMATICS] An angle equal to one sixtieth of a minute of arc.

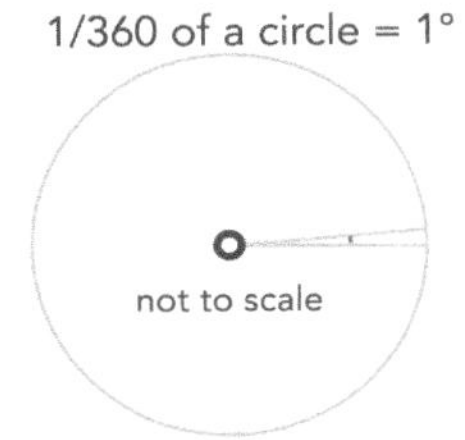

second 2

3. [PHYSICS] One sixtieth of a minute of time.

See also arcsecond, degree, degrees-minutes-seconds, minute.

second normal form [DATABASE STRUCTURES] The second level of guidelines for designing table and data structures in a relational database. The second-normal-form guideline incorporates the guidelines of first normal form; in addition, it recommends removing data that applies to multiple rows in a table into its own table and using a foreign key to create a relationship to the original table. A database that follows these guidelines is said to be in second normal form. *See also* first normal form, normal form, third normal form.

secondary reference data [GEOCODING] All material used as reference data in an address locator beyond the primary reference data. *See also* alternate name, locator, place-name alias, primary reference data, reference data.

second-order stationarity [SPATIAL STATISTICS (USE FOR GEOSTATISTICS)] In geostatistics, the assumption that a set of data comes from a random process with a constant mean, and spatial covariance that depends only on the distance and direction separating any two locations. *See also* stationarity.

section

1. [NETWORK ANALYSIS] The arcs or portions of arcs used to define a route.
2. [CADASTRAL AND LAND RECORDS] An area unit of one square mile (640 acres) in the U.S. Public Land Survey System and the equivalent 2.59 square kilometers in Canada's Dominion Land Survey. Sections are bounded by parallels and meridians and are one thirty-sixth of a township. Some sections may actually be less than 640 acres/2.59 square kilometers because of the convergence of meridians.

See also arc, DLS, PLSS, township.

section corner [SURVEYING] The point of intersection of any two survey lines, defining one corner of a section. *See also* section.

sectional aeronautical chart [NAVIGATION] A chart at 1:500,000 scale designed for visual air navigation in slow-moving to medium-speed aircraft. *See also* aeronautical chart.

segment [ESRI SOFTWARE] A geometric element from which paths are constructed. A segment consists of a start point, an endpoint, and a function that describes a straight line or curve between these two points. Curves may be circular arcs, elliptical arcs, or Bézier curves. *See also* path, Bézier curve.

segment

segmented symbol [SYMBOLOGY] A symbol in which the parts show the relative magnitudes of subcategories of attributes, such as a pie chart. *See also* symbol.

selective availability [GPS] The intentional degradation by the U.S. Department of Defense of the GPS signal for civilian receivers, which could cause errors in position of up to 100 meters. Selective availability (S/A) was removed from the civilian signal in May 2000. Since this change, position accuracy levels have improved to 20 meters or less. *See also* GPS.

self-organizing map *See* Kohonen map.

semantic search [COMPUTING] A search engine technique, augmented with a pretrained AI language model, that returns the most semantically similar results to a search query. Contrasted with a lexical search, which performs keyword matching, a semantic search retrieves results that are similar in meaning to, but do not exactly match, the original search string. Within GIS, semantic search can improve search results based on relevant spatial data. For example, when a user searches for historic landmarks, a semantic search can retrieve all relevant spatial data to which the query applies. *See also* AI, machine learning, semantics.

semantics [DATA MODELS] The study or application of attaching meaning to concepts, typically based on their relationship to other concepts within a data model. *See also* data model.

semicross variogram *See* cross variogram.

semimajor axis [GEODESY] The slightly longer equatorial radius of the ellipsoid, often referred to as "*a*." *See also* ellipsoid, radius, semiminor axis.

semimajor axis

semimajor axis

semiminor axis [GEODESY] The slightly shorter polar radius of the ellipsoid, often referred to as "*b*." *See also* ellipsoid, radius, semimajor axis.

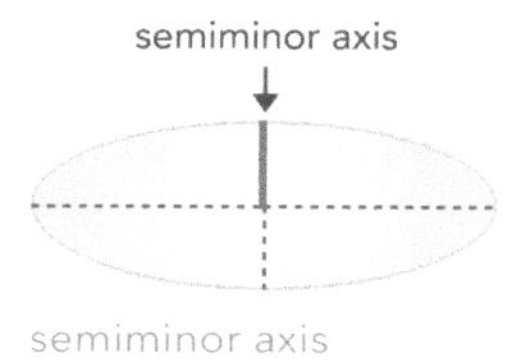

semiminor axis

semiperimeter [MATHEMATICS] Half of the perimeter of a polygon. *See also* perimeter, polygon.

semivariogram [SPATIAL STATISTICS (USE FOR GEOSTATISTICS)] The variogram divided by two. *See also* variogram.

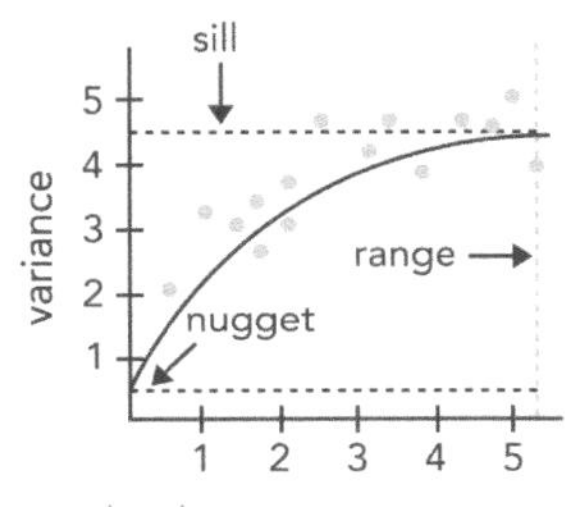

semivariogram

sense of place [COGNITION] One's perception of the essential character of a place in which one resides or has resided, stemming from a personal response to the environment. Sense of place usually refers to perceptions of a neighborhood or city but can also describe feelings about a larger region, state, or country. *See also* mental map.

sensed pixel [REMOTE SENSING] In remote sensing, a sample of sensed electromagnetic energy traceable to a single physical cell in a sensor whose ground sample distance is defined by the point spread function of the optical system and ground location. A pixel in such an image comes directly from the sensor with only sensor corrections applied. *See also* electromagnetic energy, pixel, sensor.

sensitivity analysis [STATISTICS] Analysis designed to test the robustness of a model and analytical results to ensure that small changes in model parameters or data structure do not exhibit large changes in the results. *See also* analysis, optimization.

sensor [REMOTE SENSING] An electronic device that detects electromagnetic energy and converts it into a signal that can be recorded as digital numbers (DN) and displayed as an image. *See also* passive sensors, physical sensor model, remote sensing, replacement sensor model, sensor model.

sensor model [REMOTE SENSING] In remote sensing, a mathematical solution that provides a ground-to-image transformation. Typically has a ground-to-image function, adjustable parameters, and error covariance parameters. Although physical sensor models are based on the physical design of the sensor model, replacement (theoretical) sensor models are not directly modeled. *See also* physical sensor model, replacement sensor model, RPC, sensor.

sequential analysis [STATISTICS] Analysis based on a sample of an unfixed size in which testing continues only until a trend is observed with a predefined level of certainty. *See also* analysis, trend.

sequential color scheme [SYMBOLOGY] A color scheme in which a single hue or a small range of hues that are

S

contiguous in color space vary from light to dark to show data values that range from low to high. *See also* color scheme, hue.

server-side address locator [GEOLOCATING] A locator in which processing is done on one system and made available to other systems. *See also* client-side address locator, locator.

service [COMPUTING] A persistent software process that provides a means to access resources for client applications. *See also* client, feature service, image service, map service.

service area analysis [NETWORK ANALYSIS] A type of network analysis for determining the region that encompasses all accessible streets (streets that lie within a specified impedance). For example, the 20-minute service area for a network location (such as a fire station) includes all the streets that can be reached within 20 minutes from that location. *See also* network analysis.

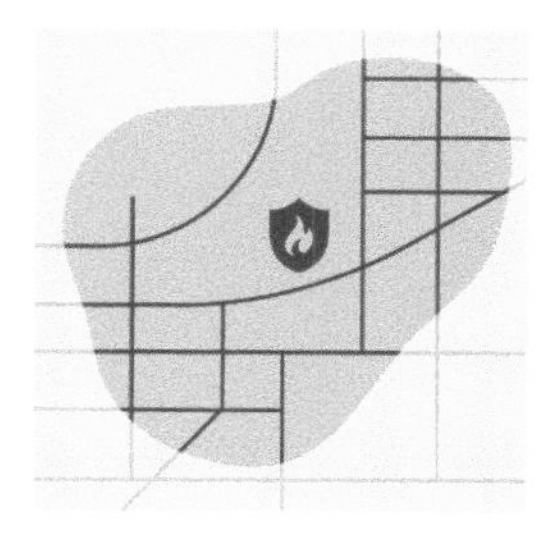

service area analysis

sextant [NAVIGATION] A handheld navigational instrument that measures, from its point of observation, the angle between a celestial body and the horizon or between two objects. The angle is measured on a graduated arc that covers one sixth of a circle (60 degrees). *See also* astrolabe, position, theodolite.

sextant

SHA1 [INTERNET] Acronym for *Secure Hash Algorithm v1.* A Federal Information Processing Standards (FIPS) method for encoding sensitive information before sending the information over the internet. The method jumbles data, such as passwords, to make it difficult to recover the original information.

shade symbol *See* fill symbol.

shaded relief image [VISUALIZATION] A raster image that shows changes in elevation using light and shadows on terrain from a given angle and altitude of the sun. *See also* hillshading.

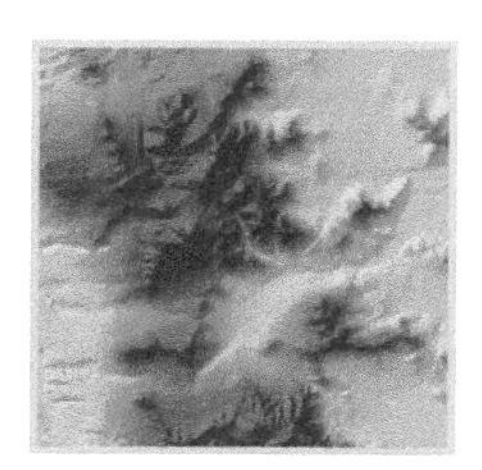

shaded relief image

shading [GRAPHICS (MAP DISPLAY)] Graphic patterns such as cross hatching, lines, or color or grayscale tones that distinguish one area from another on a map. *See also* grayscale, hillshading, relief.

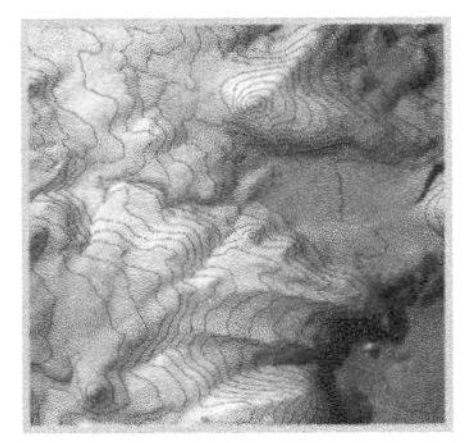

shading

shadow [DIGITAL IMAGE PROCESSING] In imagery, an area visible in the image where direct sunlight or other illumination is blocked by an object such as a tree, building, cloud, and so on. *See also* image element.

shape

1. [DATA MODELS] The characteristic appearance or visible form of a geographic object as represented on a map. A GIS uses points, lines, and polygons to represent the shapes of geographic objects.
2. [DIGITAL IMAGE PROCESSING] An image element that provides information about the form and outline of an object of interest; used to help identify the object.

See also image element, visual variable.

shape index [ONTOLOGIES] A method for describing the shape of an area feature in numeric terms. *See also* shape.

shape measure [ONTOLOGIES] A method for describing the shapes of features in numeric terms. *See also* shape.

shapefile [ESRI SOFTWARE] A vector data storage format for storing the location, shape, and attributes of geographic features. A shapefile is stored in a set of related files and contains one feature class. *See also* data source.

share [DATA SHARING, ESRI SOFTWARE] To make content available to more than one user. For ArcGIS accounts, sharing capabilities are determined by organizational account administrators. *See also* ArcGIS account, credits, group, members, organization, shared content.

shared boundary [DATA MODELS] A boundary common to two features. For example, in a parcel database, adjacent parcels share a boundary. Another example is a polygon that shares a boundary on one side with a river. The segment of the river that coincides with the polygon boundary shares coordinates with the polygon boundary. *See also* boundary, conterminous.

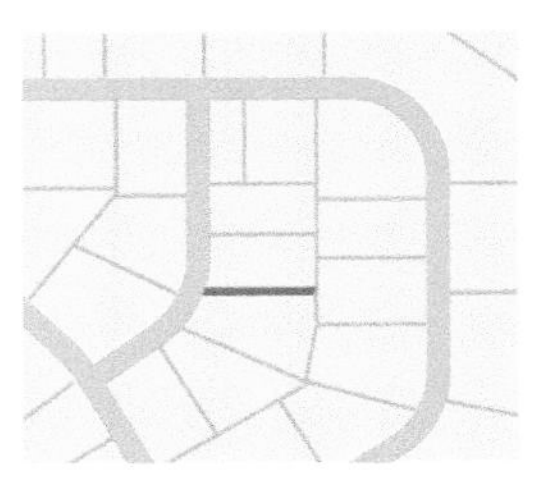

shared boundary

shared content [ESRI SOFTWARE] Data layers, services, maps, applications, and so on that are available to more than one user. *See also* ArcGIS account, organization, share.

shared vertex [DATA MODELS] A vertex common to multiple features. For example, in a parcel database, adjacent parcels share a vertex at the common corner. *See also* feature, topological association, vertex.

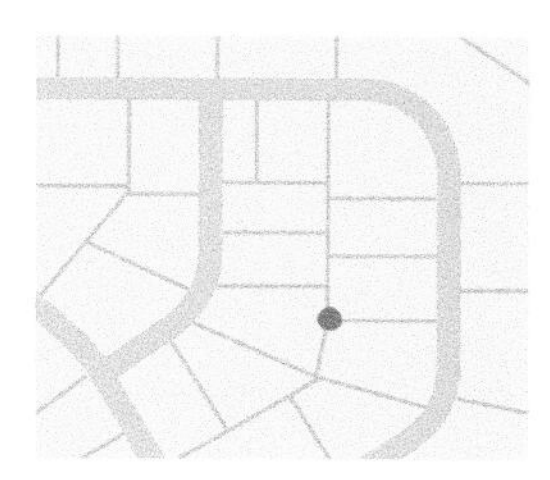

shared vertex

shield [SYMBOLOGY] A map symbol that serves as a route marker. Shields come in many varieties, but the most common shields in the United States are for interstate highways, U.S. routes, state routes, and county routes. A uniform standard exists for interstate highways, U.S. routes, and most county routes across the United States, while shields for state routes vary by state. *See also* marker symbol, route, symbol.

shift line [CARTOGRAPHY] A line between the location of individual features at two different times. *See also* spatial shift.

shortest path

1. [NETWORK ANALYSIS] The best route or the route of least impedance between two or more points, considering connectivity and travel restrictions such as one-way streets and rush-hour traffic.
2. [SPATIAL ANALYSIS, ESRI SOFTWARE] The least-cost path from a destination point to the nearest least-cost source.

See also least-cost path.

short-range variation [SPATIAL STATISTICS (USE FOR GEOSTATISTICS)] In a spatial model, fine-scale variation that is usually modeled as spatially dependent random variation. *See also* spatial modeling, variance.

shortwave infrared [PHYSICS] Also known by the abbreviation *SWIR*. A band of the electromagnetic spectrum that has wavelengths in the 1200 to 2500 nm range. In remote sensing, often used in geology for material identification and smoke penetration. *See also* band, electromagnetic spectrum, infrared radiation.

SIC codes *See* Standard Industrial Classification codes.

side offset [GEOCODING] An adjustable value that dictates how far away from either the left or right side of a line feature an address location should be placed. A side offset prevents a point feature from being placed directly over a line feature. *See also* end offset.

sidelap [REMOTE SENSING] The overlap between photos on adjacent flight lines. *See also* flight line.

signal

1. [REMOTE SENSING] Information conveyed through an electric current or electromagnetic wave.
2. [PHYSICS] The modulation of an electric current, electromagnetic wave, or other type of flow to convey information.

See also carrier, electromagnetic radiation, GPS, sensor.

signal-to-noise ratio [REMOTE SENSING] The ratio of the information content of a signal to its noninformation content (noise). *See also* attenuation, noise.

signature *See* spectral signature.

significance level [STATISTICS] In statistical testing, the probability of an incorrect rejection of the null hypothesis. *See also* confidence level, critical value.

sill [SPATIAL STATISTICS (USE FOR GEOSTATISTICS)] A parameter of a variogram or semivariogram model that represents a value that the variogram tends toward when distances become large. Under second-order stationarity, variables become uncorrelated at large distances, so the sill of the semivariogram is equal to the variance of the random variable. *See also* second-order stationarity, semivariogram, sill.

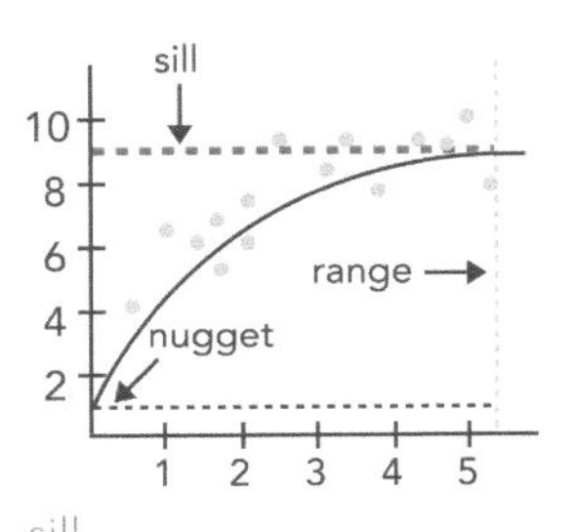

sill

simple edge feature [ESRI SOFTWARE] In a geodatabase, a line feature that corresponds to a single network element in the logical network. *See also* complex edge feature, edge, feature.

simple feature

1. [DATA MODELS] A feature that has both spatial and nonspatial attributes and is part of a simple feature class.
2. [STANDARDS] A set of standards for accessing and storing geometric data for digital use within a GIS maintained by both the Open Geospatial Consortium (OGC) and the International Organization for Standardization (ISO).

See also complex feature class, feature, Open Geospatial Consortium.

simple feature class [DATA MODELS] A point, line, polygon, or multipoint feature class that does not participate in geodatabase topology and is not part of a relationship class or a controller dataset (such as a network dataset, terrain dataset, trace network, utility network, or parcel fabric). This includes multipatch feature classes. *See also* complex feature class, feature class, geodatabase, simple feature, topology.

simple junction feature [ESRI SOFTWARE] In a geodatabase, a junction feature that corresponds to a single network element in the logical network. *See also* feature, junction.

simple kriging [SPATIAL STATISTICS (USE FOR GEOSTATISTICS)] A kriging method in which the weights of the values do not sum to unity. Simple kriging uses the average of the entire dataset, which is less accurate than ordinary kriging but produces a smoother result. *See also* kriging.

simple market area [ANALYSIS/GEOPROCESSING] An area defined by a generalized boundary drawn around the most distant set of customer points (a convex hull) that total to some value. The calculation may be unweighted (in which case every point has the same value) or weighted by a value in the underlying database, such as sales. *See also* complex market area.

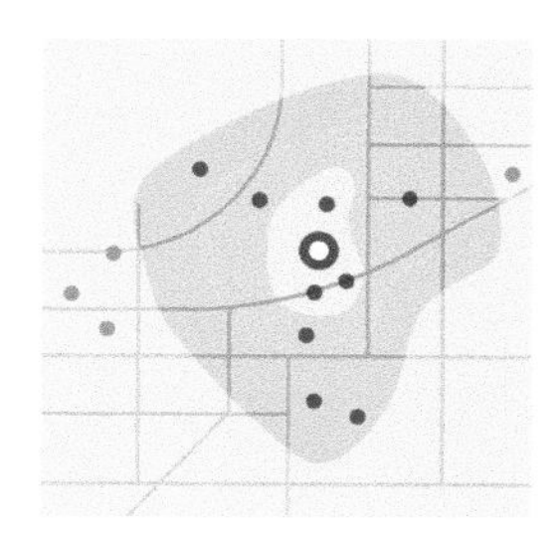

simple market area

simple measurement [SURVEYING] The simplest form in which measurements from COGO sources or TPS (Total Positioning System) sources can be stored. *See also* COGO, TPS measurement.

simple relationship [DATA STRUCTURES] A link or association between data sources that exist independently of each other. *See also* composite relationship.

simplification [MAP DESIGN] A type of cartographic generalization in which the important characteristics of features are determined, and unwanted detail is eliminated to retain clarity on a map whose scale has been reduced. *See also* generalization.

simultaneous conveyance [CADASTRAL AND LAND RECORDS] A means of defining multiple units of land in a single survey document in such a way that all their boundaries have equal legal status. A common example of simultaneous conveyance is the modern subdivision. *See also* surveying.

single precision [DATA QUALITY] A level of coordinate exactness based on the number of significant digits that can be stored for each coordinate. Single precision numbers store up to seven significant digits for each coordinate, retaining a precision of plus or minus 5 meters in an extent of 1,000,000 meters. Datasets can be stored in either single- or double-precision coordinates. *See also* double precision.

single-coordinate precision *See* single precision.

single-theme map [MAP DESIGN] A map that features one theme or variable, shown within the context of base reference information. *See also* theme.

sink [ANALYSIS/GEOPROCESSING] The location or group of locations used as the endpoint for distance analysis. *See also* source.

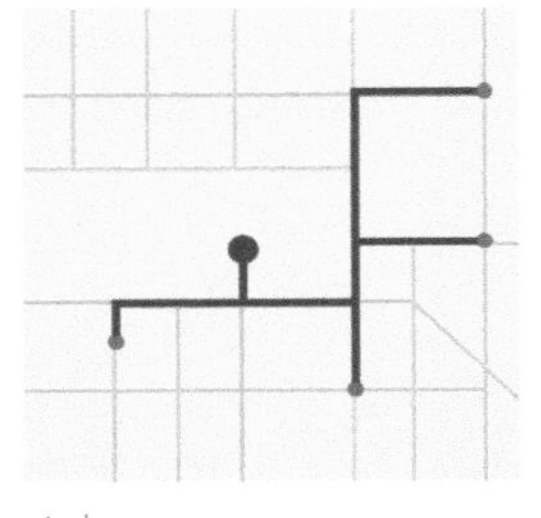

sink

sinusoidal projection [MAP PROJECTIONS] An equal-area pseudo-cylindrical projection, equidistant in the east–west direction and in the north–south direction along the central meridian, but only in these directions; developed by French hydrographer Jean Cossin in 1570. Sometimes called the Sanson-Flamsteed projection, the projection is also attributed to seventeenth-century French cartographer Nicolas Sanson and British astronomer John Flamsteed, who used the sinusoidal projection for global atlases and star maps, respectively. *See also* central meridian, pseudo-cylindrical map projection.

sinusoidal projection

site prospecting [ANALYSIS/GEOPROCESSING, BUSINESS] The process of evaluating demographic data surrounding potential locations for a business, based on a user-defined trade area or areas. *See also* demographics, market area, market penetration analysis.

size [DIGITAL IMAGE PROCESSING] An image element that provides information about the extent of an object of interest; used to help identify the object. Typically expressed in terms of ground units. *See also* image element.

sketch [VISUALIZATION, ESRI SOFTWARE] A software function used for drawing features, lines, or symbols on a web map. *See also* shape, visualization.

skew [STATISTICS] A property that indicates how the peak of a distribution differs from a normal distribution when the peak is shifted to the right or left. *See also* kurtosis.

sliver polygon [DATA MODELS] A small, narrow, polygon feature that appears along the borders of polygons following the overlay of two or more geographic datasets. Sliver polygons may indicate topology problems with the source polygon features, or they may be a legitimate result of the overlay. *See also* polygon overlay, sliver removal.

sliver polygon

sliver removal [DATA EDITING] The act of deleting unwanted sliver polygons. *See also* polygon overlay, sliver polygon.

slope

1. [EUCLIDEAN GEOMETRY] The incline, or steepness, of a surface, measured in degrees from horizontal (0–90), or percent slope (the rise divided by the run, multiplied by 100). A slope of 45 degrees equals 100 percent slope; as a slope angle approaches vertical (90 degrees), the percent slope approaches infinity. The slope of a triangulated irregular network (TIN) face is the steepest downhill slope of a plane defined by the face; the slope for a cell in a raster is the steepest slope of a plane defined by the cell and its eight surrounding neighbors.

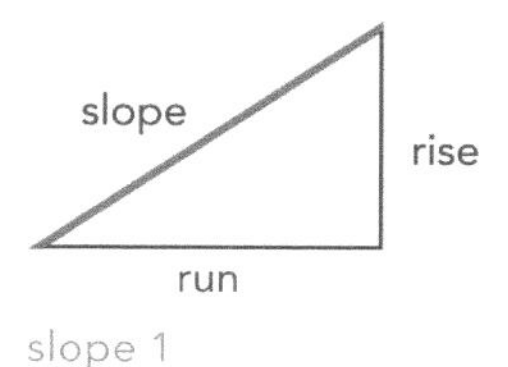

slope 1

2. [MAP DESIGN] The vertical change in the elevation of the land surface (rise), determined over a given horizontal distance (run).

See also aspect, gradient, rise, run.

slope angle [MATHEMATICS] The inverse tangent (tan–1) of the slope ratio (the angle whose tangent is the slope ratio) with values ranging from 0 to 90 degrees. *See also* slope, slope ratio.

slope curvature [EUCLIDEAN GEOMETRY] The curvature of a surface, defined as linear, convex, or concave. *See also* curvature, slope.

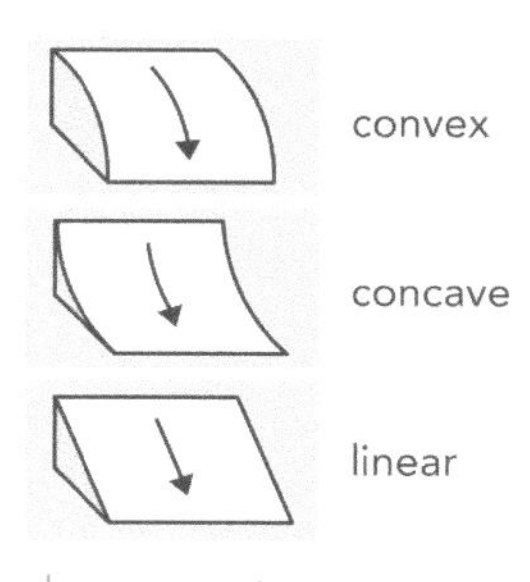

slope curvature

slope error [MAP DESIGN] The difference between the true ground and measured map distance owing to landform slope. *See also* error.

slope image *See* shaded relief image.

slope map [MAP DESIGN] A map that shows the steepness of the terrain. *See also* slope.

slope measurement error [DATA QUALITY] A type of error caused by deviation of the landform from a linear surface within the distance over which the slope is computed. *See also* error, slope.

slope ratio [MATHEMATICS] The ratio of the elevation difference (rise) and the ground distance (run) between two points. *See also* physical distance, rise, run, slope.

slope template [MAP DESIGN] A scaled template of selected contour spacing, chosen so that slope categories can be easily identified. Also known as a tick sheet. *See also* contour line, slope.

small circle [GEODESY] The circle made when a flat plane intersects a sphere anywhere but through its center. Parallels of latitude other than the equator are small circles. *See also* great circle, homolatitudes, latitude, sphere.

small circle

small multiples [GRAPHICS (MAP DISPLAY)] A series of map displays with the same graphic design but with different data depicted; used to show changes in magnitude or type from multiple to multiple (map to map). Also called constant-format displays.

small scale [MAP DESIGN] Generally, a map scale that shows a relatively large area on the ground with a low level of detail. In a two-way grouping of map scales, a map is considered small scale when the numeric value of the representative fraction 1/x is smaller than 1:500,000; in a three-way grouping of map scales, it is considered small scale when the numeric value of the representative fraction 1/x is smaller than 1:1,000,000. *See also* large scale.

smoothing

1. [CARTOGRAPHY] In cartography, reducing or removing small variations in a line or other feature to improve its appearance or simplify the feature's representation.
2. [DIGITAL IMAGE PROCESSING] In image processing, reducing or removing small variations in an image to reveal the global pattern or trend, either through interpolation or by passing a filter over the image.

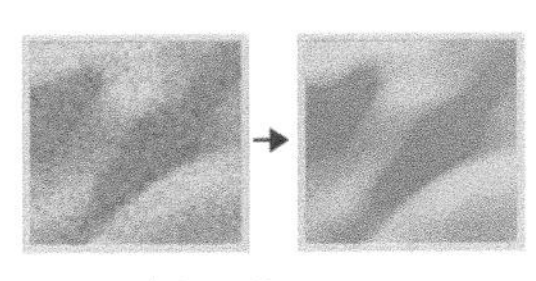

smoothing 2

See also flattening, generalization, interpolation.

smoothing error [DATA QUALITY] A type of error caused by reduction in the length of a linear feature—such as a road or river—through generalization. *See also* generalization, interpolation, smoothing.

snap extent [ESRI SOFTWARE] A geoprocessing option that snaps, or aligns, all layers to the cell registration of a specified raster. All layers will share the lower left corner and cell size of the specified raster. Snap extent is used to resample layers to the same registration and cell size in order to perform analysis. *See also* extent, snapping.

snapping [DATA EDITING] The process of moving a feature to match or coincide exactly with another point or feature's coordinates when your pointer is within a specified distance (tolerance). Commonly used to increase accuracy when using a variety of tools including editing, georeferencing, and measure tools. Snapping may also be used to adjust the extent of the cells in one raster to match the extent of the cells in another raster. *See also* coincident, tolerance.

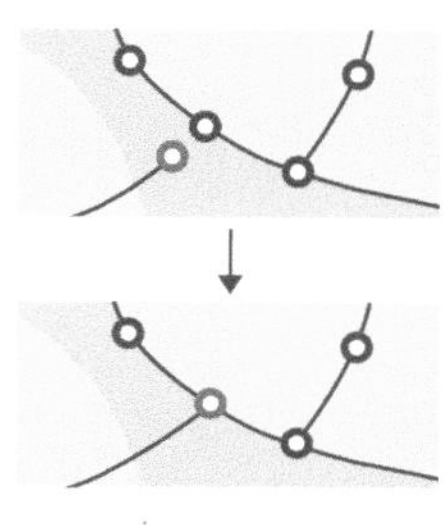

snapping

snapping tolerance [DATA EDITING] A specified distance within which points or features within are moved to match or coincide exactly with each other's coordinates. *See also* distance, snapping, tolerance.

SNR *See* signal-to-noise ratio.

software development kit *See* SDK.

solar altitude [ASTRONOMY] The angle of the sun above the horizon. *See also* altitude, hillshading, solar azimuth.

solar azimuth [ASTRONOMY] The angular direction of the sun, measured clockwise from north in degrees, from 0 to 360. *See also* azimuth, hillshading, solar altitude.

solar insolation [PHYSICS] The amount of solar radiation received in the earth's atmosphere or at its surface. Solar insolation depends primarily on the solar zenith angle, the variable distance of the earth from the sun, and scattering and absorption characteristics of the atmosphere. It is measured as the flux of solar radiation per unit of horizontal area for a given location. *See also* zenith.

solver [NETWORK ANALYSIS] A function that performs network analysis based on a set of network data. *See also* network analysis.

sonar [REMOTE SENSING] An abbreviation of *sound navigation and ranging.* A system or device that measures the time lapse between emitting a sound and receiving a returned echo to determine the location, depth, and shape of objects under water. Certain types of sonar consist only of a listening device that picks up sound emitted by underwater objects, such as submarines. *See also* bathymetry, lidar, radar.

soundex [COMPUTING] A widely used phonetic algorithm that facilitates searches and address matching. *See also* address matching, algorithm.

sounding [CARTOGRAPHY] A water depth reading. *See also* bathymetric map, depth contour.

source [ANALYSIS/GEOPROCESSING] The location or group of locations used as the starting point for distance analysis. *See also* sink.

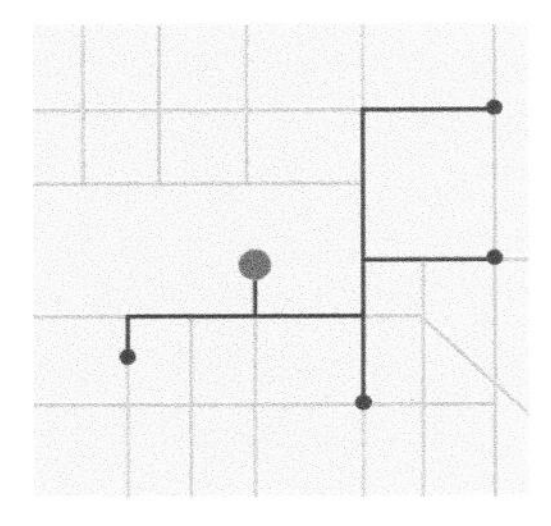

source

source scale [DATA MANAGEMENT] The scale at which information is originally derived from its source. *See also* scale.

S

south pole [NAVIGATION] The point in the southern hemisphere at which the earth's axis of rotation meets the surface of the earth. It defines latitude 90° S, as well as the direction of true south. *See also* axis, north pole.

Southern Positioning Augmentation Network *See* SouthPAN.

SouthPAN [GPS] A combined abbreviation and acronym for the *Southern Positioning Augmentation Network*. A regional satellite-based augmentation system (SBAS) that improves GNSS positioning, navigation, and timing (PNT) for Australia and New Zealand. SouthPAN is a joint initiative of the Australian and New Zealand governments and is the first SBAS in the southern hemisphere. *See also* GNSS, GPS augmentation, PNT.

space coordinate system [COORDINATE SYSTEMS] A three-dimensional, rectangular, Cartesian coordinate system that has not been adjusted for the earth's curvature. In a space coordinate system, the x- and y-axes lie in a plane tangent to the earth's surface, and the z-axis points upward. *See also* Cartesian coordinate system, x-axis, y-axis, z-axis.

space trilateration [NAVIGATION] A method of determining an object's absolute ground position using satellite signal velocity (how long it takes a signal to reach the satellite) and the travel time of the object and the satellite, calculating the distance between the GPS receiver (object on the ground) and the known locations of two or more satellites. *See also* GPS.

space-time cube [MAP DESIGN] A mapping technique introduced in 1970 by Swedish geographer Torsten Hägerstrand to study the social interaction and movement of individuals in space and time. This technique adds time as a z-axis to the familiar x,y map coordinates used by mapmakers. *See also* x-axis, y-axis, z-axis.

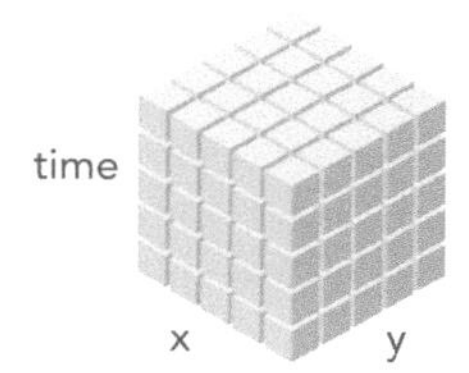

space-time cube

spacing [MAP DESIGN] The locational arrangement of objects with respect to each other rather than relative to data collection units. *See also* data collection unit.

spaghetti data [DATA CAPTURE] Vector data composed of simple lines with no topology and usually no attributes. Spaghetti lines may cross, but no intersections are created at those crossings. *See also* planar enforcement.

spaghetti digitizing [DATA CAPTURE] Digitizing that does not identify intersections as it records lines. *See also* data capture, spaghetti data.

spatial [DATA MODELS] Related to or existing within space. *See also* geographic.

spatial analysis [SPATIAL ANALYSIS]

1. The process of examining the locations, attributes, patterns, and relationships of features in spatial data to address a question or gain useful knowledge.
2. A method of advanced spatial modeling that assists with terrain modeling, finding suitable locations and routes, discovering spatial patterns, and performing hydrologic and statistical analysis.

See also spatial modeling.

spatial association [SPATIAL ANALYSIS] The degree to which two categories overlap on a map. The spatial association is positive when two sets of values increase or decrease similarly in space; the association is negative if the values of one dataset increase while those of the other decrease.

spatial autocorrelation [SPATIAL STATISTICS (USE FOR GEOSTATISTICS)] A measure of the degree to which a set of spatial features and their associated data values tend to be clustered together in space (positive spatial autocorrelation) or dispersed (negative spatial autocorrelation). *See also* autocorrelation, correlation, Tobler's First Law of Geography.

spatial cognition [COGNITION] The mental processes involved in gaining and using knowledge and beliefs about spatial environments. Spatial cognition includes issues of perception, memory, language, learning, and problem solving and is an object of study in humans, animals, and machines. *See also* mental map, spatial.

spatial configuration [DATA MANAGEMENT] The character and arrangement, position, or orientation of patches within a patch class or landscape mosaic.

spatial correspondence [SPATIAL ANALYSIS] The degree of similarity in spatial position.

spatial cross-correlation [DATA MANAGEMENT] The similarity between two different datasets for the same spatial data collection units. *See also* data collection unit, spatial autocorrelation.

spatial cross-correlation coefficient [SPATIAL ANALYSIS] Expressed as (r). A measure of the similarity between two different datasets for the same spatial data collection units; a measure of spatial association based on computing the covariance for counts of different point features within data collection units. Also known as the correlation coefficient. *See also* data collection unit, spatial autocorrelation.

spatial data
1. [DATA STRUCTURES] Information about the locations and shapes of geographic features and the relationships between them, usually stored as coordinates and topology.
2. [DATA MODELS] Any data that can be mapped to a geographic location.

See also nonspatial data, temporal data, thematic data.

S

spatial data infrastructure *See* SDI.

Spatial Data Transfer Standard *See* SDTS.

spatial database [DATABASE STRUCTURES] A structured collection of spatial data and its related attribute data, organized for efficient storage and retrieval. *See also* database, spatial, spatial data.

spatial dependence [SPATIAL ANALYSIS] A relation that is said to exist and be positive when the category ranks of two distributions increase or decrease similarly in space, as calculated using Spearman's rank correlation coefficient. If the category ranks in one distribution increase while those of the other decrease, negative spatial dependence exists. *See also* spatial autocorrelation, Spearman's rank correlation coefficient.

spatial diffusion
1. [SPATIAL ANALYSIS] The transfer or movement of things, ideas, or people from place to place.
2. [PHYSICS] The process by which features move outward from a starting position across space over time, usually continuously decreasing in density.

See also diffusion.

spatial feature *See* feature.

spatial function [SPATIAL ANALYSIS] An operation that performs specific spatial data analytics. Distance, slope, and density are examples of spatial functions. *See also* density, distance, slope.

spatial grid [DATA MODELS] A two-dimensional grid system that spans a feature class. It is used to quickly locate features in a feature class that might match the criteria of a spatial search. *See also* grid, spatial.

spatial index [ESRI SOFTWARE] In a geodatabase, a mechanism for optimizing access to data based on the spatial column of the business table. In most geodatabases, a system of grids is used for the spatial index. *See also* index, spatial.

spatial join [SPATIAL ANALYSIS] A type of table join operation in which fields from one layer's attribute table are appended to another layer's attribute table based on the relative locations of the features in the two layers. *See also* joining, spatial.

spatial modeling [MODELING] A methodology or set of analytical procedures used to derive information about spatial relationships between geographic phenomena. *See also* spatial analysis.

spatial overlay analysis [ANALYSIS/GEOPROCESSING] A type of analysis in which data is extracted from one layer (such as block groups) to an overlay layer (such as a trade area). *See also* overlay.

spatial pattern [GEOGRAPHY] The arrangement or placement of features on the earth, as well as the space between them. The repetitive or structured arrangement of features on the ground or map.

spatial pattern analysis [SPATIAL ANALYSIS] The process of identifying and quantifying geographic patterns or the configuration of features and objects. Used to understand the behavior of geographic phenomena and compare patterns of different distributions or time periods. Spatial pattern analysis tools incorporate space (proximity, area, connectivity, and other spatial relationships) directly into the statistical analysis of spatial distributions and patterns. *See also* geographic phenomena, pattern.

spatial query [SPATIAL ANALYSIS] A statement or logical expression that selects geographic features based on location or spatial relationship. For example, a spatial query might find which points are contained within a polygon or set of polygons, find features within a specified distance of a feature, or find features that are adjacent to each other. *See also* proximity query.

spatial reference [COORDINATE SYSTEMS] A set of parameters that define the coordinate system and spatial properties for geographic data. A spatial reference is used to provide a framework when measuring and representing geographic features from different sources. Each spatial reference is defined either by a WKID—also called a spatial reference ID (SRID)—or by a WKT. Common spatial references include WGS84, used by GPS systems, and Web Mercator projections, used in most web maps. *See also* coordinate system, GPS, Web Mercator projection, WGS84, WKID, WKT.

spatial reference system *See* coordinate system.

spatial resolution [PHOTOGRAMMETRY] The dimensions of the area on the ground represented by a single cell in a raster or pixel in an image. The size of a pixel, or its spatial resolution, affects the level of detail represented in an image.

spatial shift [CARTOGRAPHY] The difference in location for individual features between two time periods. *See also* temporal data.

spatial statistics [STATISTICS] Statistical methods that use space and spatial relationships (such as distance, area, volume, length, height, orientation, centrality, or other spatial characteristics of data) in mathematical computations. Spatial statistics are used for a variety of analysis types, including pattern analysis, shape analysis, surface modeling and surface prediction, spatial regression, statistical comparisons of spatial datasets, statistical modeling, and prediction of spatial interaction. The types of spatial statistics include descriptive, inferential, exploratory, geostatistical, and econometric statistics. *See also* geostatistics.

spatial weights matrix [SPATIAL STATISTICS (USE FOR GEOSTATISTICS)] A file that quantifies spatial relationships among a set of features. Typical examples of such relationships are inverse distance, contiguity, travel time, and fixed distance. *See also* contiguous, inverse distance.

spatialization [COGNITION] The transformation of complex, multivariate, nonspatial data into a spatial representation located in an information space. The relative positioning of data elements within the spatial representation shows relationships between them. Spatialization is used to allow exploration of nonspatial data using spatial metaphors and spatial analysis. *See also* nonspatial data, spatial analysis.

Spearman's rank correlation coefficient [SPATIAL ANALYSIS] Expressed as (p). A nonparametric measure of correlation that shows the degree to which two datasets are spatially independent, without having to assume that each dataset has a normal distribution of values. Also known as Spearman's rho. *See also* correlation, distribution.

specialty [NETWORK ANALYSIS] An object used in vehicle routing problem (VRP) analysis. A specialty is used to represent a specific route or vehicle capability. Only items that match the specialty can be paired. For example, when an order requires an electrician, it can only be serviced by a route that supports the electrician specialty. *See also* vehicle routing problem.

species distribution map [MAP DESIGN] A map that shows the distribution of plants and animals. *See also* distribution.

species range map [MAP DESIGN] A map that shows the range of one or more species. *See also* range.

spectral band *See* band.

spectral resolution [SATELLITE IMAGING] The range of wavelengths that a single band in an imaging system can detect. Sensors are characterized by their spectral resolution, including both the number of bands and the wavelength range detected by each band. *See also* band.

spectral signature [PHYSICS] The pattern of electromagnetic radiation that identifies a chemical or compound. Unique differences in the reflectance characteristics of materials can be distinguished from one another by examining which portions of the spectrum they reflect and absorb. *See also* anisotropy, reflectance, wavelength.

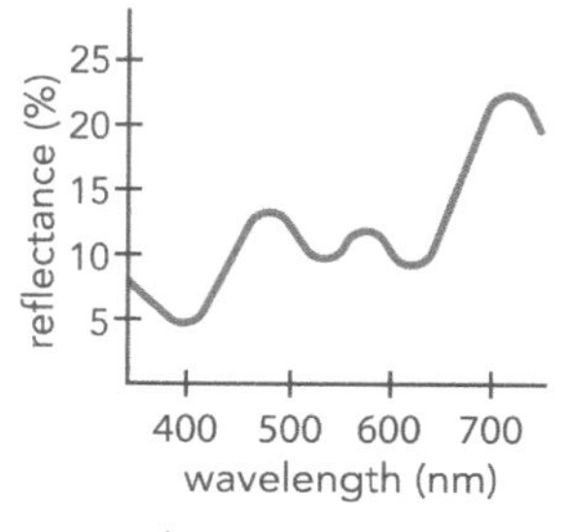

spectral signature

spectrometer *See* spectrophotometer.

spectrophotometer [PHYSICS] A photometer that measures the intensity of electromagnetic radiation as a function of its frequency. Spectrophotometers are usually used for measuring the visible portion of the spectrum. *See also* radiometer, photometer.

spectroscopy [PHYSICS] The scientific study of how chemicals and other substances absorb and reflect various parts of the electromagnetic spectrum.

spectrum *See* electromagnetic spectrum.

spelling sensitivity [GEOLOCATING] In geocoding, the degree to which the spelling variation of a street name is allowed during a search for likely match candidates. The lower the value, the more likely it is that additional candidates will be retrieved, and vice versa. *See also* locator.

sphere [EUCLIDEAN GEOMETRY] A three-dimensional solid in which all points on the surface are the same distance from the center; used to approximate the true shape and size of the earth or other celestial bodies for mapping purposes. *See also* ellipsoid, globe, hemisphere, radius, spheroid.

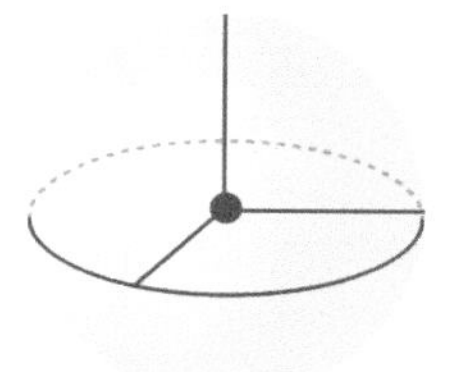

sphere

spherical coordinate method [EUCLIDEAN GEOMETRY] A method of area measurement based on the use of geographic coordinates (latitude and longitude) in calculating the area of spherical triangles. *See also* measurement.

spherical coordinate system [COORDINATE SYSTEMS] A reference system using positions of latitude and longitude to define the locations of points on the surface of a sphere or spheroid. *See also* geographic coordinate system.

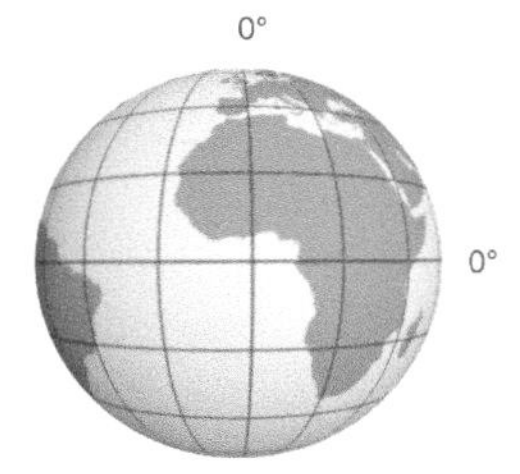

spherical coordinate system

spheroid

1. [EUCLIDEAN GEOMETRY] A three-dimensional shape obtained by rotating an ellipse about its minor axis, resulting in an oblate spheroid, or about its major axis, resulting in a prolate spheroid.
2. [GEODESY] When used to represent the earth, a three-dimensional shape obtained by rotating an ellipse about its minor axis, with dimensions that either approximate the earth as a whole, or with a part that approximates the corresponding portion of the geoid.

See also ellipsoid, geoid.

spider diagram *See* desire-line analysis.

spike

1. [STATISTICS] An anomalous data point that protrudes above or below an interpolated surface.

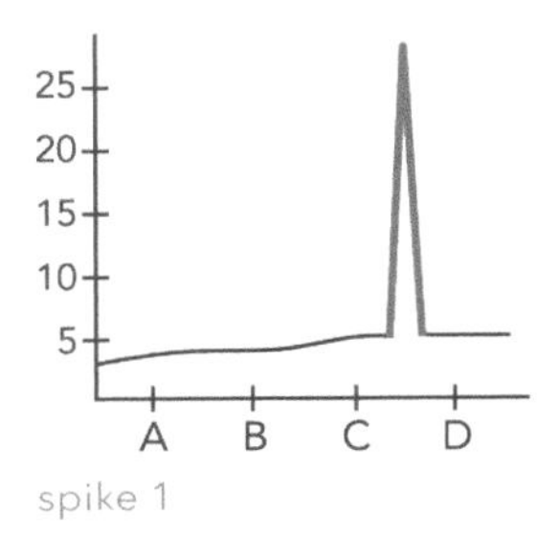

spike 1

2. [DATA CAPTURE] An overshoot line created erroneously by a scanner and its rasterizing software.

See also outlier.

spline [MATHEMATICS] In mathematics, a piecewise polynomial function used to approximate a smooth curve in a line or surface. *See also* spline interpolation, densify.

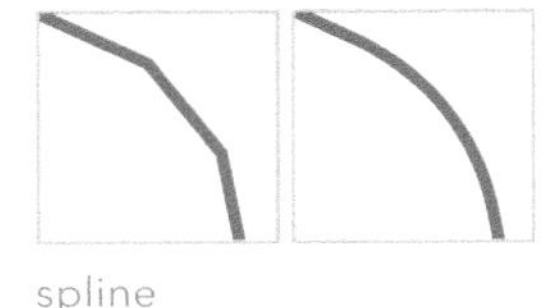
spline

spline interpolation [SPATIAL STATISTICS (USE FOR GEOSTATISTICS)] An interpolation method in which cell values are estimated using a mathematical function that minimizes overall surface curvature, resulting in a smooth surface that passes exactly through the input points. *See also* spline, interpolation.

split character [ESRI SOFTWARE] A user-designated character that divides long labels into two or more lines during the labeling process. There can be more than one split character. *See also* label.

SPOT [SATELLITE IMAGING] Acronym for *Satellite Pour l'Observation de la Terre.* Earth observation satellites developed by Centre National d'Etudes Spatiales (CNES), the space agency of France. The SPOT satellites gather high-resolution imagery used in natural resource management, climatology, oceanography, environmental monitoring, and the monitoring of human activities. *See also* Argos, DORIS, sensor, satellite constellation.

spot elevation [SURVEYING] An elevation measurement taken at a single location. *See also* contour line, elevation.

spot height *See* spot elevation.

spurious polygon *See* sliver polygon.

stacked scale bar [SYMBOLOGY] Two scale bars that use different distance units that are stacked one on top of the other. *See also* distance unit, scale bar.

staging geodatabase *See* CAD staging geodatabase.

standard annotation [ESRI SOFTWARE] Annotation that is stored in the geodatabase, consisting of geographically placed text strings that are not associated with features in the geodatabase. *See also* annotation, feature-linked annotation.

standard deviation [STATISTICS] A statistic used to measure the dispersion or variation of values in a distribution. Equal to the square root of the variance. Often expressed as *SD* or σ. *See also* calculated average mean, distribution, mean, normal distribution, quality control, variance.

standard deviation classification [STATISTICS] A data classification method that finds the mean value, then places class breaks above and below the mean at intervals of either .25, .5, or 1 standard deviation until all the data values are contained within the classes. Values that are beyond three standard deviations from the mean are aggregated into two classes, greater than three standard deviations above the mean and less than three standard deviations below the mean. *See also* classification.

standard deviational ellipse [UNCERTAINTY] An ellipse formed when the standard deviation of the x- and y-coordinates from their average values is used to define the major and minor axes of the ellipse. *See also* ellipse, standard deviation.

standard distance [SPATIAL STATISTICS (USE FOR GEOSTATISTICS)] A measure of the compactness of a spatial distribution of features around its mean center. Standard distance (or standard distance deviation) is usually represented as a circle where the radius of the circle is the standard distance. *See also* mean center.

standard error [ACCURACY] A measure of a map's accuracy; sums the major diagonal of the error matrix and divides by the total number of samples.

Standard Industrial Classification codes [STANDARDS] Also known by the acronym *SIC* codes. The federal U.S. standard prior to 1997 for classifying establishments by their primary type of business activity. SIC codes are used as an identification system in business directories, publications, and statistical sources. The classification system was officially replaced by NAICS in 1997 but is still used by some organizations outside the federal government. *See also* North American Industry Classification System.

standard line [MAP PROJECTIONS] A line on a sphere or spheroid that has no length compression or expansion after being projected, usually a standard parallel or central meridian. *See also* central meridian, line, sphere.

standard parallel [MAP PROJECTIONS] The line of latitude in a conic or cylindrical projection in normal aspect where the projection surface touches the globe. A tangent conic or cylindrical projection has one standard parallel, whereas a secant conic or cylindrical projection has two. At the standard parallel, the projection shows no distortion. *See also* parallel.

standard parallel

standardization [STATISTICS] A process in which a range of data values is normalized to measure how many standard deviations each value is from its range mean. Although this term is often used interchangeably with normalization, standardization is a special case of normalization. *See also* normalization.

star chart [ASTRONOMY] A chart used to plot stellar and planetary positions throughout the year.

star diagram [CARTOGRAPHY] A type of diagram that consists of a central point from which lines radiate outward. The central point typically represents a geographic location whereas the length of each line represents an attribute value or ratio. The direction of the line may represent a compass direction, a specific time frame, or some other attribute classification. A wind rose is a common example of a star diagram. *See also* chart, wind rose.

state [GOVERNMENT] An autonomous political and administrative division of geography. The United States is composed of 50 states.

state grid [COORDINATE SYSTEMS] A special grid coordinate system for select U.S. states that fall into two UTM zones. Created by shifting the central meridian of a UTM zone to the center of the state. Examples include the Oregon Lambert system and the Wisconsin transverse Mercator (WTM) system. *See also* Oregon Lambert system, UTM, WTM.

state plane coordinate system [COORDINATE SYSTEMS] A group of planar coordinate systems based on the division of the United States into more than 130 zones to minimize distortion caused by map projections while defining property boundaries in a way that simplifies calculating land parcel perimeters and areas. Each zone has its own map projection and parameters and uses either the NAD27 or NAD83 horizontal datum. The Lambert conformal conic projection is used for states that extend mostly east–west, whereas transverse Mercator is used for those that extend mostly north–south. The oblique Mercator projection is used for the panhandle of Alaska. *See also* Lambert conformal conic projection, planar coordinate system, transverse Mercator projection.

static [PHYSICS] Lacking in movement, action, or change.

static positioning [GPS] Determining a position on the earth by averaging the readings taken by a stationary antenna over a specific time. *See also* kinematic positioning.

stationarity [SPATIAL STATISTICS (USE FOR GEOSTATISTICS)] In geostatistics, a property of a spatial process in which all statistical properties of an attribute depend only on the relative locations of attribute values. *See also* intrinsic stationarity, second-order stationarity.

stationary front [METEOROLOGY] The boundary between cold and warm air masses that show little or no movement. *See also* front.

stationing [LINEAR REFERENCING] In the pipeline industry, another name for linear referencing. Stationing allows any point along a line feature representing a pipeline to be uniquely identified by its relative position along the line feature. *See also* linear referencing.

statistic F *See* F statistic.

statistical surface [STATISTICS] Ordinal, interval, or ratio data represented as a surface in which the height of each area is proportional to a numerical value. *See also* representation.

steepest path [NETWORK ANALYSIS] A line that follows the steepest downhill direction on a surface. Paths terminate at the surface perimeter or in surface concavities or pits. *See also* path, pit, slope.

stepped-surface map [CARTOGRAPHY] A map made by dividing the region into data collection areas and showing the magnitude values by proportionately varying the heights of the areas. Also called a prism map. *See also* data collection unit, mapping methods.

steradian [EUCLIDEAN GEOMETRY] The solid (conical) angle subtended at the center of a sphere of radius *r* by a bounded region on the surface of the sphere having an area *r* squared. There are 4π steradians in a sphere. *See also* radian.

stereocompilation [MAP DESIGN] A map produced with a stereoscopic plotter using aerial photographs and geodetic control data. *See also* stereoplotter.

stereogrammatic organization *See* visual hierarchy.

stereographic projection [MAP PROJECTIONS]

1. A planar conformal projection based on a light source on the surface of the generating globe opposite the point of tangency.
2. A secant planar projection that views the earth from a point on the globe opposite the center of the projection.

See also point of tangency, projection.

stereometer [PHOTOGRAMMETRY] A stereoscope containing a micrometer for measuring the effects of parallax in a stereoscopic image. *See also* micrometer, stereoscope.

stereomodel

1. [BIOLOGY] A three-dimensional image formed when rays from points in the images of a stereoscopic pair intersect. In vision, this is represented by binocular (two eye) sight.
2. [REMOTE SENSING] In remote sensing, the region of overlap in two oriented images through which a 3D relief model is observed, analyzed, and exploited with point intersection techniques.

See also remote sensing, stereopair.

stereopair [PHOTOGRAMMETRY] Two aerial photographs of the same area taken from slightly different angles that when viewed together through a stereoscope produce a three-dimensional image. *See also* stereoscope.

stereopair

stereoplotter [PHOTOGRAMMETRY] An instrument that projects a stereoscopic image from aerial photographs, converts the locations of objects and landforms on the image to x-, y-, and z-coordinates, and plots these coordinates as a drawing or map. *See also* stereocompilation.

stereoscope [PHOTOGRAMMETRY] A binocular device that produces the impression of a three-dimensional image from photographic stereopairs. *See also* stereopair.

stereoscopic pair *See* stereopair.

stereoscopic polarization [PHOTOGRAMMETRY] A technically complex way to display the terrain stereoscopically using a special computer monitor that alternates two images at least 30 times a second. The first image is displayed with horizontally polarized light, the second with vertically polarized light. Stereoscopic analysts wear special goggles with polarizing filters that allow the right eye to see only the horizontally polarized image and the left eye the vertically polarized image, so that the terrain is seen stereoscopically in the mind. Also called stereoscopic projection. *See also* stereopair, stereographic projection.

stereoscopy [GRAPHICS (MAP DISPLAY)] Any technique capable of creating the illusion of depth in a photograph, map, or other two-dimensional image. *See also* stereocompilation, stereomodel, stereopair, stereoscope.

stochastic model

1. [MODELING] A model that includes a random component. The random component can be a model variable, or it can be added to existing input data or model parameters.
2. [SURVEYING] A model that describes the expected error distribution of the measurements.

See also deterministic model, model, Monte Carlo method.

stop [NETWORK ANALYSIS] A network location used to determine a route in route analysis. Users can specify multiple stops, of which two must be used to represent an origin and a destination. Stops in between (known as intermediary stops) are visited en route from the first to the last stop. *See also* destination, origin, route analysis.

stop impedance [NETWORK ANALYSIS] The time it takes for a stop to occur, used to compute the impedance of a path or tour. For example, when a school bus drops children off or picks them up at their homes, the stop impedance might be 2 minutes at each stop. *See also* impedance, stop.

store market analysis [ANALYSIS/GEOPROCESSING, BUSINESS] A type of business analysis that uses mostly data about a store or stores rather than about customers. Examples include ring studies and analyses of equal competition areas and drive-time areas. *See also* customer market analysis.

store prospecting [ANALYSIS/GEOPROCESSING] A type of business analysis that assesses the potential of a site by performing simple ring or drive-time analysis. *See also* site prospecting.

story [VISUALIZATION, ESRI SOFTWARE] A web page containing some combination of interactive maps, text, images, and other multimedia to tell a story. *See also* interactive map.

straight-line allocation [SPATIAL ANALYSIS, ESRI SOFTWARE] A spatial analysis function that identifies which cells belong to which source, based on closest proximity in a straight line. *See also* allocation, spatial analysis.

straight-line direction [ESRI SOFTWARE] A spatial analysis function that identifies the azimuth direction from each cell to the nearest source. *See also* direction, spatial analysis.

straight-line distance [SPATIAL ANALYSIS, ESRI SOFTWARE] A spatial analysis function that calculates the distance in a straight line from every cell to the nearest source. *See also* distance, Euclidean distance, spatial analysis.

stream digitizing *See* stream mode digitizing.

stream flow [NATURAL RESOURCE MANAGEMENT] The volume of water at a given time in a river, stream, or other runoff source. *See also* streamgage.

stream mode digitizing [DATA CAPTURE] A method of digitizing in which, as the cursor is moved, points are recorded automatically at preset intervals of either distance or time. *See also* point mode digitizing.

S

stream tolerance [DATA CAPTURE] During stream mode digitizing, the minimum interval between vertices. Stream tolerance is measured in map units. *See also* map unit, stream mode digitizing, tolerance.

streamgage [NATURAL RESOURCE MANAGEMENT] A monitoring station that continuously collects data about stream flow. *See also* stream flow.

streaming [IMPORT/EXPORT] A technique for transferring data over the internet in real time, allowing large multimedia files to be viewed. *See also* live feed, real time.

street network [NETWORK ANALYSIS] A system of interconnecting lines and points that represent a system of roads for a given area. A street network provides the foundation for network analysis; for example, finding the best route or creating service areas. *See also* connectivity, network, route.

street segment [NAVIGATION] The portion of a street between two adjacent intersections, or, for a dead-end street, between an intersection and the endpoint of the dead-end street.

street-based mapping [ADDRESS MATCHING] A form of digital mapping that links information to geographic locations and displays address locations as point features on a map.

stretch [VISUALIZATION] A display technique applied to the histogram of raster datasets, most often used to increase the visual contrast between cells. *See also* histogram.

stretch

string [DATA STRUCTURES] A set of coordinates that defines a group of linked line segments. *See also* expression, line.

structure *See* drift.

structure line [3D ANALYSIS] A line feature constraint in a TIN. There are two types of structure lines: hard and soft. Hard structure lines, also known as breaklines, represent interruptions in the slope of the surface. Soft structure lines are used to add information about the surface without implying a change in the surface behavior across the line. *See also* breakline, surface constraint, TIN.

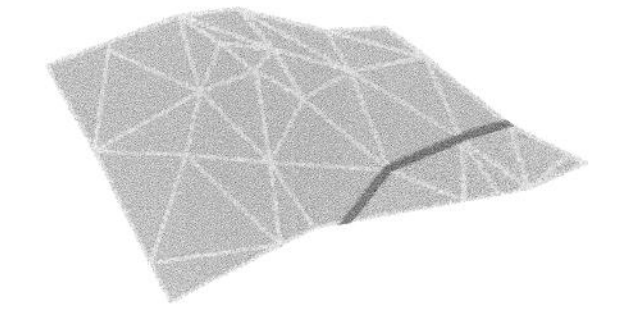

structure line

study area [ANALYSIS/GEOPROCESSING] The geographic area treated in an analysis.

style

1. [CARTOGRAPHY] An organized collection of predefined colors, symbols, properties of symbols, and map elements. Styles promote standardization and consistency in mapping products.
2. [GRAPHICS (MAP DISPLAY)] Techniques applied to enhance imagery, feature, and raster data to reveal, emphasize, and convey information in the data, depending on the purpose or application.

See also symbology.

subdivision [CADASTRAL AND LAND RECORDS] A piece of land divided from a larger area. *See also* lot, parcel.

sublayer [ESRI SOFTWARE] One of several layers that are part of a group layer in a map document. *See also* group layer.

submerged contour [CARTOGRAPHY] A contour that depicts variations in the land beneath the surface of water. Also called an underwater contour. *See also* bathymetry, contour, depth contour.

subtractive color [GRAPHICS (COMPUTING)] The color space defined by the subtractive properties of cyan, magenta, and yellow, the primary colors used in the printing process. Because these colors reflect light, the absence of them allows the viewer to see the white substrate of print; when these three are combined, the print result is effectively black. *See also* CMYK.

subtype [DATABASE STRUCTURES] In a geodatabase, a subset of features in a feature class or objects in a table that share the same attributes. For example, the streets in a street feature class can be divided into three subtypes: local streets, collector streets, and arterial streets. Each subtype has unique attribute values or behavior and shares the values and behavior of the street feature class. Subtypes are used to simplify data editing and improve the geodatabase's performance. *See also* feature class.

suitability model [MODELING] A model that weights locations relative to each other based on given criteria. Suitability models might aid in finding a favorable location for a new facility, road, or habitat for a species of bird. *See also* location, model, weight.

sun-synchronous orbit [REMOTE SENSING] A special case of the near-polar orbit in which a satellite passes over the same part of the earth at roughly the same local time each day. *See also* near-polar orbit.

superimposed map [MAP DESIGN] A technique that shows change by overlaying one map over another.

supplemental contour [CARTOGRAPHY] A contour placed in an area in which elevation change is minimal. Supplemental contours are added to help delineate small features that otherwise would be missed by the contour interval used for the rest of the map. *See also* contour line.

surface [DATA MODELS]

1. A geographic phenomenon represented as a set of continuous data (such as elevation, geological boundaries, or air pollution).

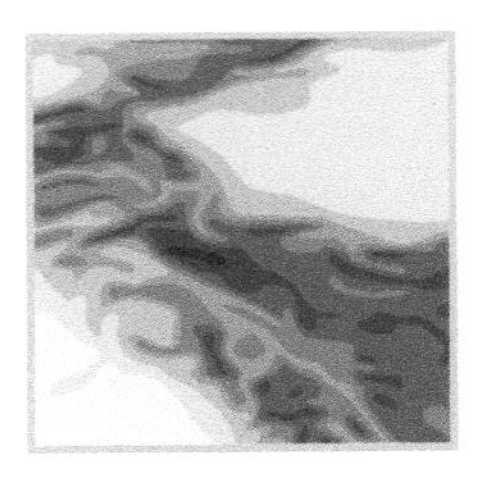

surface 1

2. A spatial distribution that associates a single value with each position in a plane, usually associated with continuous attributes.

See also distribution, geographic phenomena.

surface constraint [DATA MODELS] A point, line, or polygon feature used to reinforce a behavior in a surface. Structure lines and breaklines are examples of surface constraints. *See also* breakline, constraint, structure line.

surface fitting [SPATIAL STATISTICS (USE FOR GEOSTATISTICS)] A generated statistical surface that approximates the values of a set of known x,y,z points. *See also* interpolation, statistical surface, surface.

surface model [MODELING] A representation of a geographic feature or phenomenon that can be measured continuously across some part of the earth's surface (for example, elevation). A surface model approximates a surface, generalized from sample data. Surface models are stored and displayed as rasters, TINs, or terrains. *See also* geographic phenomena.

surficial geology map [GEOLOGY] A map that shows bands of differing rock types at the earth's surface. *See also* geologic cross section.

surficial geology map

surround element *See* map element.

survey

1. [SURVEYING] The use of precise principles, instruments, and methods to measure physical or geometric characteristics of the earth. Surveys are classified by the data studied or by the methods used. Examples include geologic, soil, engineering, land parcels, and more.
2. [CARTOGRAPHY] To use linear and angular measurements and apply principles of geometry and trigonometry to determine the exact boundaries, position, or extent of a tract of land.

See also boundary monument, COGO, parcel.

survey marker *See* survey monument.

survey monument [SURVEYING] An object, such as a metal disk, permanently mounted in the landscape to denote a survey station. *See also* surveying.

survey township [SURVEYING] A 6-by-6-mile block of land arranged in rows and columns, bound by township lines on the north and south and range lines on the east and west; used in the U.S. Public Land Survey System and Canada's Dominion Land Survey. Sometimes called a congressional township to distinguish it from political townships. Sometimes shortened to "township," though not to be confused with the distance measure of township. *See also* DLS, PLSS, township.

surveying [SURVEYING] Measuring physical or geometric characteristics of the earth. Surveys are often classified by the type of data studied or by the instruments or methods used. Examples include geodetic, geologic, topographic, hydrographic, land, geophysical, soil, mine, and engineering surveys.

surveyor's compass [SURVEYING] A compass with a bearing card that shows quadrants from 0 to 90 degrees, with the north and south directions identified as 0 degrees and the east and west directions labeled as 90 degrees. *See also* compass, quadrant.

SVG [GRAPHICS (COMPUTING)] Acronym for *scalable vector graphics.* An XML-based graphics file format that describes two-dimensional vector images, including animation. SVG images scale to fit the display window without compromising quality. *See also* graphic, image, vector.

symbol [SYMBOLOGY] A graphic used to represent a geographic feature or class of features. Symbols can look like what they represent (trees, railroads, houses), or they can be abstract shapes (points, lines, polygons) or characters. Symbols are typically explained in a map legend. *See also* glyph, legend, representation.

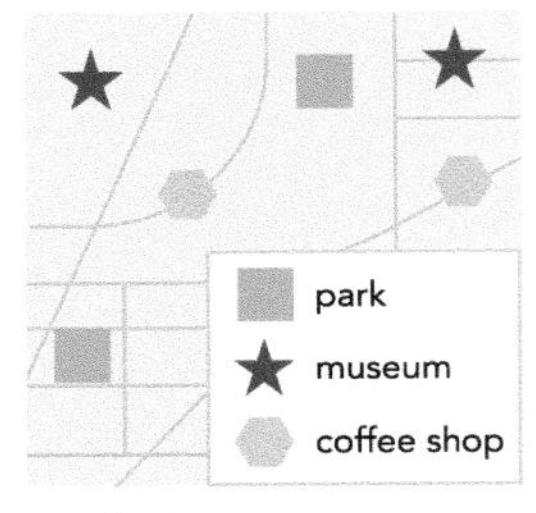

symbol

Symbol ID code [DEFENSE] A 15-character identifier that provides the information necessary to display or transmit a tactical symbol between MIL-STD-2525B-compliant systems. *See also* identifier, symbol.

symbol level drawing [ESRI SOFTWARE] A setting that determines the drawing order of features based on their symbol. When symbols have more than one layer, symbol level drawing can be used to specify the order in which each layer of the symbol is drawn. *See also* symbol.

symbolization [SYMBOLOGY] The use of visual variables to represent data attributes. *See also* visual variable.

symbology [CARTOGRAPHY] The set of conventions, rules, or encoding systems that define how geographic features are represented with symbols on a map. A characteristic of a map feature may influence the size, color, and shape of the symbol used. *See also* style.

synchronization [DATABASE STRUCTURES] The process of automatically updating certain elements of a metadata file.

synchronous

1. [PHYSICS] Occurring together, or at the same time.
2. [DATA TRANSFER] In data transmission, precisely timed and steady transmission of information that allows for higher rates of data exchange.

See also asynchronous.

System for Differential Corrections and Monitoring *See* SDCM.

table [DATA MODELS] A set of data elements arranged in rows and columns. Each row represents a single record. Each column represents a field of the record. Rows and columns intersect to form cells, which contain a specific value for one field in a record. *See also* cell; column; row; row, column.

FID	Species	Color
1	oak	brown
2	pine	green
3	fir	green

table

tabular data [DATA STORAGE] Descriptive information, usually alphanumeric, that is stored in rows and columns in a database and can be linked to spatial data. *See also* table.

tactical symbol *See* control measure.

tag [DATA MANAGEMENT] A keyword used to describe and index an online resource. *See also* label, annotation.

tangent projection [MAP PROJECTIONS] A projection whose surface touches the globe's surface without piercing it. A tangent planar projection touches the generating globe at a point of tangency for planar projections or along a line of tangency for conic and cylindrical projections. At the point or line of tangency, the projection is free from distortion. *See also* secant projection.

tangent projection

tangential lens distortion [PHOTOGRAMMETRY] An occurrence when the lens and the image plane are not parallel, creating a linear displacement of image points. This distortion is typically the result of a compound lens misalignment and considered to be of minimal concern due to modern lens production quality. *See also* distortion.

tasseled cap transformation [DIGITAL IMAGE PROCESSING] An image transformation method that uses principal component analysis and a linear rotation of multispectral image bands to yield transformed bands. The index is commonly defined by categories of brightness, greenness, and wetness, and subsequent noise bands, such as clouds and haze. It is used especially in the analysis of vegetation growth cycles, soil characterization, and

T

urban development. Also known as the Kauth-Thomas transformation. *See also* image transformation, index, principal component analysis.

TAT *See* text attribute table.

TDI [REMOTE SENSING] Acronym for *time delay integration.* In remote sensing, a method used to gather energy in a pixel of a focal plane using electronic shuttering. As the sensor moves, the pixel receives information for a fixed period (dwell time) and is then read out and reset. Carefully synchronized with the speed of the sensor, the ground sample distance (GSD) of the pixel, and the exposure required to form a good signal. *See also* focal plane, pixel.

TDOP *See* dilution of precision.

telemetry [DATA TRANSFER] An automated communication process through wireless data transfer; may also use radio, telephone, or wire communications.

temperature gradient [CLIMATOLOGY] A visual representation of the difference in temperature between different locations.

T

temperature map [CLIMATOLOGY] A map that shows variations in surface temperature across an area. *See also* temperature gradient.

temporal data [DATA STRUCTURES] Data that specifically refers to times or dates. Temporal data may refer to discrete events, such as lightning strikes; moving objects, such as trains; or repeated observations, such as counts from traffic sensors. *See also* continuous data, discrete data, spatial data, thematic data.

temporal GIS [DATA MODELS] The integration of temporal data with location and attribute data. *See also* temporal data.

temporal resolution [PHOTOGRAMMETRY] The frequency or rate at which images are captured over the same geographic location.

terrain

1. [GEOLOGY] The ground surface.
2. [GEOGRAPHY] An area of land having a particular characteristic, such as sandy terrain or mountainous terrain.

See also surface.

terrain dataset [3D GIS] A multiresolution, TIN-based surface built from measurements stored as features in a geodatabase. Associated and supporting rules help organize the data and control how features are used to define the surface. Terrain datasets are typically derived from sources such as lidar, sonar, and photogrammetric data. *See also* terrain, TIN.

terrain dataset pyramid [3D GIS] A data structure associated with a terrain dataset used to define a multiresolution surface, which organizes data into different levels of detail, or pyramid levels, and serves to improve performance by enabling the terrain to access only the data required for a particular display or analysis function. Data that is oversampled or redundant can be avoided. Two pyramid types can be used to build a terrain: z-tolerance and window size. Z-tolerance pyramids filter data into different pyramid levels based on their vertical significance. Window size pyramids filter the data through localized selection biased to max, min, or mean height. *See also* pyramid.

terrain model [MAP DESIGN] A portion of the surface of a raised-relief globe. *See also* raised-relief globe.

terrain pyramid group [3D GIS] In a terrain dataset, a collection of line or polygon feature classes used to represent those geographic features at different levels of detail. *See also* pyramid, terrain.

terrain pyramid level [3D GIS] An individual level of detail in a terrain dataset pyramid. Each level has a pair of properties: resolution and scale. The resolution defines the amount of detail represented by the level, and the scale is a threshold that indicates at which display scale the level becomes active.

terrain pyramid resolution bounds [3D GIS] The range of pyramid levels for which polygon or polyline features will be enforced in the surface of a terrain dataset. *See also* terrain pyramid level.

terrain surface [CARTOGRAPHY] A three-dimensional representation of elevation data.

terrain tiles [3D GIS] A spatially coherent organization of terrain data facilitating efficient retrieval and editing. Tile definition is based on the average point spacing of the source data. *See also* average point spacing.

tessellation [DATA STRUCTURES] The division of a two-dimensional area into polygonal tiles, or a three-dimensional area into polyhedral blocks, in such a way that no figures overlap and there are no gaps. *See also* Thiessen polygon.

text attribute table [ESRI SOFTWARE] A table containing text attributes, such as color, font, size, location, and placement angle, for an annotation subclass in a coverage. In addition to user-defined attributes, the text attribute table contains a sequence number and text feature identifier. *See also* attribute table.

text element [SYMBOLOGY] The characteristics of text on a map, which can include font, font size, color, and symbols. *See also* font, image element, point size, text symbol.

text label [CARTOGRAPHY] The words and numbers on a map that identify or describe features. For example, a text label that delineates the extent of an area feature.

text modifier *See* attribute.

T

text symbol [SYMBOLOGY] A text style defined by font, size, character spacing, color, and so on, used to label maps and geographic features.

texture

1. [3D GIS] A digital representation of the surface of a feature.
2. [PHOTOGRAMMETRY] In imagery, a series of metrics resulting from image processing techniques used to characterize and identify an area of interest. It defines a pattern based on variability and other statistical measures across an object or feature. These are used in traditional pattern recognition techniques or machine learning algorithms, including convolutional neural networks (CNN).

See also machine learning.

thematic accuracy [ACCURACY] A measure of the accuracy of an object or feature's semantic label (such as water, commercial building, secondary road, and so on) in a map layer. Typically stated as a percentage.

thematic data [DATA STRUCTURES] Features of one type that are generally placed together in a single layer. *See also* coverage.

thematic map [MAP DESIGN] A map that focuses on a specific subject and is organized so that the subject stands out above the geographic setting; an incomplete list of examples includes maps of zoning classes, land cover, vegetation zones, species ranges, population density, soils, per capita income, and race and ethnicity. *See also* choropleth map.

theme [VISUALIZATION, ESRI SOFTWARE] A set of related geographic features such as streets, parcels, or rivers, along with their attributes. All features in a theme share the same coordinate system, are located within a common geographic extent, and have the same attributes. Themes are similar to layers. *See also* layer.

theme table [VISUALIZATION, ESRI SOFTWARE] A document object linked to the set of features in a theme. It serves as an interface to the underlying database and is a mechanism for manipulating data.

theodolite [SURVEYING] A surveying instrument for measuring vertical and horizontal angles, consisting of an alidade, a telescope, and graduated circles mounted vertically and horizontally. *See also* alidade, sextant.

thermal radiation [PHYSICS] Electromagnetic radiation of any frequency created from the motion of particles in matter. *See also* electromagnetic spectrum.

Thiessen polygon [EUCLIDEAN GEOMETRY] Polygons generated from a set of sample points. Each Thiessen polygon defines an area of influence around its sample point, so that any location inside the polygon is closer to that point than any of the other sample points. Named for the American meteorologist Alfred H. Thiessen (1872–1931), Thiessen polygons are more commonly known as Voronoi diagrams, other than when used to analyze spatially distributed data in geophysics and meteorology. *See also* Delaunay triangles, Voronoi diagram.

thinning *See* weeding.

third normal form [DATABASE STRUCTURES] The third level of guidelines for designing table and data structures in a relational database. The third-normal-form guideline incorporates the guidelines of first and second normal form; in addition, it recommends removing from a table those columns that do not depend on the table's primary key. A database that follows these guidelines is said to be in third normal form. *See also* normal form.

three-dimensional shape *See* 3D shape.

threshold ring analysis [ANALYSIS/GEOPROCESSING, BUSINESS] In business analysis, an operation that creates rings that contain a given population around a store or stores on a map. *See also* analysis, ring study, store market analysis, store prospecting.

thumbnail [ESRI SOFTWARE] A summary of the geographic data contained in a data source or layer, or a map layout. A thumbnail might provide an overview of all the features in a feature class or a detailed view of the features in, and the symbology of, a layer. Thumbnails are not updated automatically; they become outdated if features are added to a data source or if the layer symbology changes. *See also* data source, layer.

tic [ESRI SOFTWARE] A registration or geographic control point for a coverage representing a known location on the earth's surface. Tics allow all coverage features to be recorded in a common coordinate system. Tics are used to register map sheets when they are mounted on a digitizer. They are also used to transform the coordinates of a coverage, for example, from digitizer units (inches) to the appropriate values for a particular coordinate system. *See also* coverage, register.

tick marks

1. [SYMBOLOGY] Graphics that mark divisions of measurement on a scale bar.
2. [GRAPHICS (MAP DISPLAY)] Short, regularly spaced lines along the edge of an image or neatline that indicate intervals of distance, such as the intersection of longitude and latitude lines to denote the graticule.

See also graticule, neatline, scale bar.

tidal datum [GEODESY] A vertical datum in which zero height is defined by a particular tidal surface, often mean sea level. Examples of tidal surfaces include mean sea level, mean low water springs, and mean lower low water. Most traditional vertical geodetic datums are tidal datums. *See also* datum, vertical geodetic datum, local datum.

tie point

1. [SURVEYING] A point whose location is determined by a tie survey.
2. [PHOTOGRAMMETRY] A point in a digital image or aerial photograph that represents the same location in an adjacent image or aerial photograph. Usually expressed as a pair, tie points can be used to link images and create mosaics.

See also tie survey.

tie survey [SURVEYING] A survey that uses a point of known location on the ground to determine the location of a second point. *See also* tie point.

tied candidates [GEOCODING] In geocoding, two or more records that yield the same score when matching an address. *See also* geocoding.

TIGER [DATABASE STRUCTURES] Acronym for *Topologically Integrated Geographic Encoding and Referencing.* The United States digital database developed for the 1990 U.S. Census, succeeding the DIME format. TIGER files contain street address ranges, census tracts, and block boundaries. *See also* Dual Independent Map Encoding, GBF/DIME, TIGER.

TIGER/Line files [DATA STORAGE] A digital database of geographic features, covering the entire United States and its territories, that provides a topological description of the geographic structure of these areas. The files are a public product created from the U.S. Census Bureau Topologically Integrated Geographic Encoding and Referencing (TIGER) database. TIGER/Line files define the locations and spatial relationships of streets, rivers, railroads, and other features to each other and to the numerous geographic entities for which the Census Bureau tabulates data from its censuses and sample surveys. *See also* TIGER.

tight coupling [MODELING] A high or complex degree of interconnections between the components within a program or between programs that requires substantial overlap between methods, ontologies, class definitions, and so on. *See also* loose coupling.

tile [DATA STRUCTURES] Each image (such as a .jpg) that is part of a set of tiles that compose a tile layer. Tiles are stored in a tile cache and have a configuration determined by a tiling scheme. *See also* tiling, tiling scheme.

tile cache [DATA STRUCTURES] Spatial dataset or imagery files that are processed, compressed, and stored, or "cached" into a defined tiling scheme that enables efficient access. Multiple tiles are combined to create a map. How each tile cache fits into the map, along with its metadata and spatial reference, is defined in a tiling scheme. *See also* tiling, tiling scheme.

tile layer [DATA STRUCTURES] A data layer that displays either raster image or vector tiles, pregenerated as a tile cache. Tiled layers are often used for basemaps. A tile layer is distinguishable from a map image layer, which is rendered dynamically in response to user actions. *See also* map image layer, tile, tile cache, tiling scheme.

tile service [DATA MANAGEMENT, PROGRAMMING] A persistent software process that provides access to static, prerendered image tiles—typically as an image tile layer. Most often used to generate basemaps and elevation layers. *See also* image tile layer.

tiling [DATA STRUCTURES] An organization method for a spatial dataset (specifically a raster) into rectangular blocks of pixels, typically used to improve data access for large datasets. *See also* tessellation, tile cache, tiling scheme.

tiling scheme [DATA STRUCTURES] The information that instructs how to reference the tiles in a tile cache. Defines spatial reference, naming convention, tiling grid, size of tiles in pixels, and scale levels at which the cache has tiles. Based on this schema, a system can determine what tiles are required to cover a map at a suitable resolution for the screen or output device. *See also* tile cache, tiling.

time line [VISUALIZATION] A graphic, linear representation of a time range.

time window [NETWORK ANALYSIS] In routing and network analysis, the time during which a stop can be visited. For example, on a bus route, each stop may have a time window of 15 minutes. If the bus arrives before its 15-minute time slot, it will wait until the appropriate time before proceeding. If a bus arrives after its 15-minute time slot, the stop will display a symbol to denote a time violation. *See also* stop, path.

TIN [DATA STRUCTURES] Acronym for *triangulated irregular network.* A vector data structure that partitions geographic space into contiguous, nonoverlapping triangles. The vertices of each triangle are sample data points with x-, y-, and z-values. These sample points are connected by lines to form Delaunay triangles. TINs are used to store and display surface models. *See also* Delaunay triangulation.

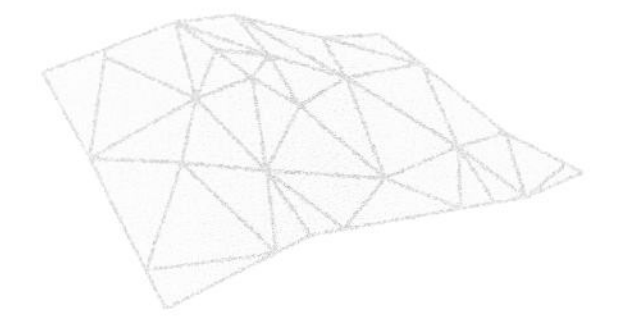

TIN

TIN dataset [DATA STRUCTURES] A dataset containing a triangulated irregular network (TIN). The TIN dataset includes topological relationships between points and neighboring triangles. *See also* TIN.

TIN layer [ESRI SOFTWARE] A layer that references a set of TIN data. TIN data contains a triangulated irregular network (TIN) and includes topological relationships between points and neighboring triangles. *See also* layer.

TIN line type [DATA STRUCTURES] One of four types of edges that may be found in a TIN: regular lines, hard breaklines, soft breaklines, and outside lines. Regular lines define the TIN's basic structure, connecting triangle nodes. Hard breaklines represent features that mark pronounced changes in slope, like roads or rivers. Soft breaklines mark milder changes in slope and sometimes artificial boundaries, such as the border of a study area. Outside lines designate parts of a TIN structure that lie beyond the TIN's zone of interpolation. Every TIN contains regular lines; other line types may or may not be present. *See also* breakline, edge, TIN.

Tissot's indicatrix [MAP PROJECTIONS] A graphical representation of the spatial distortion at a particular map location. The indicatrix is the figure that results when a circle on the earth's surface is plotted to the corresponding point on a map. The shape, size, and orientation of an indicatrix at any given point depend on the map projection used. In conformal (shape-preserving) projections, the indicatrix is a circle; in nonconformal projections, it is an ellipse at most locations. As a visual aid, indicatrices convey a general impression of distortion; as mathematical tools, they can be used to quantify distortion of scale and angle precisely. Tissot's indicatrix was devised by French cartographer Nicolas-Auguste Tissot (1824–1907). *See also* distortion, projection.

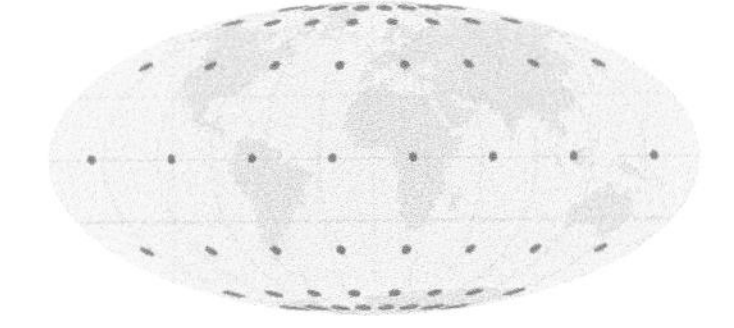

Tissot's indicatrix

TLM [CARTOGRAPHY] Acronym for *topographic line map*. A map that uses line contours to show elevations and depressions of the earth's surface. Topographic line maps may be used to portray topography, elevations, infrastructure, hydrography, and vegetation. *See also* topographic map.

TLM

Tobler's First Law of Geography [GEOGRAPHY] A formulation of the concept of spatial autocorrelation by the American-Swiss geographer and cartographer Waldo Rudolph Tobler (1930–2018), which states "Everything is related to everything else, but near things are more related than distant things." *See also* autocorrelation.

tolerance [DATA EDITING] The minimum or maximum variation allowed when processing or editing a geographic feature's coordinates. For example, during editing, if a second point is placed within the snapping tolerance distance of an existing point, the second point will be snapped to the existing point. *See also* comparison threshold, error, precision.

tone

1. [AERIAL PHOTOGRAPHY] In photography, the intensity of an object in a panchromatic image.
2. [PHOTOGRAMMETRY] An image element derived from the spectral response in each band of an image that is used to help identify an object of interest. Related to hue, saturation, and intensity, or other measures of chromaticity.

See also image element.

to-node [DATA STRUCTURES] One of an arc's two endpoints. From- and to-nodes give an arc a sense of direction. *See also* from-node.

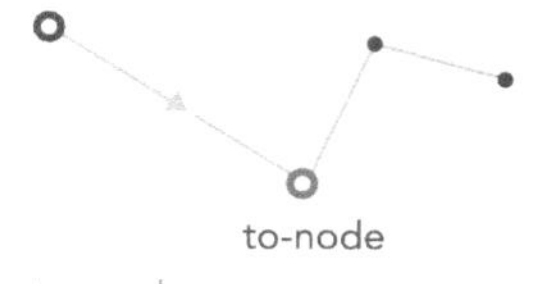

to-node

topo sheet *See* quadrangle.

topographic line map *See* TLM.

topographic map [CARTOGRAPHY] A map that represents the three-dimensional nature of the terrain surface (elevation and landforms) as well as natural and cultural features in the environment. Topographic maps typically show relief, such as contour lines, hypsometric tints, and hillshading. *See also* cultural feature, natural feature, planimetric map, TLM, topography.

topographic profile [MAP DESIGN] A profile based on contours. Also known as a terrain profile. *See also* contour, profile.

topography [CARTOGRAPHY] A record and measurements of the terrain, relief, and landforms on the surface of the earth, planets, and moons. *See also* bathymetric map, elevation guide, hypsometric map, Tissot's indicatrix, topographic map.

topological association [ESRI SOFTWARE] The spatial relationship between features that share geometry such as boundaries and vertices. *See also* shared boundary, shared vertex, topology.

topological error [ACCURACY] A type of error that occurs when multiple layers of maps on which features are not aligned are overlaid, resulting in small lines or polygons that are identified as unique features rather than part of larger adjacent features. *See also* error.

topological feature [ESRI SOFTWARE] A feature that supports network connectivity that is established and maintained based on geometric coincidence. *See also* feature, topology.

topological overlay *See* overlay.

topological transformation [SPATIAL ANALYSIS] The process of modifying the topological relationships between spatial features, such as adjacency, containment, and connectivity. Topological transformations are used to modify or update the topology of spatial data to ensure data integrity, consistency, and accuracy. Topological transformations are typically applied to vector data, such as points, lines, and polygons. *See also* adjacency, connectivity, containment.

topology

1. [EUCLIDEAN GEOMETRY] The study of the properties and spatial relationships of figures that remain unchanged even when they are bent, stretched, or otherwise distorted.
2. [ESRI SOFTWARE] In geodatabases, the arrangement that constrains how point, line, and polygon features share geometry. For example, street centerlines and census blocks share geometry, and adjacent soil polygons share geometry. Geodatabase topology defines and enforces data integrity rules (for example, there should be no gaps between polygons), and topological relationship queries, navigation, editing, and feature construction.

See also arc-node topology, polygon-arc topology.

topology error *See* error.

topology rule [ESRI SOFTWARE] An instruction to the geodatabase defining the permissible relationships of features within a given feature class or between features in two different feature classes. *See also* topology.

toponym [GEOGRAPHY] A place-name.

tour [NETWORK ANALYSIS] A path through a network that visits each stop in the network only once and then returns to its point of origin. The path is determined based on a user-specified set of criteria (such as shortest distance or fastest time). *See also* cycle, Hamiltonian path, network, path.

township

1. [CADASTRAL AND LAND RECORDS] A measure of distance north or south of the baseline (in units of 6 miles) in the U.S. Public Land Survey System (PLSS).

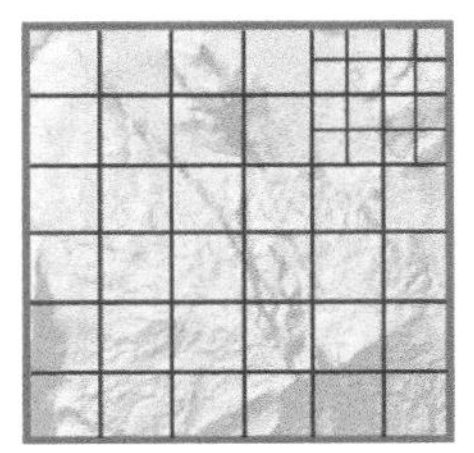

township 1

2. [LOCAL GOVERNMENT] A governmental subdivision, which may vary from the standard size and shape.

See also meridian, parallel, PLSS, quadrangle, section.

township line [SURVEYING] A parallel surveyed at a 6-mile interval north or south of the baseline or another township line in the U.S. Public Land Survey System and Canada's Dominion Land Survey. *See also* baseline, DLS, PLSS.

TPS measurement [SURVEYING] TPS is an acronym for *Total Positioning System*. An entry in an electronic or paper field book that represents observations from a theodolite. A slope distance, vertical angle, horizontal angle, and a height of target define a single TPS measurement. *See also* simple measurement, theodolite.

tracking data *See* temporal data.

tract *See* census tract. *See also* parcel.

traffic flow map [NETWORK ANALYSIS] A map created from counts taken at automatic traffic recorder (ATR) stations to show traffic volumes for the region. *See also* ATR station.

training [PHOTOGRAMMETRY] In image classification and deep learning, the process of deep model learning from training samples that represent features of interest. Training serves to develop a prediction model for the identification of features and objects in imagery. *See also* deep learning, inferencing, training samples.

training samples [PHOTOGRAMMETRY] In supervised image classification and deep learning, a representative sample of specific features and objects used to train the classifier and a deep learning model. *See also* deep learning, inferencing, training.

transect line [BIOLOGY] A line segment along which statistical samples are collected.

transform events [LINEAR REFERENCING] In linear referencing, an operation that produces a new table by copying and transforming events from one route reference to another. This allows the events to be used with a route reference having different route identifiers and/or measures. *See also* linear referencing, route.

transformation [DATA CONVERSION]

1. The process of converting or altering data within a GIS to integrate differing data types, improve visualization, or provide an analytical framework. Transformation may refer to coordinate transformation, geometric transformation, topological transformation, image transformation, attribute transformation, or network transformation.
2. The process of converting the coordinates of a map or an image from one system to another, typically by shifting, rotating, rescaling, skewing, or projecting them.

T

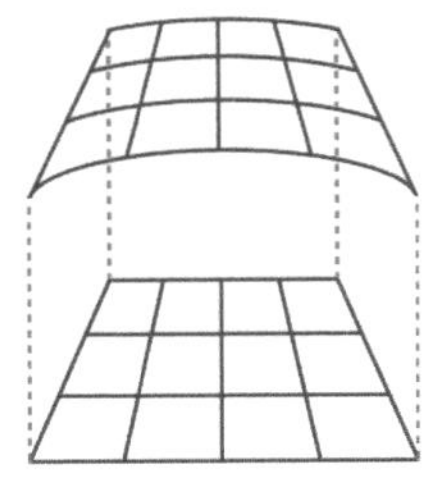

transformation 2

See also attribute transformation, coordinate transformation, geometric transformation, image transformation, network transformation, topological transformation.

transformed band [PHOTOGRAMMETRY] An output image having a different form or characteristics from the input band. *See also* band, transformation.

transit map [MAP DESIGN] A map used to navigate a transit system, such as a subway or bus system. Routes are commonly simplified and stretched to show the sequence of stops clearly. *See also* route.

transit rule [SURVEYING] A rule for adjusting the closure error in a traverse. The transit rule distributes the closure error by changing the northings and eastings of each traverse point in proportion to the northing and easting differences in each course. More specifically, a correction is computed for each northing coordinate as the difference in the course's northings divided by the sum of all the courses' northing differences. Similarly, a correction is computed for each easting coordinate using the easting coordinate differences. The corrections are applied additively to each successive coordinate pair, until the final coordinate pair is adjusted by the whole closure error amount. The transit rule assumes that course directions are measured with a higher degree of precision than the distances. Usually, observed angles are balanced for angular misclosure prior to applying a transit rule adjustment, and corrections are proportional to the x and y components of the measured line. The transit rule is used infrequently since it is only valid in cases in which the measured lines are approximately parallel to the grid of the coordinate system in which the traverse is computed. *See also* closure error, easting, northing.

translate [COORDINATE SYSTEMS] The movement or shifting of a dataset horizontally or vertically to a new location. *See also* registration, rescale, rotate.

translation [DATA EDITING] Adding a constant value to a coordinate. *See also* geometric transformation, transformation.

transmittance [PHOTOGRAMMETRY] In imagery, the ratio of light that is passed through a medium divided by the amount of light that is incident on that medium. Value ranges between 0 and 1.

transparency

1. [GRAPHICS (MAP DISPLAY)] A technique that combines datasets with varying degrees of opaqueness to reveal,

emphasize, and convey information in the data.
2. [VISUALIZATION] In GIS, transparency and its opposite, opacity, are measures of how see-through a feature or layer appears. For example, increasing a layer's transparency allows you to see more of the layers underneath.

See also style, symbology.

transverse aspect [MAP PROJECTIONS] A cylindrical map projection with the line or lines of tangency oriented along a meridian rather than along the equator. *See also* cylindrical projection, line of tangency, meridian.

transverse Mercator projection
[MAP PROJECTIONS] A conformal cylindrical projection—invented by Swiss polymath Johann Heinrich Lambert in 1772—that rotates the Mercator projection 90 degrees so that the line of tangency is a pair of meridians. Similar to the Gauss-Krüger projection. *See also* Gauss-Krüger projection, line of tangency, Mercator projection, Web Mercator projection.

trap feature [SYMBOLOGY] An element or feature that is added to or removed from a map for copyright concerns.

travel time [NAVIGATION] The amount of time spent in transit between two locations in a street network. Travel time is typically calculated using standard transportation modes, such as automobile, public transit, or walking.

traveling salesperson problem
[NETWORK ANALYSIS] A Hamiltonian circuit problem in which a traveler must find the most efficient way to visit a series of stops, then return to the starting location. In the original version of the problem, each stop may be visited only once. *See also* Hamiltonian path.

traverse
1. [SURVEYING] A predefined path or route across or over a set of geometric coordinates.
2. [SURVEYING]A method of surveying in which lengths and directions of lines between points on the earth are obtained by or from field measurements across terrain or a digital elevation model.
3. [COORDINATE GEOMETRY (COGO)] A sequence of connected two-point line features that are created by entering the coordinate geometry (COGO) dimensions for each line. If the lines are COGO-enabled, the entered dimension values are stored in COGO attribute fields. The traverse is provided with a start location for the first line and may optionally also be provided with a closing location. A traverse provided with a closing location is called a closed traverse. A traverse that is not given a closing location is called an open traverse. When the closing location is the same as the start location, the traverse is called a closed loop traverse.

See also closed loop traverse, DEM.

trend [SPATIAL STATISTICS (USE FOR GEOSTATISTICS)] In a spatial model, nonrandom variation in the value of a variable that can be described by a mathematical function such as a polynomial. *See also* short-range variation.

trend surface analysis [MATHEMATICS] A surface interpolation method that fits a polynomial surface by least-squares regression through the sample data points. This method results in a surface that minimizes the variance of the surface in relation to the input values. The resulting surface rarely goes through the sample data points. This is the simplest method for describing large variations, but the trend surface is susceptible to outliers in the data. Trend surface analysis is used to find general tendencies of the sample data, rather than to model a surface precisely. *See also* interpolation.

triangle [EUCLIDEAN GEOMETRY] A face on a TIN surface. Each triangle on a TIN surface is defined by three edges and three nodes and is adjacent to one to three other triangles on the surface. TIN triangles can be used to derive aspect and slope information and may be attributed with tag values. *See also* edge, TIN.

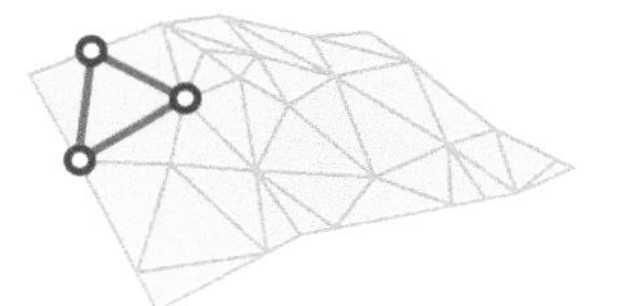

triangle

triangulated irregular network *See* TIN.

triangulation [SURVEYING] Locating positions on the earth's surface using the principle that if the measures of one side and the two adjacent angles of a triangle are known, the other dimensions of the triangle can be determined. Surveyors begin with a known length, or baseline, and from each end use a theodolite to measure the angle to a distant point, forming a triangle. Once the lengths of the two sides and the other angle are known, a network of triangles can be extended from the first. *See also* Delaunay triangulation, natural neighbors, trilateration.

trilateration [SURVEYING] Determining the position of a point on the earth's surface with respect to two other points by measuring the distances between all three points. *See also* triangulation.

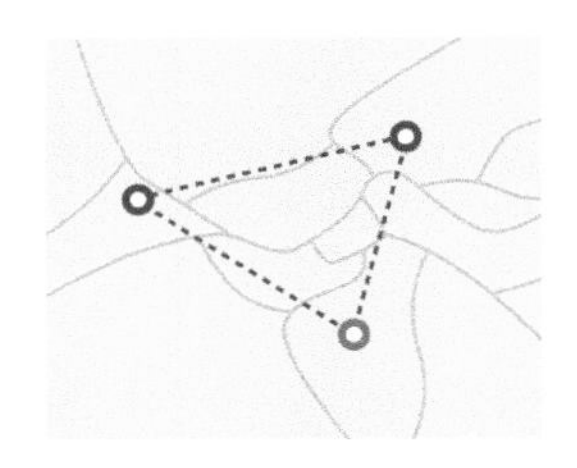

trilateration

tripel [SEMANTICS] A German word meaning a combination of three elements; used to indicate that the Winkel Tripel projection minimizes the geometric distortion of area, shape, and distance. *See also* geometric distortion, Winkel Tripel projection.

true azimuth [NAVIGATION] The angle clockwise from a true north reference line to a direction line. *See also* direction line, reference line, true north.

true bearing [NAVIGATION] A bearing measured relative to true north. *See also* bearing, navigation, true north.

true curve *See* parametric curve.

true north [GEOGRAPHY] The direction from any point on the earth's surface to the geographic north pole. *See also* grid north, magnetic north, north pole.

true north reference line [GEOGRAPHY] A line from any point on the earth to the north pole. *See also* north pole, true north.

true-direction projection
See azimuthal projection.

TSP *See* traveling salesperson problem.

TSPI [REMOTE SENSING] Acronym for *time-space-position-information.* In remote sensing, the position and attitude information provided about an airborne or space-based platform by the onboard use of an Inertial Measurement Unit (IMU) integrated with a Global Positioning System (GPS) and potentially other sensors and computational algorithms. Commonly used to provide the exterior orientation of a remote sensor, either in real time or as a post-processed result. *See also* GPS, inertial measurement unit, sensor.

tuple [DATABASE STRUCTURES] An individual row or record in a database table. Each tuple records the values for the columns defined in the table. *See also* record.

turn [NETWORK ANALYSIS] A movement that explicitly models transitions between edge elements during navigation. *See also* network element, turn feature class, turn table.

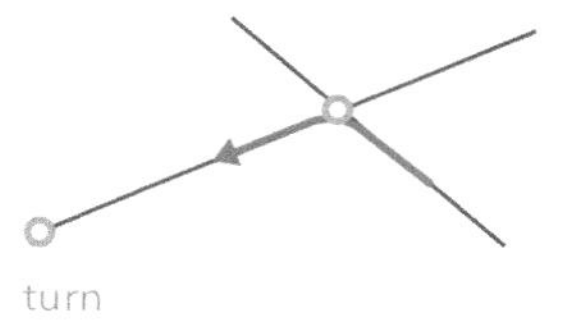

turn

turn feature class [NETWORK ANALYSIS, ESRI SOFTWARE] A specialized feature class that defines turn movements in a network dataset. Turn features explicitly model subsets of possible transitions between edge elements during navigation and may also store the turn impedance. *See also* turn, turn table.

turn impedance [NETWORK ANALYSIS] The cost of making a turn at a network node. The impedance for making a left turn, for example, can be different from the impedance for making a right turn or a U-turn at the same place. *See also* impedance, network, network analysis, turn, turn feature class.

turn table [NETWORK ANALYSIS] A table that stores information about the cost of making each turn movement in a network. A turn table identifies the edge from which the turn movement originated, the junction where the turn occurs, and the edge that it turns onto. *See also* turn, turn feature class.

turn-by-turn maps [MAP DESIGN] A series of small maps detailing where route segments meet.

tween [GRAPHICS (COMPUTING)] In animated applications, the process of creating intermediate frames between two images so it seems like the first image merges smoothly into the second image. *See also* animation.

T

UAS [AERIAL PHOTOGRAPHY] Acronym for *uncrewed aerial system.* Aircraft and related equipment piloted by remote control or onboard computers. Also known as uncrewed reconnaissance vehicles or, more commonly, drones. *See also* drone imagery.

UAV *See* UAS.

ultraviolet [PHYSICS] Also known by the abbreviation *UV.* The wavelength of energy in the electromagnetic spectrum just shorter than visible light. *See also* band, electromagnetic spectrum.

uncertainty [DATA QUALITY] The degree to which the measured value of some quantity is estimated to vary from its true value. Uncertainty can arise from a variety of sources, including limitations to the precision or accuracy of a measuring instrument or system; measurement error; the integration of data that uses different scales or that describes phenomena differently; conflicting representations of the same phenomena; the variable, unquantifiable, or indefinite nature of the phenomena being measured; or the limits of human knowledge. Uncertainty is often used to describe the degree of accuracy of a measurement. *See also* error, vagueness.

uncrewed aerial system *See* UAS.

undershoot [DATA CAPTURE] A line that falls short of another line that it should intersect. *See also* dangling arc.

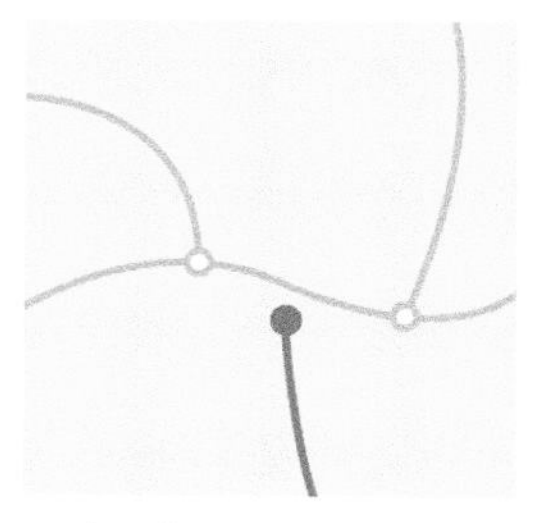

undershoot

undevelopable surface [MAP PROJECTIONS] A surface, such as the earth's, that cannot be flattened into a map without stretching, tearing, or squeezing it. To produce a flat map of the round earth, its three-dimensional surface must be projected onto a developable shape, such as a plane, cone, or cylinder. *See also* developable surface, projection.

undirected network flow [NETWORK ANALYSIS] A network state in which each edge may or may not have an associated direction of flow. In an undirected network flow, the resource that traverses a network's components can decide which direction to take, such as traffic in transportation systems. *See also* directed network flow.

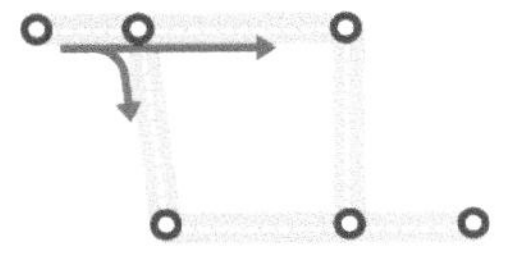

undirected network flow

uninitialized flow direction [ESRI SOFTWARE] A condition that occurs in a network when an edge feature is not connected through the network to sources and sinks or if the edge feature is only connected to sources and sinks through disabled features. *See also* edge, flow direction, sink, source.

union [ANALYSIS/GEOPROCESSING] The overlay of two or more spatial polygon datasets that preserves the features that fall within the spatial extent of either input dataset. In a union, all features from both datasets are retained and extracted into a new polygon dataset. *See also* identity, intersect.

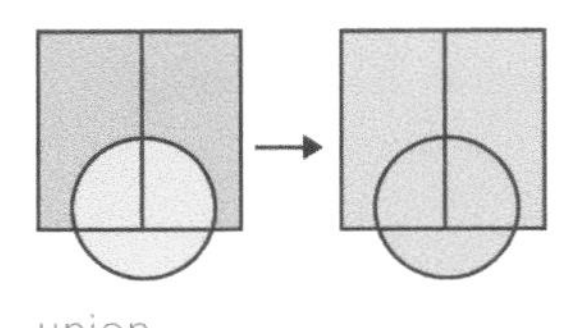

union

unit of measure [STANDARDS] A standard quantity used for measurements, such as length, area, and height. *See also* distance unit, English unit, linear unit, map unit, metric system.

United States Geological Survey *See* U.S. Geological Survey.

univariate analysis [STATISTICS] Any statistical method for evaluating a single variable, rather than the relationship between two or more variables. *See also* analysis, multivariate analysis.

univariate distribution [STATISTICS] A function for a single variable that gives the probabilities that the variable will take a given value. *See also* distribution.

universal kriging [SPATIAL STATISTICS (USE FOR GEOSTATISTICS)] A kriging method often used on data with a significant spatial trend, such as a sloping surface. In universal kriging, the expected values of the sampled points are modeled as a polynomial trend. Kriging is carried out on the difference between this trend and the values of the sampled points. *See also* kriging.

universal polar stereographic [COORDINATE SYSTEMS] A projected coordinate system that covers all regions not included in the UTM coordinate system; that is, regions above 84 degrees north and below 80 degrees south. Its central point is either the north or south pole. *See also* coordinate system, projection, UTM.

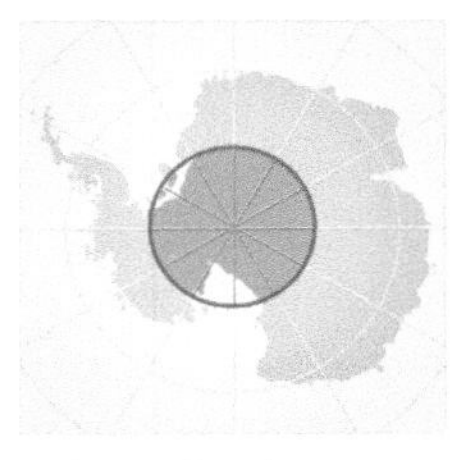

universal polar stereographic

Universal Soil Loss Equation [ENVIRONMENTAL GIS] An erosion model developed by the Agricultural Research Service of the United States Department of Agriculture that computes average annual soil loss caused by rainfall and associated overland flow. Factors used in the equation include rainfall, soil characteristics, topography, and land use and land cover. Each major factor is divided into numerous subfactors.

universal time [ASTRONOMY] A timekeeping system that defines local time throughout the world by relating it to time at the prime meridian. Universal time is based on the average speed at which the earth rotates on its axis. For official purposes, universal time has been replaced by coordinated universal time; universal time is, however, still used in navigation and astronomy. Different versions of universal time correct for irregularities in the earth's rotation and orbit. *See also* coordinated universal time, Greenwich mean time, prime meridian.

universal transverse Mercator *See* UTM.

universe polygon [DATA MODELS] In coverages, the first record in a polygon attribute table, representing the area beyond the outer boundary of the coverage. *See also* boundary, coverage, polygon.

UNIX time [ASTRONOMY] The number of seconds, in coordinated universal time format, since January 1, 1970 (the start of the UNIX system). *See also* coordinated universal time.

unknown points [SURVEYING] In surveying, previously uncoordinated points. *See also* point, surveying.

unmanaged raster catalog [DATA STRUCTURES] A raster catalog in which the raster datasets are not copied or altered by the geodatabase and there will only be a pointer connecting the raster catalog row to the raster datasets. Deleting a row in an unmanaged raster catalog will not delete the raster dataset from its storage location. *See also* managed raster catalog.

unprojected coordinates *See* geographic coordinates.

UPS *See* universal polar stereographic.

upstream [NETWORK ANALYSIS] In network tracing, the direction along a line or edge that opposes the direction of flow. *See also* downstream, directed network flow.

urban geography [GEOGRAPHY] The field of geography concerning the spatial and cultural patterns and processes of cities and neighborhoods. *See also* geography, infrastructure.

Urban Vector Map *See* UVMap.

U.S. Geological Survey [FEDERAL GOVERNMENT] A multidisciplinary scientific agency of the United States government, part of the U.S.

Department of the Interior, focused on biology, geography, geology, geospatial information, and water. The U.S. Geological Survey is the primary civilian mapping agency in the United States. It produces digital and paper map products; aerial photography; and remotely sensed data for land cover, hydrology, geology, biology, and geography; it is also the agency responsible for the national topographic map series. *See also* digital elevation model, National Map Accuracy Standards.

U.S. National Geodetic Survey [FEDERAL GOVERNMENT] The U.S. government agency responsible for maintaining the National Spatial Reference System (NSRS) of the United States. It is a part of the National Oceanic and Atmospheric Administration (NOAA). *See also* National Spatial Data Infrastructure, National Spatial Reference System, NOAA.

USGS *See* U.S. Geological Survey.

USLE *See* Universal Soil Loss Equation.

UT *See* universal time.

UTC *See* coordinated universal time.

U

utility network [NETWORK ANALYSIS] A framework that models and manages a utility infrastructure, such as water, electricity, gas, or telecommunication. Typically composed of nodes, edges, a subnetwork, and flow or connectivity management systems. *See also* edge, node.

UTM [COORDINATE SYSTEMS] Acronym for *universal transverse Mercator.* A projected coordinate system that extends around the world, from 84° N to 80° S. The UTM system has 60 north–south zones, each 6 degrees in longitude. *See also* digital raster graphic, Gauss-Krüger projection, projection, universal polar stereographic.

UV map [3D GIS] A 2D vertex texture that stores the horizontal (u) and vertical (v) coordinates of a 3D model. Additional x,y,z, and w (quaternion or rotation) coordinates may also be used. Typically associated with 3D mesh or texture mapping. *See also* 3D mesh.

UVMap [CARTOGRAPHY] UV is an acronym for *Urban Vector.* A vector-based data product in vector product format (VPF), typically at larger scales ranging from 1:2,000 to 1:25,000. UVMap data is typically collected over densely populated urban areas. *See also* map, vector, VMap.

UVMap

vagueness [UNCERTAINTY] In GIS, a state of uncertainty in data classification that exists when an attribute applies to an indeterminate quality of an object or describes an indefinite quantity. For example, the classification of an area of land as the range of golden-winged warblers (a rare species of bird) is vague for two reasons. The area populated by the birds is indefinite: It is changing constantly and can never be precisely defined. The term "range" is also somewhat vague since the birds migrate and occupy the territory for only part of the year. *See also* ambiguity, fuzzy classification, uncertainty.

valency [ESRI SOFTWARE] In coverages, the number of arcs that begin or end at a node. *See also* valency table.

valency table [ESRI SOFTWARE] A table that lists the nodes in a data layer along with their valencies. *See also* valency.

validation

1. [DATA QUALITY] The process, using formal methods, of evaluating the integrity and correctness of data or a measurement.
2. [MODELING] In modeling, the evaluation of a method to show whether it is assessing the parameter of interest rather than something else.
3. [DATA QUALITY] The process of comparing the topology rules against the features in a dataset. Features that violate the rules are marked as error features. Topology validation is typically performed after the initial topology rules have been defined, after the feature classes have been modified, or if additional feature classes or rules have been added to the map topology.

See also cross validation.

validation rule [DATA QUALITY] A rule applied to objects in the geodatabase to ensure that their state is consistent with the system that the database is modeling. The geodatabase supports attribute, connectivity, relationship, and custom validation rules. *See also* validation.

value

1. [MATHEMATICS] A measurable quantity that may be passed to a function. Values are either assigned or determined by calculation.
2. [GRAPHICS (COMPUTING)] The lightness or darkness of a color.
3. [PHYSICS] The brightness of a color or how much light it reflects, for instance, blue, light blue, dark blue.

See also hue, saturation.

value attribute table *See* VAT.

vanishing point [PHOTOGRAMMETRY] In imagery, a distant point on an image plane where the two-dimensional perspective projections of mutually parallel lines in three-dimensional space appear to converge in an image.

variable

1. [MATHEMATICS] A symbol or placeholder that represents a changeable value or a value that has not yet been assigned.

$$a^2 + b^2 = c^2$$

variable 1

2. [COMPUTING] A symbol or quantity that can represent any value or set of values, such as a text string or number. Variables may change depending on how they are used and applied.

See also attribute, expression, symbol.

variable contour density [MAP DESIGN] A mapping technique in which different contour intervals are used on different parts of the same map; most commonly used when the terrain changes markedly from one region on the map to another. *See also* contour interval.

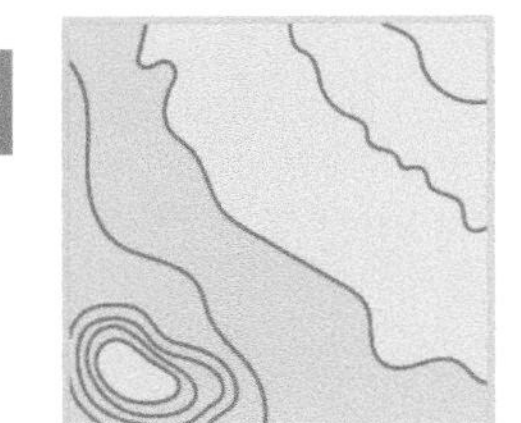

variable contour density

variable depth masking [SPATIAL ANALYSIS] A drawing technique for hiding part of one layer using another set of features. Variable depth masking allows a layer to be drawn with gaps at specific locations without affecting other layers at these locations. *See also* mask.

variance [STATISTICS] A numeric description of how values in a distribution vary or deviate from the mean. The larger the variance, the greater the dispersion of values around the mean. The standard deviation for a distribution is the square root of the variance. *See also* mean, standard deviation.

variance-covariance matrix [SURVEYING] In surveying, the symmetric 3×3 matrix that mathematically expresses the correlation between errors in coordinates x, y, and z. *See also* covariance, error, matrix, variance, x,y,z coordinates.

variant [DATA STRUCTURES] A data type that can contain any kind of data. *See also* data.

variogram [SPATIAL STATISTICS (USE FOR GEOSTATISTICS)] A function of the distance and direction separating two locations that is used to quantify dependence. The variogram is defined as the variance of the difference between two variables at two locations. The variogram generally increases with distance and is described by nugget, sill, and range parameters. If the data is stationary, then the variogram and the covariance are theoretically related to each other. *See also* covariance, nugget, range, semivariogram, sill.

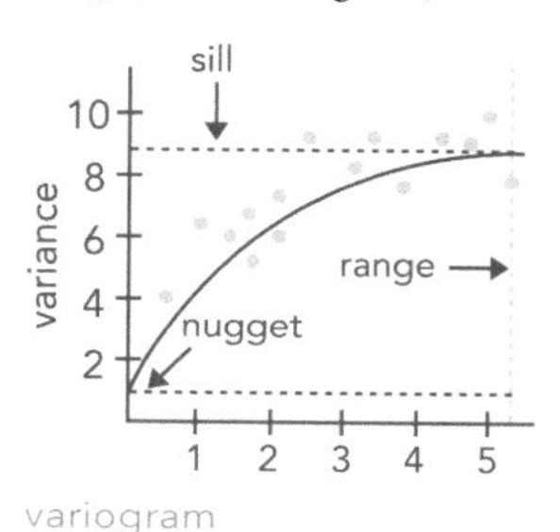

variogram

variography [SPATIAL STATISTICS (USE FOR GEOSTATISTICS)] The process of examining spatial dependence using a variogram. The procedures used to interpret variograms. *See also* variogram.

VAT [ESRI SOFTWARE] Acronym for *value attribute table.* A table containing attributes for a grid, including user-defined attributes, the values assigned to cells in the grid, and a count of the cells with those values. *See also* attribute, grid.

VDOP *See* dilution of precision.

vector

1. [DATA MODELS] A coordinate-based data model that represents geographic features as points, lines, and polygons. Each point feature is represented as a single coordinate pair, while line and polygon features are represented as ordered lists of vertices. Attributes are associated with each vector feature, as opposed to a raster data model, which associates attributes with grid cells.

vector 1

2. [MATHEMATICS] Any quantity that has both magnitude and direction.

See also line, point, polygon, raster.

vector data [MODELING] Data in point, line, or area (polygon) format. *See also* line, point, polygon.

vector data model [DATA MODELS] A representation of the world using points, lines, and polygons. Vector models are useful for storing data that has discrete boundaries, such as country borders, land parcels, and streets. *See also* line, point, polygon, raster data model.

vector flow map [MAP DESIGN] A flow map that uses a regular grid of vectors that are identical in shape but with arrows that vary in orientation to show the direction of significantly high waves, or incidents, at a particular time. *See also* flow map.

vector model *See* vector data model.

V

vector product format [DATA STRUCTURES] Also known by the acronym *VPF.* A vendor-neutral data format used to structure, store, and access geographic data according to a defined standard. *See also* UVMap, VMap.

vectorization [DATA CONVERSION] The conversion of raster data (an array of cell values) to vector data (a series of points, lines, and polygons). *See also* batch vectorization, centerline vectorization, interactive vectorization, raster data, rasterization, vector data.

vegetation zone [CLIMATOLOGY] Areas of land that would have a specific plant cover in the absence of human activity.

vehicle navigation system [NAVIGATION] A GPS- or radio-enabled system that provides route driving information. The system can be installed (interfaced with a vehicle's own technology) or portable (on handheld GPS receivers or smart devices). *See also* GPS, shortest path.

vehicle routing problem [NETWORK ANALYSIS] A type of network analysis that involves routing a fleet of vehicles to known specifications with the goal to (a) achieve a known objective, for example, to lower operating costs, while (b) simultaneously satisfying certain constraints, such as achieving optimal time windows, multiple route capacities, travel duration constraints, and so on. *See also* network analysis, route renewal, time window.

verbal scale [MAP DESIGN] A way to express map scale in commonly used map and ground units through a descriptive expression; for example, "so many centimeters to a kilometer" or "one inch represents 20 miles." *See also* scale bar, representative fraction.

vernier scale [AERIAL PHOTOGRAPHY] A counting dial on a polar planimeter that has a graduated scale calibrated to indicate fractional parts of the subdivisions of the larger scale; allows partial revolutions to be determined with greater precision. Also called a vernier unit. *See also* counting dial, polar planimeter.

vertex

1. [EUCLIDEAN GEOMETRY] One of a set of ordered x,y coordinate pairs that defines the shape of a line or polygon feature.

vertex 1

2. [EUCLIDEAN GEOMETRY] The junction of lines that form an angle.
3. [GEODESY] The highest point of a feature.

See also node.

vertical control datum *See* vertical geodetic datum.

vertical coordinate system [COORDINATE SYSTEMS] A reference system that defines the location of z-values relative to a surface. The surface may be gravity related, such as a geoid, or a more regular surface like a spheroid or sphere. *See also* coordinate system.

vertical datum [COORDINATE SYSTEMS] A surface, or the set of locations that define a surface, used to determine heights or depths on, above, or below the earth's surface. *See also* vertical geodetic datum.

vertical exaggeration [CARTOGRAPHY] A multiplier applied uniformly to the z-values of a three-dimensional model to enhance the natural variations of its surface. Scenes may appear too flat when the range of x- and y-values is much larger than the z-values. Setting vertical exaggeration can compensate for this apparent flattening by increasing relief. *See also* 3D model, elevation, z-value.

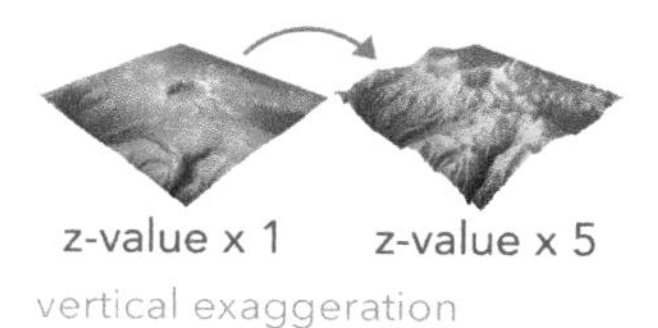

vertical exaggeration

vertical geodetic datum [GEODESY] A geodetic datum for any extensive measurement system of heights on, above, or below the earth's surface. Traditionally, a vertical geodetic datum defines zero height as the mean sea level at a particular location or set of locations; other heights are measured relative to a level surface passing through this point. Examples include the North American Vertical Datum of 1988; the Ordnance Datum Newlyn (used in Great Britain); and the Australian Height Datum. *See also* benchmark, datum, elevation, geodetic datum.

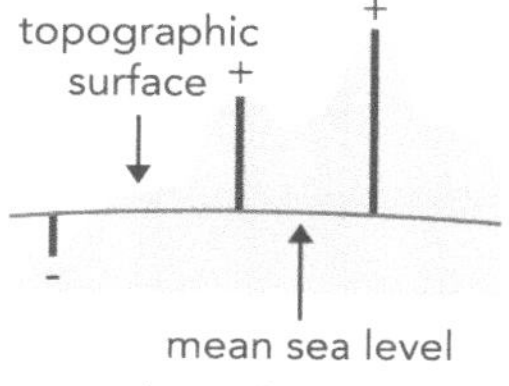

vertical geodetic datum

vertical image [REMOTE SENSING] An image collected with the sensor axis vertical (sensor pointing straight down) or as nearly vertical as possible. The practical consideration for being considered true vertical is that the principal axis remains within ± 3 degrees of the nadir direction. *See also* sensor, vertical photograph.

vertical line *See* plumb line.

vertical photograph [AERIAL PHOTOGRAPHY] An aerial photograph taken with the camera lens pointed straight down. *See also* aerial photograph.

vertical photograph

vertical shift [ESRI SOFTWARE] A parameter that offsets the z-origin from the surface of a vertical coordinate system. The vertical shift is similar in effect to the false easting or false northing parameters of a projected coordinate system. *See also* false easting, false northing, vertical coordinate system.

VFR chart [MAP DESIGN] VFR is an acronym for *visual flight rules.* An aeronautical chart that is based on a set of regulations under which a pilot operates an aircraft in weather conditions generally clear enough to allow the pilot to see where the aircraft is going. Specifically, the pilot must be able to see outside the cockpit, control the aircraft's altitude, navigate, and avoid obstacles and other aircraft. *See also* aeronautical chart.

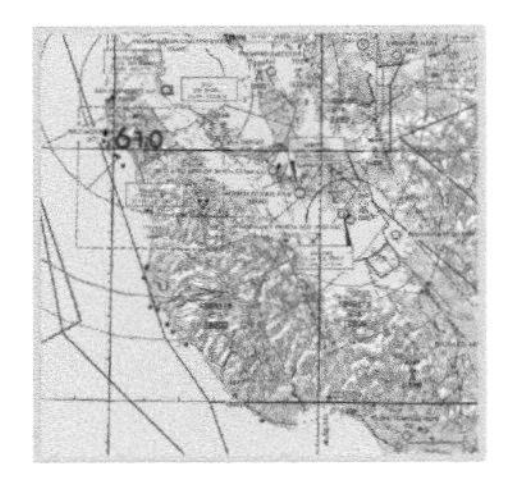

VFR chart

view [ESRI SOFTWARE]

1. In a database, a stored query used to display data from specified columns from one or multiple tables. Views can be used by the database administrator to restrict access to data by displaying only certain columns or to join information from two or more tables and display them in one table. In geodatabases, views can also include the spatial column of a feature class. These spatial views can be used as feature classes in client applications.
2. The user interface component that controls the extent, layers, and features in a web map or web scene. A view typically contains the extent's properties and the position and zoom level or scale of the map or scene.

See also extent, feature, layer, map view, scale, web map, zoom.

viewing mirror [NAVIGATION] The part of a compass that is used to help determine angles more accurately. *See also* compass.

viewpoint [VISUALIZATION] The point from which a line of sight or viewshed is determined. Also known as an observer point. *See also* line of sight, viewshed.

viewshed [GEODESY] The field of view from a given vantage point. Viewshed maps are useful for applications such as finding well-exposed places for communication towers. *See also* field of view.

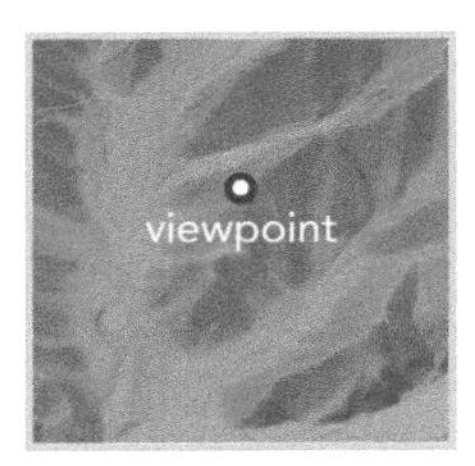

viewshed

virtual table [COMPUTING] A logical table in a database that stores a pointer to the data rather than the data itself. *See also* table.

virtually lossless compression *See* lossless compression.

visibility *See* line of sight.

visible satellite image [SATELLITE IMAGING] An image that shows the amount of visible sunlight reflected to a satellite sensor by clouds, the land surface, and water bodies.

visible scale range [ESRI SOFTWARE] A minimum and maximum value that a map scale must fall between so that the map layers will display. *See also* display scale, range, scale.

visual balance [MAP DESIGN] An impression of ordered distribution of the elements placed on the map influenced by the organization of the mapped area and marginalia. *See also* map marginalia.

visual center [MAP DESIGN] The point on a rectangular map or image to which the eye is drawn. The visual center lies slightly (about 5 percent of the total height) above the geometric center of the page.

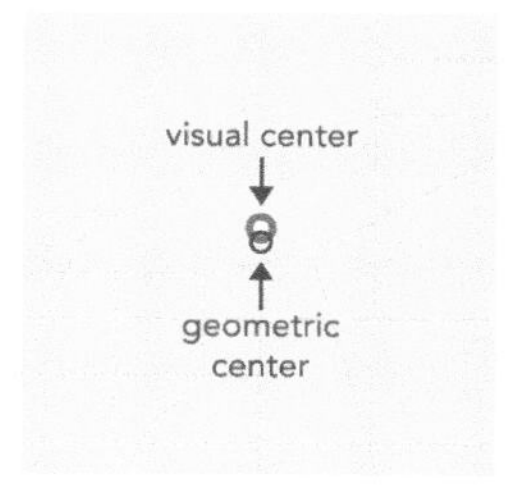

visual center

visual field effect [SYMBOLOGY] The effect that occurs when a map symbol's appearance is modified by nearby symbols.

visual hierarchy [MAP DESIGN] The presentation of features on a map in a way that implies order, priority, or relative importance, usually achieved with visual contrast. *See also* contrast, visual balance.

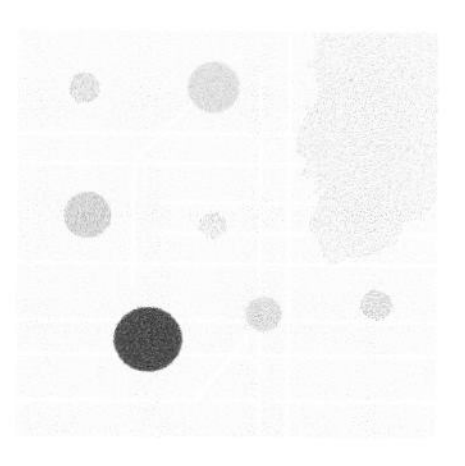

visual hierarchy

visual variable [SYMBOLOGY] A set of parameters for a renderer symbol that can be changed to provide qualitative or quantitative information to the map reader. Visual variable parameters include color settings (hue, value, and saturation), size, shape, orientation, and pattern details. *See also* image element.

visualization [VISUALIZATION] The representation of data in a viewable medium or format. In GIS, visualization is used to organize spatial data and related information into layers that can be analyzed or displayed as maps, three-dimensional scenes, summary charts, tables, time-based views, and schematics. *See also* animation, hillshading, representation, spatial data.

VMap [STANDARDS] An abbreviation of *vector map,* also known as a vector smart map. A vector-based GIS data output of the earth in vector product format (VPF), produced at several scales or levels. For example, VMap Level 1 scale is 1:250,000, and VMap Level 2 scale is 1:50,000. *See also* UVMap.

VMR [SPATIAL ANALYSIS] Acronym for *variance-to-mean ratio.* A mathematical measure that indicates whether features are arranged spatially in a clustered, random, or regular manner. This measure is used only for features, not their attributes.

volume
1. [DATA ANALYSIS] In a TIN, the space (measured in cubic units) between a TIN surface and a plane at a specified elevation. Volume may be calculated above or below the plane.
2. [PHYSICS] The amount of space occupied by a three-dimensional object, calculated as area times height or depth, and usually expressed in cubic units.

volume 2

See also TIN.

VOR beacon [NAVIGATION] VOR is an abbreviation for *VHF omnidirectional range.* A very high frequency beacon used for air navigation, usually located at or near airports and frequently coupled with a distance-measuring equipment (DME) ground station. The compass rose on an aeronautical chart is often centered on this type of beacon. *See also* aeronautical chart, compass rose.

Voronoi diagram [EUCLIDEAN GEOMETRY] A partition of space into areas, or cells, that surround a set of geometric objects (usually points). These cells, or polygons, must satisfy the criteria for Delaunay triangles. All locations within an area are closer to the object it surrounds than to any other object in the set. Voronoi diagrams are often used to delineate areas of influence around geographic features. Voronoi diagrams are named for the Ukrainian mathematician Georgy Feodosevich Voronyi (1868–1908). *See also* Delaunay triangles, Thiessen polygon.

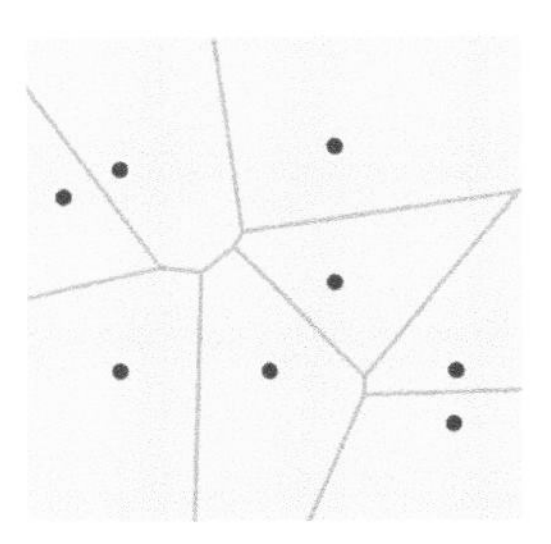

Voronoi diagram

voxel [GRAPHICS (MAP DISPLAY)] A three-dimensional pixel used to display and rotate three-dimensional images. Typically represents a specific cell value in three-dimensional space. *See also* pixel.

voxel

VPF *See* vector product format.

VRP *See* vehicle routing problem.

WAAS [GPS] Acronym for *Wide Area Augmentation System.* A regional satellite-based augmentation system (SBAS) used to augment GPS, with the goal to improve its accuracy, integrity, and availability. This system was developed and is operated by the U.S. Federal Aviation Administration (FAA) for use in North American airspace. *See also* GPS, GPS augmentation.

WAMI [PHOTOGRAMMETRY] Acronym for *wide area motion imagery.* In imagery, a large-format, sequential imagery (typically greater than 10k × 10k pixels per frame) that provides persistent surveillance for large regions of interest. Commonly used to produce observations of moving objects or to track observations.

warping *See* rubber sheeting.

watershed [HYDROLOGY] A basinlike terrestrial region consisting of all the land that drains water into a common terminus. Also known as a catchment. *See also* source, sink.

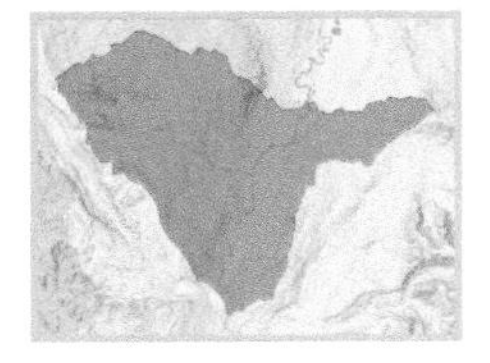

watershed

wavelength [PHYSICS] The distance between two successive crests on a wave, calculated as the velocity of the wave divided by its frequency. *See also* electromagnetic spectrum, spectral resolution.

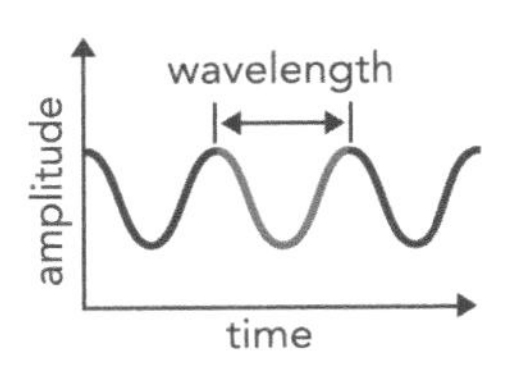

wavelength

wavelet compression [DATA STORAGE] A lossy method of data compression that uses mathematical functions and is best used in image or sound compression. *See also* compression, lossy compression.

wavelet transform [DIGITAL IMAGE PROCESSING] In imagery, a method of reducing the size of a stored image through analysis and removal of high frequency, nonessential components, specifically those considered to be noise. *See also* noise.

wayfinding

1. [GEOGRAPHY] The mental activities engaged in by a person trying to reach a destination, usually an unfamiliar one, in real or virtual space. Wayfinding consists of

W

acquiring information that is relevant to choosing a route, or a segment of a route, and of evaluating that information during travel so the route can be changed as needed. Wayfinding is the cognitive component of navigation.

2. [GEOGRAPHY] The academic study of wayfinding behavior; also, the scientific art of designing real or virtual environments to make wayfinding easier.
3. [NAVIGATION] Long-distance, open-sea navigation without instruments. Historical wayfinding typically refers to the navigation techniques used by Indigenous people from the greater Pacific Islands, also known as Polynesian wayfinding.

See also dead reckoning, locomotion, navigation.

waypoint [GPS] A specified geographic location, destination, or point on a route—defined by longitude and latitude—that is used with GPS for navigation purposes. *See also* GPS, latitude-longitude, location.

weather map [METEOROLOGY] A map that shows short-term, current atmospheric conditions and predictions.

web feature server specification *See* Web Feature Service.

Web Feature Service [INTERNET] Also known by the acronym *WFS*. A set of interface specifications that standardizes the methods used to display and manipulate geographic features and data in web maps. The WFS specification is the result of a collaborative effort assembled by the Open Geospatial Consortium (OGC). *See also* Open Geospatial Consortium, Web Map Service.

web map [ESRI SOFTWARE, SOFTWARE] A digital map file that contains a map's layers, styles, and data; end users can use services to call the web map and can display and query the web map's layers. *See also* interactive map.

web map server specification *See* Web Map Service.

Web Map Service [INTERNET] Also known by the acronym *WMS*. A set of interface specifications that standardizes how maps are rendered by servers to web clients. The WMS specification is the result of a collaborative effort assembled by the Open Geospatial Consortium (OGC). *See also* Open Geospatial Consortium, Web Feature Service.

Web Mercator projection [MAP PROJECTIONS] A cylindrical map projection that assumes a spherical earth, commonly used for online maps. *See also* Mercator projection, transverse Mercator projection, WTM.

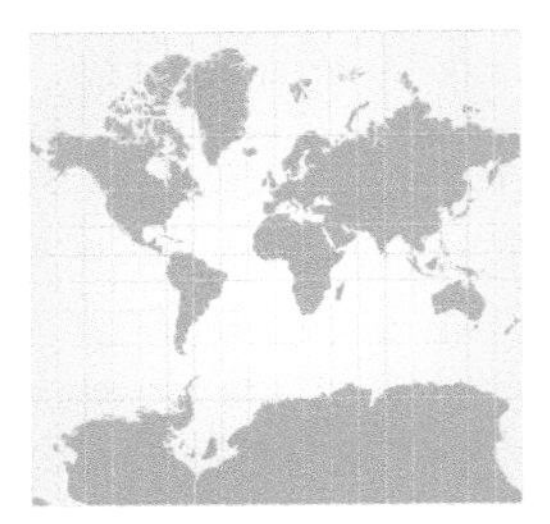
Web Mercator projection

web service [INTERNET] A persistent software process that provides access to web content for client applications. Web content can include data, programs, processing templates, and ancillary information related to GIS material. *See also* service.

weed tolerance [DATA CAPTURE] The minimum distance allowed between any two vertices along a line, set before digitizing. When new lines are added, vertices that fall within that distance of the last vertex are ignored. Weed tolerance applies only to vertices, not to nodes. *See also* weeding.

weeding [DATA CAPTURE] The act of reducing the number of points that define a line while preserving its essential shape. *See also* line smoothing.

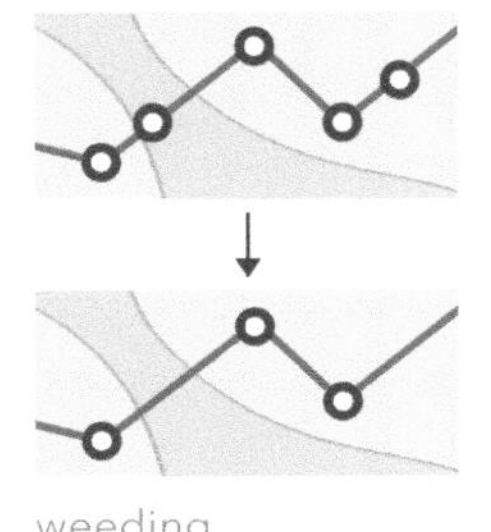
weeding

weight

1. [MATHEMATICS] A number that indicates the importance of a variable or measure for a particular calculation. The larger the weight assigned, the more the variable will influence the outcome.
2. [MODELING] The process of assigning a value to a feature according to its importance.
3. [MATHEMATICS] A measure of the relative contribution or influence that one attribute makes to the whole.

See also cost, kriging.

weighted mean center [SPATIAL STATISTICS (USE FOR GEOSTATISTICS)] The geographic center of a set of points as adjusted for the influence of a value associated with each point. For example, although the mean center of a group of grocery stores would be the location obtained by averaging the stores' x,y coordinates, the weighted mean center would be shifted closer to stores with higher sales, more square footage, or a greater quantity of some other specified attribute. *See also* mean, mean center, weight.

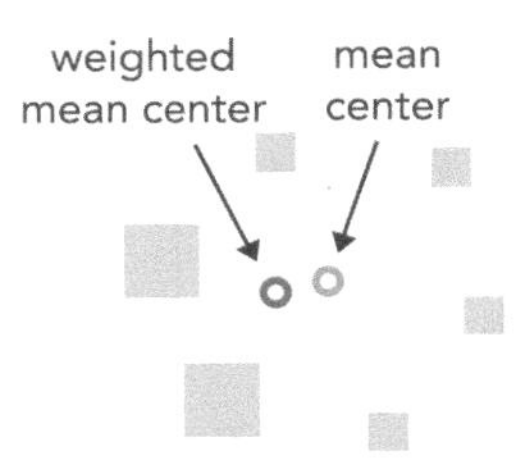

weighted mean center

weighted moving average [SPATIAL STATISTICS (USE FOR GEOSTATISTICS)] The value of a point's attribute computed by averaging the values of its surrounding points, considering their importance or distance from the point. *See also* weight.

weighted overlay [DATA ANALYSIS] A technique for combining multiple rasters by applying a common measurement scale of values to each raster, weighting each according to its importance, and adding them together to create an integrated analysis. *See also* weight.

weighted standard deviational ellipse [DATA QUALITY] A standard deviational ellipse calculated using the locations influenced by a weight that indicates which points are more important than others. *See also* standard deviational ellipse, weight.

weird polygon *See* nonsimple polygon.

well-known ID *See* WKID.

well-known text *See* WKT.

WFS *See* Web Feature Service.

WGS 1972 *See* WGS72.

WGS 1984 *See* WGS84.

WGS72 [GEODESY] Acronym for *World Geodetic System 1972.* A geocentric datum and coordinate system designed by the U.S. Department of Defense. No longer in use. *See also* WGS84.

WGS84 [GEODESY] Acronym for *World Geodetic System 1984.* The most widely used geocentric datum and geographic coordinate system globally. The WGS84 coordinate system has its origin at the earth's center of mass. The geoid is based on the Earth Gravitational Model 1986 (updated in 2004). Designed by the U.S. Department of Defense to replace WGS72, the elliptical shape of the system mirrors the earth's actual shape. GPS measurements are based on WGS84. *See also* geocentric datum, GPS.

whisk broom scanner *See* across-track scanner.

Wide Area Augmentation System *See* WAAS.

wide area motion imagery *See* WAMI.

widget [ESRI SOFTWARE] An interactive graphic component of a user interface (such as a button, scroll bar, or menu bar), its controlling program, or the combination of both the component and program.

wind barb [METEOROLOGY] A map symbol that shows wind direction and speed in knots. *See also* symbol.

wind rose [METEOROLOGY] A diagram showing, for a given place and time, how much of the time the wind blows from each direction. Wind roses have many variations, but in the typical pattern, a number of wedges (usually

8, 12, or 16) radiate from the center of a circle. The width and orientation of a wedge represent the direction from which the wind blows; the length of a wedge represents the percentage of time the wind blows from that direction. More complex wind roses use color schemes and other graphic devices to represent wind speed and related information. *See also* chart.

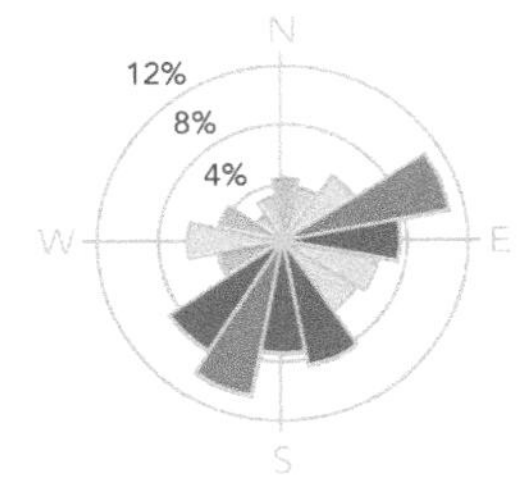

wind rose

wind vector [CLIMATOLOGY] A 2D vector that describes the wind magnitude and direction at a point. *See also* vector.

windowing [GRAPHICS (MAP DISPLAY)] The process of limiting the viewable extent of a map or data by panning and zooming. *See also* extent, pan, zoom.

Winkel Tripel projection [MAP PROJECTIONS] A compromise projection that is not equal area, conformal, nor equidistant, but rather minimizes all three forms of geometric distortion. Constructed as a mathematical combination of two existing projections; proposed by German cartographer Oswald Winkel (1874–1953) in 1921. *See also* conformal projection, equal-area projection, equidistant projection, tripel.

wireframe [GRAPHICS (COMPUTING)] A 3D picture of an object, composed entirely of lines (wires). The lines represent the edges or surface contours, including those that would otherwise be hidden by a solid view. Wireframes facilitate faster editing speeds. *See also* line, representation.

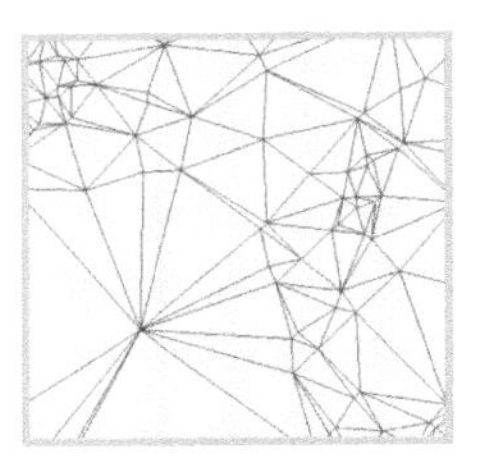
wireframe

WKID [COORDINATE SYSTEMS] Acronym for *well-known ID.* A unique number used to identify a predefined coordinate reference system or spatial reference system within a GIS. Each WKID defines the coordinate system with all parameters needed to represent spatial data. The source, origin, or authority of the number is either Esri or the former European Petroleum Survey Group (EPSG). Also known as a spatial reference ID (SRID). *See also* coordinate reference system, EPSG code, WKT.

WKT

1. [COORDINATE SYSTEMS] Acronym for *well-known text.* A standard text representation format that is used to provide a complete definition of

a coordinate reference system or a spatial reference system. A WKT follows a specific syntax and includes information such as the coordinate system, datum, WKID and authority, and units. Although a WKID is used to identify known, predefined coordinate reference systems, a WKT can be used to share customized coordinate reference systems within a GIS.

2. [SPATIAL ANALYSIS] Acronym for *well-known text.* A standard text representation format used to describe spatial objects or geometries. A WKT provides a way to define points, lines, polygons, and other geometric shapes with their coordinates and associated attributes.

See also coordinate reference system, datum, EPSG code, WKID.

WMS *See* Web Map Service.

workflow [ORGANIZATIONAL ISSUES] A set of tasks carried out in a certain order to achieve a goal.

world file [DATA STORAGE] A text file containing information about where an image should be displayed in real-world coordinates. When an image has a properly configured world file, GIS software can use the information (a total of six values, including the starting coordinates, the cell size in both x- and y-dimensions, and any rotation and scaling information) to accurately overlay the image with any other data already in a projected or geographic coordinate system. *See also* coordinate system, image, overlay, real world.

World Geodetic System 1972 *See* WGS72.

World Geodetic System 1984 *See* WGS84.

WorldView-3 [REMOTE SENSING] A commercial earth observation satellite owned by DigitalGlobe and launched in 2014 to provide panchromatic imagery, eight-band multispectral imagery, shortwave infrared imagery, and CAVIS (Clouds, Aerosols, Vapors, Ice, and Snow) data. *See also* satellite constellation, sensor.

WTM [COORDINATE SYSTEMS] Acronym for *Wisconsin transverse Mercator.* A special grid coordinate system for Wisconsin, USA, formed by shifting the central meridian of a UTM zone to the center of the state. *See also* Mercator projection, transverse Mercator projection, UTM, Web Mercator projection.

WYSIWYG [COMPUTING] Acronym for *what you see is what you get.* In software, this term refers to the editing tools that provide visual feedback of an editing operation while the process is still under way. For cartographic representations, this references the ability to see a symbolized representation of a feature instead of a wireframe of its underlying geometry during editing.

x-axis

1. [COORDINATE SYSTEMS] In a planar coordinate system, the horizontal line that runs right and left (east and west) of the origin (0,0). Numbers east of the origin are positive, and numbers west of it are negative.

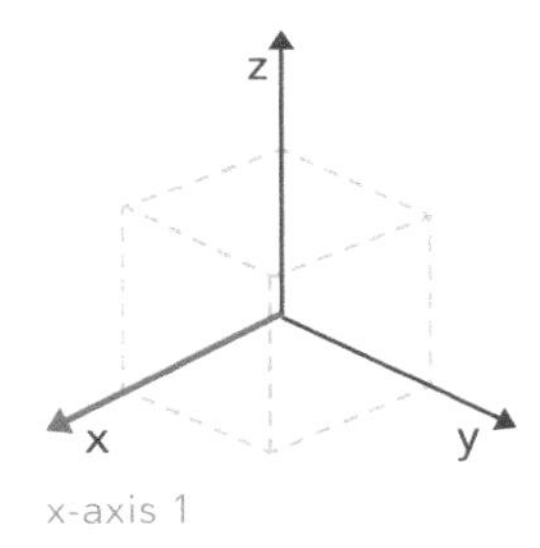

x-axis 1

2. [COORDINATE SYSTEMS] In a spherical coordinate system, a line in the equatorial plane that passes through 0 degrees longitude.
3. [MATHEMATICS] On a chart, the horizontal axis.

See also Cartesian coordinate system, y-axis, z-axis.

xeric [ENVIRONMENTAL GIS] A desert and dry shrubland biome characterized by—or requiring—limited moisture, as in drought-tolerant plants and animals.

x-slope [CARTOGRAPHY] The slope in the x (easting) direction when determining the gradient. *See also* easting, gradient.

XTE *See* cross-track error.

x,y coordinates [COORDINATE SYSTEMS] A pair of values that represents the distance from an origin (0,0) along two axes, a horizontal axis (x), and a vertical axis (y). On a map, x,y coordinates are used to represent features at the location they are found on the earth's spherical surface. *See also* x-axis, y-axis.

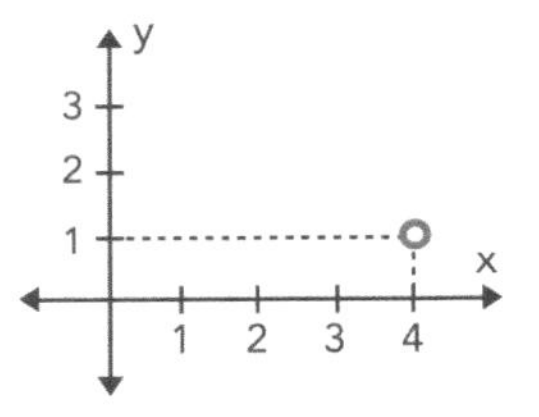

x,y coordinates

x,y event [MATHEMATICS] A simple coordinate pair that describes the location of a feature, such as a set of latitude and longitude degrees. *See also* coordinates, event, event table, x,y coordinates.

x,y values *See* x,y coordinates.

x,y,z coordinates [COORDINATE SYSTEMS] In a planar coordinate system, three coordinates that locate a point by its distance from an origin (0,0,0) where three orthogonal axes cross. Usually, the x-coordinate is measured along the east–west axis, the y-coordinate is measured along the north–south axis, and the z-coordinate measures height or elevation. *See also* x-axis, y-axis, z-axis.

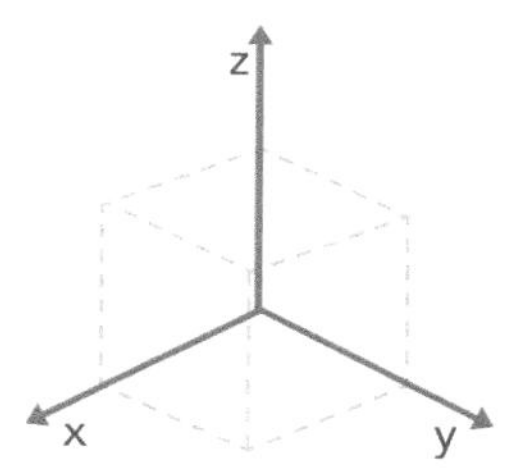

x,y,z coordinates

yard [MATHEMATICS] A measurement in English units of distance equal to three feet (.91 meters); decreed by King Henry I of England to be the distance from the tip of his nose to the end of his thumb with his arm outstretched. *See also* English unit.

y-axis

1. [COORDINATE SYSTEMS] In a planar coordinate system, the vertical line that runs above and below (north and south of) the origin (0,0). Numbers north of the origin are positive, and numbers south of it are negative.

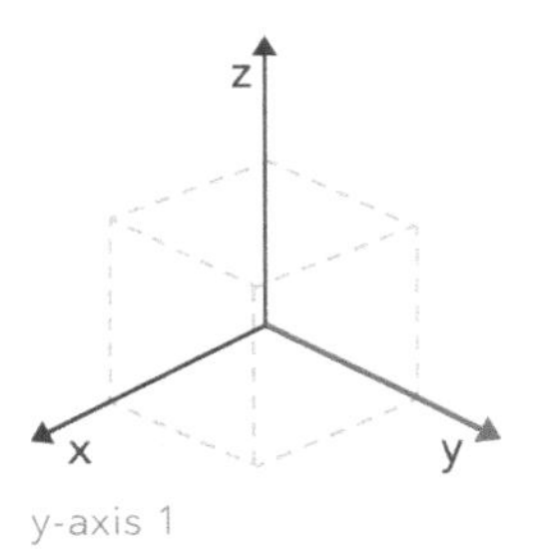

y-axis 1

2. [COORDINATE SYSTEMS] In a spherical coordinate system, a line in the equatorial plane that passes through 90 degrees east longitude.
3. [MATHEMATICS] On a chart, the vertical axis.

See also Cartesian coordinate system, x-axis, z-axis.

y-slope [CARTOGRAPHY] The slope in the y (northing) direction when determining the gradient. *See also* northing, slope.

Y

Z

z-axis [COORDINATE SYSTEMS] In a spherical coordinate system, the vertical line that runs parallel to the earth's rotation, passing through 90 degrees north latitude, and perpendicular to the equatorial plane, where it crosses the x- and y-axes at the origin (0,0,0). *See also* x-axis, y-axis.

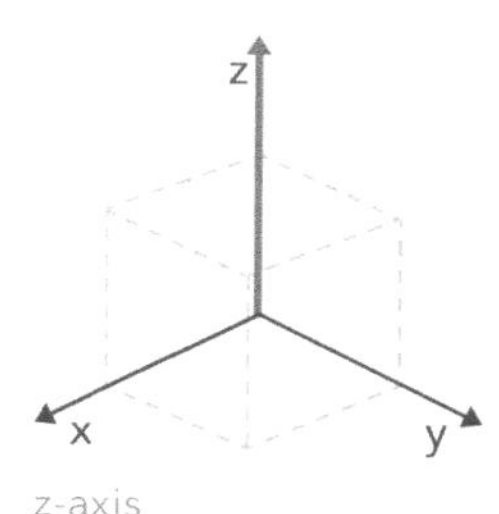

z-axis

z-coordinate *See* z-value.

zenith [ASTRONOMY] In astronomy, the point on the celestial sphere directly above an observer. Both the zenith and nadir lie on the observer's meridian; the zenith lies 180 degrees from the nadir and is observable. *See also* celestial sphere, nadir.

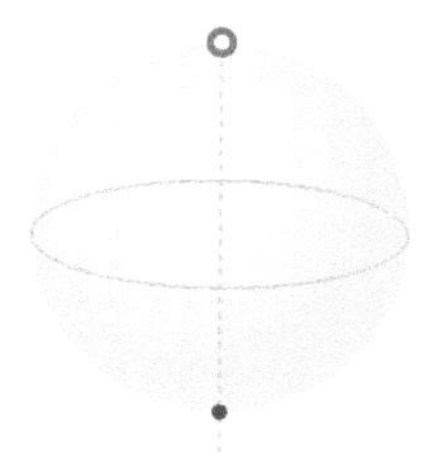

zenith

zenith angle [SURVEYING] A vertical angle that is formed by the intersection of two lines in a vertical plane. Zenith angles are observed on the vertical circle of a TPS instrument. *See also* surveying, zenith.

zenithal projection *See* azimuthal projection.

zero contour [CARTOGRAPHY] The contour that corresponds to mean sea level. Also called the datum contour. *See also* contour, mean sea level.

zero-length line event [LINEAR REFERENCING] A line event where the from-measure is equal to the to-measure. A zero-length line may occur, for example, along routes when a polygon touches a route but does not overlap it. *See also* event, line event, linear referencing.

z-factor [COORDINATE SYSTEMS] A conversion factor used to adjust vertical and horizontal measurements into the same unit of measure. Specifically, the number of vertical units (z-units) in each horizontal unit. For example, if a surface's horizontal units are meters and its elevation (z) is measured in feet, the z-factor is 0.3048 (the number of meters in a foot). *See also* data conversion, unit of measure.

ZIP code [FEDERAL GOVERNMENT] ZIP is an acronym for *zone improvement plan*. A five-digit code, developed by the U.S. Postal Service, which identifies the United States geographic delivery area served by an individual post office or metropolitan area delivery station. *See also* ZIP+4 code.

ZIP+4 code [FEDERAL GOVERNMENT] An enhanced U.S. ZIP code that consists of the five-digit ZIP code plus four additional digits that identify a specific geographic segment within the five-digit delivery area, such as a city block, office building, or other unit. *See also* ZIP code.

zonal analysis [SPATIAL ANALYSIS] The creation of an output raster in which the desired function is computed on the cell values from the input value raster that intersect or fall within each zone of a specified input zone dataset. The input zone dataset is used to define only the size, shape, and location of each zone, whereas the value raster identifies the values to be used in the evaluations within the zones. *See also* analysis, raster, zone.

zonal functions *See* zonal analysis.

zonal statistics [SPATIAL ANALYSIS] The calculation of a statistic for each zone of a zone dataset based on values from another dataset, a value raster. A single output value is computed for each cell in each zone defined by the input zone dataset. *See also* spatial analysis, zone.

zone

1. [ANALYSIS/GEOPROCESSING] All cells or pixels in a raster image with the same value, regardless of whether they are contiguous.

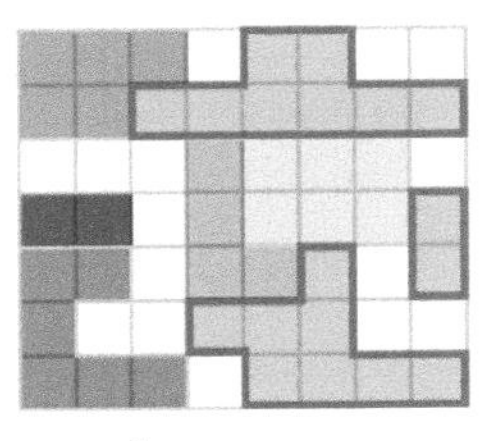

zone 1

2. [CARTOGRAPHY] A well-defined region on the earth between two meridians or parallels.

See also cell, meridian, pixel.

zone of indifference weighting [MATHEMATICS] A composite distance weighting method that combines a fixed-distance component with an inverse-distance component, in which features within a critical distance of a feature are given a constant weight. Once the critical distance is exceeded, the level of impact declines. *See also* distance weighting, weight.

zone of interpolation [SURVEYING] The area in a TIN layer for which values (elevation, slope, and aspect) are calculated. When a TIN layer is clipped to a smaller size to create a more focused study area, the parts that lie outside the study area remain triangulated and are represented as outside lines, but they have no values. These parts are said to be outside the zone of

Z

interpolation. *See also* interpolation, surveying, TIN, zone.

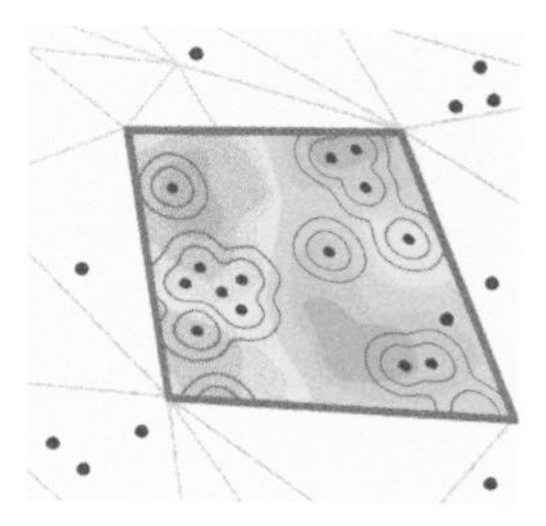

zone of interpolation

zoning [LOCAL GOVERNMENT] The application of local government regulations that permit certain land uses within geographic areas under the government's jurisdiction. Zoning regulations typically set a broad category of land use permissible in an area, such as residential, commercial, agricultural, or industrial. *See also* land use, zoning code.

zoning code [REGIONAL PLANNING] An ordinance used by local governments to specify the type of use to which privately owned land may be put in specific areas.

zoning districts [REGIONAL PLANNING] A section of a city or town restricted by law for a particular use, such as for homes, parks, or businesses.

zoom [GRAPHICS (MAP DISPLAY)] To change the scale of a web map, or how large or small items appear. Zoom in shows more detail in a smaller area; zoom out shows less detail and a larger area. *See also* pan, web map.

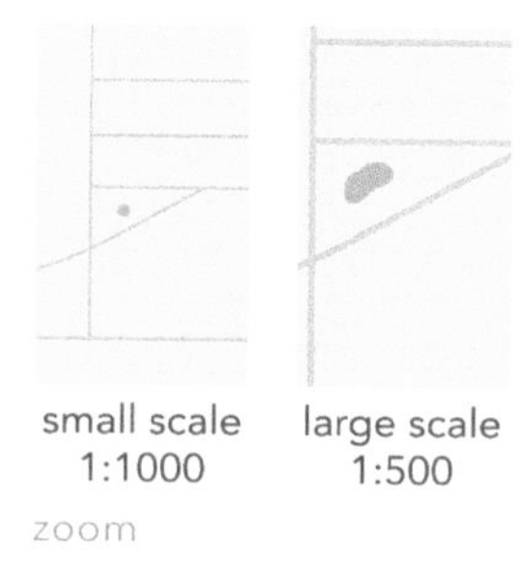

zoom

zoom level [SOFTWARE] A value that sets the scale for a view. Lower values are the farthest from the earth and the smallest in scale; the higher values are the closest to the earth and the largest in scale. The higher the zoom level value, the more geographic details are present. *See also* scale, view.

z-score [STATISTICS] A statistical measure of the spread of values from their mean, expressed in standard deviation units, in which the z-score of the mean value is zero and the standard deviation is one. In a normal distribution, 68 percent of the values have a z-score of plus or minus 1, meaning they are within one standard deviation of the mean. Ninety-five percent of the values have a z-score of plus or minus 1.96, meaning they are within two standard deviations of the mean; 99 percent of the values have a z-score of plus or minus 2.58. Z-scores are a common scale on which different distributions, with different means and standard deviations, can be compared. *See also* normal distribution, standard deviation.

z-tolerance [COORDINATE SYSTEMS] In raster-to-TIN conversion, the maximum allowed difference between the z-value of the input raster cell and the z-value of the output TIN at the location corresponding to the raster cell center.

z-value [COORDINATE SYSTEMS] The value for a given surface location that represents an attribute other than position. In an elevation or terrain model, the z-value represents elevation; in other kinds of surface models, it represents the density or quantity of a particular attribute. *See also* altitude, breakline, digital elevation model, drift, elevation, height, vertical coordinate system, vertical exaggeration.

z-value

About Esri Press

Esri Press is an American book publisher and part of Esri, the global leader in geographic information system (GIS) software, location intelligence, and mapping. Since 1969, Esri has supported customers with geographic science and geospatial analytics, what we call The Science of Where. We take a geographic approach to problem-solving, brought to life by modern GIS technology, and are committed to using science and technology to build a sustainable world.

At Esri Press, our mission is to inform, inspire, and teach professionals, students, educators, and the public about GIS by developing print and digital publications. Our goal is to increase the adoption of ArcGIS and to support the vision and brand of Esri. We strive to be the leader in publishing great GIS books, and we are dedicated to improving the work and lives of our global community of users, authors, and colleagues.

Related titles

The Power of Where: A Geographic Approach to the World's Greatest Challenges

Jack Dangermond, Esri, and the GIS community

9781589486065

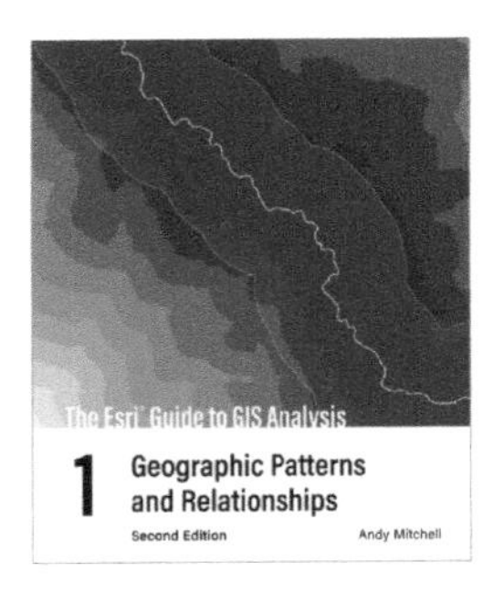

The Esri Guide to GIS Analysis, Volume 1: Geographic Patterns and Relationships, second edition

Andy Mitchell

9781589485792

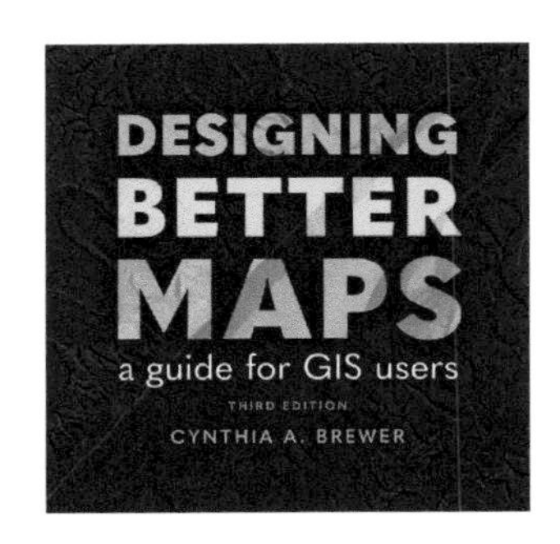

Designing Better Maps: A Guide for GIS Users, third edition

Cynthia A. Brewer

9781589487826

Top 20 Essential Skills for ArcGIS Online

Craig Carpenter, Jian Lange, and Bern Szukalski

9781589487802

For more information about Esri Press books and resources, or to sign up for our newsletter, visit

esripress.com.